VIRTUAL REALITY IN PSYCHOLOGICAL, MEDICAL AND PEDAGOGICAL APPLICATIONS

Edited by **Christiane Eichenberg**

Virtual Reality in Psychological, Medical and Pedagogical Applications
http://dx.doi.org/10.5772/2607
Edited by Christiane Eichenberg

Contributors

Giuseppe Riva, Fabrizia Mantovani, Christiane Eichenberg, Carolin Wolters, Birgit U. Stetina, Anna Felnhofer, Oswald D. Kothgassner, Mario Lehenbauer, Stéphane Bouchard, Geneviève Robillard, Serge Larouche, Claudie Loranger, Annie Aimé, Karine Cotton, Tanya Guitard, Stéphane Bouchard, Linda Garcia, Adi Kartolo, Eric Méthot-Curtis, C.E. Buckley, E. Nugent, D. Ryan, P.C. Neary, Bernadette McElhinney, Angela Beard, Krishnan Karthigasu, Roger Hart, Kristiina M. Valter McConville, Ying-hui Chou, Carol P. Weingarten, David J. Madden, Allen W. Song, Nan-kuei Chen, José Luis Mosso, Gregorio T. Obrador, Brenda Wiederhold, Mark Wiederhold, Verónica Lara, Amador Santander, Claudio Kirner, Christopher Shneider Cerqueira, Tereza Gonçalves Kirner, Jamshid Beheshti

Published by InTech

Janeza Trdine 9, 51000 Rijeka, Croatia

Copyright © 2012 InTech

Notice

Publishing Process Manager Mirna Cvijic
Technical Editor InTech Prepress, Novi Sad
Cover InTech Design team

Additional hard copies can be obtained from orders@intechopen.com

Virtual Reality in Psychological, Medical and Pedagogical Applications,
Edited by Christiane Eichenberg
p. cm.
ISBN 978-953-51-0732-3

For EU product safety concerns:
IN TECH d.o.o., Prolaz Marije Krucifikse Kozulić 3, 51000 Rijeka, Croatia, info@intechopen.com or visit our website at intechopen.com.

Meet the editor

Dr Christiane Eichenberg received 2001 Diploma in Psychology (University of Cologne, Germany) and she obtained PhD in Psychology (University of Cologne, Germany) in 2006. 2010 follows venia legendi in Psychology (Ilmenau University of Technology, Germany) She became certifiticated psychotherapist recently, in the year 2012. She currently works as lecturer at Sigmund Freud University, Vienna in Austria. Some research topics which are of her great interest are: E-Mental-Health; Psychotraumatology; Research of psychotherapy. You can find more information about her numerous publications and scientific lectures on her personal page: www.christianeeichenberg.de

Contents

Preface

Virtual reality applications (VR) allows creation of computer-based models of the real world, which interacte by using the human-machine interfaces. This research field developed in the last 20 years for potential use also in the *human sciences*. While first applications were implemented mostly in industry, (e.g. virtual prototyping and assembly training in the automotive industry, ergonomic assessments of the operation of cockpits through the simulation of different body sizes), in the military (e.g. operational training, using simulated parachute jumps), in geology and architecture (e.g. spatial studies, or archaeological reconstructions) and last but not least in the entertainment area (especially games development), the scientific and practical field of application extended to VR in recent years. The *medical field* has used VR for example in surgery, for laparoscopic surgery and applications have been developed for rehabilitation, which enabled mute patients to communicate verbally, by capturing mute gestures with the aid of a data glove, interpreted by the computer and routed to a voice system, which translated it into synthetic speech.

This book has an aim to present latest applications, trends and developments of virtual reality technologies in three humanities disciplines: in *medicine, psychology* and *pedagogy*.

In the *medical field* the potential uses in the various specialized medical areas are wide spread and go far beyond surgical options. VR is successfully applied e.g. in anatomical education, in functional diagnostics and before and during interventions for pain and anxiety reduction. The same is true for simulations of the musculoskeletal system, for example in the diagnosis and training of patients whose posture balance is impaired due to osteoporosis. VR also shows great promise as a new paradigm of imaging in the neurosciences.

In *psychology*, VR is utilized in addition to fundamental research (especially in general psychology and cognitive and social psychology), which has in recent years very greatly contributed to our understanding of presence- experience in virtual environments, in particular in the application in *clinical psychology* and *psychotherapy*. The observation that virtual stimuli cause real fear led to VR to be integrated into the spectrum of therapeutic interventions. While initially VR was rather intuitively integrated into the behavioral treatment of mostly specific phobias, there meanwhile exist evaluated treatment manuals for a broader spectrum disorder. Thus, the benefits to body image disorders, as well as in the neuropsychological contexts such as the treatment of dementia, were examined. On the other hand meanwhile, even if only hesitantly, other therapeutic schools which traditionally are more restrictive in the flexibility of the therapeutic setting - such as psychoanalysis, are open to the integration of VR.

Other constructive uses of VR have been recognized for other target groups, e.g. for children and young people for whom VR-based games can also have both therapeutic and pedagogical value (so-called *serious games*). This relativizes the often single-sided debate about the harmful effects of computer games. In *pedagogy* the term *edutainment* is applied to such applications that provide entertainment-oriented knowledge. Utilizing the affinity of young people to modern technologies in the field of education, different VR-based scenarios were developed for integration into elaborate didactic concepts (e.g. so-called integrated *augmented reality*, which incorporate virtual objects into the real world, in geometry or biology lessons).

Studies show that people in both educational as well as in the medical therapeutic range expect more and more that modern media are included in the corresponding demand and supply structures. For the Internet and various mobile media, associated research and application projects now have fixed key words such as "E-learning" and "E-Mental Health" or "M-Learning", "M-Mental Health". This book aims to contribute to the current state of the corresponding efforts in the area of further promising technology - the Virtual Reality - designed to give an overview and secondly to provide a stimulus on specific projects, associated with the hope of giving to scientists and practitioners from the humanities an impulse for their own (further-) development, evaluation and implementation of various VR scenarios in the education and health sectors.

Dr. Christiane Eichenberg
Department of Psychology
Sigmund Freud University Vienna, Vienna,
Austria

Section 1

VR in Psychological Applications

Chapter 1

Being There: Understanding the Feeling of Presence in a Synthetic Environment and Its Potential for Clinical Change

Giuseppe Riva and Fabrizia Mantovani

Additional information is available at the end of the chapter

http://dx.doi.org/10.5772/46411

1. Introduction

Virtual Reality (VR) has been usually described as a collection of technological devices: a computer capable of interactive 3D visualization, a head-mounted display and data gloves equipped with one or more position trackers [1]. The trackers sense the position and orientation of the user and report that information to the computer which updates the images for display in real time.

However, in the behavioral sciences, VR is usually described as [2] "an advanced form of human-computer interface that allows the user to interact with and become immersed in a computer-generated environment in a naturalistic fashion" (p. 82).

This feature transforms VR in an "empowering environment", a special, sheltered setting where patients can start to explore and act without feeling of being threatened [3]. Nothing the patients fear can "really" happen to them in VR. With such assurance, they can freely explore, experiment, and experience feelings and/or thoughts. VR thus becomes a very useful intermediate step between the therapist's office and the real world [4; 5]. In other words, the key feature of VR for clinical goals is that it offers an effective support to the activity of the subject by activating the feeling of "presence", the feeling of being *inside* the virtual world.

But what is presence? In this chapter we will use the following three research outcomes emerging from the recent work of cognitive sciences to build a cognitive theory of presence:

1. *Cognitive processes can be either rational or intuitive*: we will argue that presence is an intuitive feeling that is the outcome of an experience-based metacognitive judgment;
2. *Skills become intuitive when our brain is able to simulate their outcome*: we will show argue that presence monitors intuitively our activity processes using embodied simulations;

3. *Space is perceived in terms of the actions we could take towards them*: we will argue that the feeling of Presence in a real or virtual space is directly correlated to the outcome of the actions the subject can enact in it;

In sum, the feeling of presence can be described as the product of an intuitive experience-based metacognitive judgment related to the enaction of our intentions: We are present in an environment - real and/or synthetic - when we are able, inside it, to intuitively transform our intentions in actions. The consequences of this claim for the development of clinical virtual environments are presented and discussed.

2. Virtual reality: From technology to experience

Since 1986, when Jaron Lamier used the term for the first time, VR has been usually described as a collection of technological devices. In general, a VR system is the combination of the hardware and software that enables developers to create VR applications [6]. The hardware components receive input from user-controlled devices and convey multi-sensory output to create the illusion of a virtual world. The software component of a VR system manages the hardware that makes up VR system. This software is not necessarily responsible for actually creating the virtual world. Instead, a separate piece of software (the VR application) creates the virtual world by making use of the VR software system.

Typically, a VR system is composed by [6]:

- the *output tools* (visual, aural and haptic), that immerse the user in the virtual environment;
- the *input tools* (trackers, gloves or mice) that continually reports the position and movements of the users;
- the *graphic rendering system* that generates the virtual environment;
- the *database construction and virtual object modeling software* for building and maintaining detailed and realistic models of the virtual world. In particular, the software handles the geometry, texture, intelligent behavior, and physical modeling of hardness, inertia, and surface plasticity of any object included in the virtual world.

However, as we have seen in the introduction VR can be described, too, as an advanced form of human-computer interface. Specifically, what distinguishes VR from other media or communication systems is the sense of *presence*. VR can be considered the leading edge of a general evolution of present communication interfaces such as television, computer and telephone whose ultimate goal is the full immersion of the human sensorimotor channels into a vivid and interactive communication experience. But what is presence?

The term "*Presence*" entered the general scientific debate in 1992 when Sheridan and Furness used it in the title of a new journal dedicated to the study of virtual reality systems and teleoperations: *Presence, Teleoperators and Virtual Environments.* In the first issue, Sheridan clearly refers to presence as an experience elicited by technology use [7]: the effect felt when controlling real world objects remotely as well as the effect people feel when they interact with and immerse themselves in virtual environments.

This vision describes presence as "Media Presence", a function of our experience of a given medium [7-10]. The main outcome of this approach is the *"perceptual illusion of non-mediation"* [10] definition of presence. Following it, presence is produced by means of the disappearance of the medium from the conscious attention of the subject. The main advantage of this approach is its predictive value: the level of presence is reduced by the experience of mediation during the action. The main limitation of this vision is what is not said. What is presence for? Is it a specific cognitive process? What is its role in our daily experience?

To address these questions, a second group of researchers considers presence as "Inner Presence", the feeling of being located in a perceived external world around the self [11-13]. In this view presence is broad psychological phenomenon, not necessarily linked to the experience of a medium, whose goal is the control of the individual and social activity. In the next paragraphs we will justify this statement using the recent work of cognitive sciences.

3. The first feature of presence: it is an intuitive process

A first problem related to the research about presence is its role in cognitive science: what is its foundation in terms of the cognitive processes involved in it? Stanovich & West, [14] noted that in the last forty years, different authors from different disciplines suggested a two-process theory of reasoning based on Intutive and Rational processes. Even if the details and specific features of these theories do not always match perfectly, nevertheless they share the following properties:

- Intuitive operations are faster, automatic, effortless, associative, and difficult to control or modify.
- Rational operations, instead, are slower, serial, effortful, and consciously controlled.

One of the theories based on this distinction is the cognitive-experiential self-theory (CEST). As explained by Epstein [15]:

"A fundamental assumption in CEST is that people operate by two cognitive systems: an "experiential system", which is a nonverbal automatic learning system, and a "rational system," which is a verbal reasoning system. The experiential system operates in a manner that is preconscious, automatic, nonverbal, imagistic, associative... and its schemas are primarily generalizations from emotionally significant intense or repetitive experience... In contrast to the automatic learning of the experiential system, the rational system is a reasoning system that operates in a manner that is conscious, verbal, abstract, analytical, affect free, effortful, and highly demanding of cognitive resources. It acquires its beliefs by conscious learning from books, lectures and other explicit sources of information, and from logical inference; and it has a very brief evolutionary history." (pp. 24-25).

The differences between the two systems are described in Table 1. An interesting feature of this approach is that intuition is not only innate. As demonstrated by the research on perceptual-cognitive and motor skills, these skills are automatized through experience and thus rendered intuitive [16]. In the case of motor skill learning, the process is initially rational and controlled by consciousness, as shown, for example, by the novice driver's rehearsal of the steps involved in parking a car: check the mirrors and blind spots; signal to the side of the space; position the car beside the vehicle I'm parking behind, etc.

	Experiential/Intuitive System	*Rational System*
Main Features	• **Intuitive**: Preconscious, automatic, and intimately associated with affect • **Concrete**: Encodes reality in images, metaphors, and narratives • **Associative**: Connections by similarity and contiguity • **Rapid processing**: Oriented toward immediate action • **Resistant to change**: Changes with repetitive or intense experience • **Differentiated**: Broad generalization gradient; categorical thinking • **Integrated**: Situationally specific; organized in part by cognitive-affective modules • Experienced passively and preconsciously: We are seized by our emotions • **Self-evidently valid:** "Experiencing is believing"	• **Rational**: Conscious, deliberative and affect-free • **Abstract**: Encodes reality in symbols, words, and numbers • **Analytic**: Connections by cause-and-effect relations • **Slower processing**: Capable of long delayed action • **Less resistant to change**: Can change with speed of thought • More highly differentiated: nuanced thinking • **More highly integrated**: Organized in part by cross-situational principles • **Experienced actively and consciously**: We believe we are in control of our thoughts • **Not Self-evident:** Requires justification via logic and evidence
How it works	• Operates by **hedonic principle** (what feels good) • Acquires its schemas by **learning from experience** • Outcome oriented • Behavior mediated by "vibes" from past experience	• Operates by **reality principle** (what is logical and supported by evidence) • Acquires its beliefs by conscious learning and logical inference • More process oriented • Behavior mediated by conscious appraisal of events

Table 1. Differences between the Intuitive and Rational system according to the cognitive-experiential self-theory

However, later the skill becomes intuitive and consciously inaccessible by virtue of practice, as shown, for example, by the difficulty of expert drivers to describe how to perform a complex maneuver to others, and by the fact that conscious attention to it actually interferes with their driving performance.

In sum, perceptual-motor skills that are not innate – e.g. driving a car - may become automatic through practice, and their operations thereby rendered intuitive. Using a metaphor derived from computer science, this process can be described as "*knowledge compilation*" [16]: a knowledge given in a general representation format (linguistic-semantic) is translated into a different one, more usable and less computationally demanding (perceptual-motor).

Are presence and telepresence intuitive or rational cognitive processes? On one side, it is evident that presence is the *outcome* of an intuitive cognitive process: no rational effort is required to experience a feeling of presence. On the other side, however, presence is *different* from an acquired motor skill or a behavioral disposition.

A possible path to find a better answer comes from the concept of metacognition. Koriat [17] defines "metacognition" as "the processes by which people self-reflect on their own cognitive and memory processes (monitoring) and how they put their metaknowledge to use in regulating their information processing and behavior (control)." (p. 289). Following the distinction between Intuition and Reasoning, researchers in this area distinguish between *information-based* (or theory-based) and *experience-based* metacognitive judgments [17].

Information-based metacognitive judgments are based on a deliberate use of one's beliefs and theories to reach an evaluation about one's competence and cognitions: they are deliberate and largely conscious, and draw on the contents of declarative information in long term memory. By contrast, experience-based metacognitive judgments are subjective feelings that are product of an inferential intuitive process: they operate unconsciously and give rise to a "sheer subjective experience". An example of these metacognitive judgment are [18]: the "*feeling of knowing*" (knowing that we are able to recognize the correct answer to a question that we cannot currently recall), or the "*feeling of familiarity*" (knowing that we have encountered a given situation before, even if we don't have an explicit memory of it).

In conclusion, we may describe presence as the sheer subjective experience of being in a given environment (the feeling of "being there") that is the product of an intuitive experience-based metacognitive judgment.

4. The second feature of presence: it is the outcome of a simulation

A second critical question is "What is intuitively judged by Presence?". Different authors have suggested a role of presence in the monitoring of action. For example, Zahoric and Jenison [19] underlined that ''presence is tantamount to successfully supported action in the environment'' (p. 87); Riva and colleagues [13]: suggested that "…the evolutionary role of presence is the control of agency" (p. 24); finally, Slater and colleagues [20] argued that "humans have a propensity to find correlations between their activity and internal state and their sense perceptions of what is going on out there" (p. 208). But, how may this work? And how this process is related to intuition? As suggested by Reber [21]:

"To have an intuitive sense of what is right and proper, to have a vague feeling of the goal of an extended process of thought, to "get the point" without really being able to verbalize what it is that one has gotten, is to have gone through an implicit learning experience and have built up the requisite representative knowledge base to allow for such judgment." (p. 233).

In simpler words, through implicit learning the subject is able to represent complex actions using perceptual-motor data and enact/monitor them intuitively. An empirical proof of this hypothesis is the recent discovery of neuronal resonance processes activated by the simple observation of others. Rizzolatti and colleagues found that a functional cluster of premotor neurons (F5c-PF) contains *"mirror neurons"*, a class of neurons that are activated both during the execution of purposeful, goal-related hand actions, and during the observation of similar actions performed by another individual [22].

The general framework outlined by the discovery of neuronal resonance processes was used by Simulation Theorists – for example, Lawrence Barsalou, Vittorio Gallese, Alvin Goldman, Jane Heal, Susan Hurley, Marc Jeannerod, Guenter Knoblich and Margaret Wilson – to support the following view: the mirror system instantiates simulation of transitive actions used to map the goals and purposes of others' actions [23; 24]. As clearly explained by Wilson and Knoblich [25] this is the outcome of an implicit/covert, subpersonal process:

"The various brain areas involved in translating perceived human movement into corresponding motor programs collectively act as an emulator, internally simulating the ongoing perceived movement… The present proposal suggests that, in tasks requiring fast action coordination, the emulator derives predictions about the future course of others' actions, which could be integrated with the actions one is currently planning." (pp. 468-469).

According to this approach, action and perception are more closely linked than has traditionally been assumed. Specifically, for the *Common Coding Theory* [26], the cognitive representations for perceived events (perception) and intended or to-be generated events (action) are formed by a common representational domain: actions are coded in terms of the perceivable effects they should generate. For this reason, when an effect is intended, the movement that produces this effect as perceptual input is automatically activated, because actions and their effects are stored in a common representational domain.

In simpler words, the brain has its own virtual reality system that is used in both action planning and action understanding. If this is true, how we can distinguish between the virtual action planning and the real action? The answer is easy: using presence. In his book "Inner Presence" Revuonso [12] clearly states:

"To be conscious is to have the sense of presence in a world… To have contents of consciousness is to have patterns of phenomenological experience present… In the philosophy of presence, consciousness is an organized whole of transparent surrogates of virtual objects that are immediately present for us in the here-and-now of subjective experience." (pp. 126-129).

In this view, to be directly present right here or for an object to be directly present for me require some form of "acquaintance": a direct awareness (intuition) based on a non propositional knowledge or nonconceptual content [27]. This view is surprisingly near to the vision of presence as *"perceptual illusion of non-mediation"* [10] introduced before. In both cases, presence is related to a direct experience.

However, if in the Lombard and Ditton definition the mediation is given by the used medium (virtual reality) in the Revuonso view [12], *the mediation is given by the body*: the experience of the body is our first virtual reality system. This vision is shared by many cognitive scientists. For instance Andy Clark [28] underlines that:

"The infant, like the VR-exploring adult, must learn how to use initially unresponsive hands, arms, and legs to obtain its goals… With time and practice enough bodily fluency is achieved to make the wider world itself directly available as a kind of unmediated arena for embodied action… At such moments the body has become "transparent equipment"… that is not the focus of attention in use." (p. 10).

More, different neurological disorders clearly support this view, showing how the direct experience of presence in our body is the result of different and separable subcomponents that can be altered in some way [29]: *agency, ownership and location*.

- *Autopagnosia (agency)*: it is a neurological disease characterized by the inability to recognize or to orient any part of one's own body, caused by a parietal lobe lesion [30]: a patient with Autopagnosia is not present in his/her body;
- *Anarchic Hand (ownership)*: it is a neurological disease in which patients are aware of the actions of their anarchic hand but do not attribute its intentional behavior to themselves (it is not "owned" by them) (Della Sala 2006): the anarchic hand is not present to the patient who owns it;
- *Hemispatial Neglect (location)*: it is a neurological disease characterized by a deficit in attention to and awareness of one side of space. For example, a stroke affecting the right parietal lobe of the brain can lead to neglect for the left side of the visual field, causing a patient with neglect to behave as if the left side of sensory space is nonexistent: a patient with left neglect will not be present in the left part of a room.

Recently, different authors showed that is possible to induce an illusory perception of a fake limb [31] as part of our own body, by altering the normal association between touch and its visual correlate. It is even possible to generate a body transfer illusion [31]: Slater and colleagues substituted the experience of male subjects' own bodies with a life-sized virtual human female body. This was demonstrated subjectively by questionnaire and physiologically through heart-rate deceleration in response to a threat to the virtual body [31].

5. The third feature of presence: we use it to monitor our actions

As we have seen before, Lombard and Ditton defined presence as the *"perceptual illusion of non-mediation"* [10] linking it to the experience of a medium:

"An illusion of nonmediation occurs when a person fails to perceive or acknowledge the existence of a medium in his/her communication environment and responds as he/she would if the medium were not there. ... Presence in this view cannot occur unless a person is using a medium."

However, in the previous paragraph we suggested that the outcome of many recent neurological studies considers the body as the first medium, through which we articulate ourselves and engage with others. More, recent studies on peripersonal space demonstrated that tool-mediated actions modify the multisensory coding of near peripersonal space [32]: the active use of a tool for physically and effectively interact with objects in the distant space appears to produce a spatial extension of the multisensory peri-hand space corresponding to the whole length of the tool. *In other words, through the successful enaction of the subject's intentions using the tool, he/she becomes physically present in the tool [33].*

These studies confirm that the subject locates himself/herself in an external space according to the action he/she can do in it. As suggested by Zahoric and Jenison [19]: *"presence is tantamount to successfully supported action in the environment"* (p. 87, italics in the original). In sum, the subject is *"present"* in a space if he/she can act in it. More, the subject is *"present"* in the space – real or virtual – where he/she can act in. Interestingly, what we need for presence are both the affordance for action (the possibility of acting) and its enaction (the possibility of successfully acting).

The first suggestion this framework offers to the developers of virtual worlds, is that for presence action is more important than perception [34]: I'm more present in a perceptually poor virtual environment (e.g. a textual MUD) where I can act in many different ways than in a real-like virtual environment where I cannot do anything.

Another consequence of this framework is the need to understand more what "acting successfully" means. We can start from the definition of "Agency": "the power to alter at will one's perceptual inputs" [35]. But how can we define our will? A simple answer to this question is: through intentions. Following this line of reasoning *Presence can be defined as "the non mediated (prereflexive) perception of using the body to successfully transforming intentions in action (enaction)"*

A possible criticism to this definition is the following: "I may be asked to repair a computer, and I may be unable to fix it. This does not mean that I am not present in the environment (real or virtual) where the computer and I are." This objection makes sense if we use the folk psychology definition of intention: the intention of an agent performing an action is his/her specific purpose in doing so. However, the latest cognitive studies clearly show that any behavior is the result of a complex intentional chain that cannot be analyzed at a single level [36].

According to the *Dynamic Theory of Intentions* presented by Pacherie [36; 37] and to the *Activity Theory* introduced by Leont'ev and disseminated by Kaptelinin, & Nardi [38], repairing a computer is driven by an above objective (e.g., obtaining the money for paying a new car) and is the result of lower-level operations (e.g., removing the hard disk or the CPU, cleaning them, etc.) each driven by specific purposes. So, for an intention that failed (repairing the computer) many others were successful (removing the hard disk, cleaning it, etc.) inducing Presence [33; 39].

Specifically, the *Dynamic Theory of Intentions* identifies three different "levels" or "forms" of intentions (Figure 2), characterized by different roles and contents: distal intentions (D-intentions), proximal intentions (P-intentions) and motor intentions (M-intentions):

- *D-intentions (Future-directed intentions).* These high-level intentions act both as intra- and interpersonal coordinators, and as prompters of practical reasoning about means and plans: in the activity "obtaining a Ph.D. in psychology" described in Figure 2, "helping others to solve problems" is a D-intention, the object that drives the activity of the subject.
- *P-intentions (Present-directed intentions).* These intentions are responsible for high-level (conscious) forms of guidance and monitoring. They have to ensure that the imagined actions become current through situational control of their unfolding: in the activity described in Figure 1, "preparing the dissertation" is a P-intention.
- *M-intentions (Motor intentions).* These intentions are responsible for low-level (unconscious and intuitive) forms of guidance and monitoring: we may not be aware of them and have only partial access to their content. Further, their contents are not propositional: in the activity described in Figure 2, the motor representations required to write using the keyboard are M-intentions.

Any intentional level has its own role: the rational (D-intentions), situational (P-Intention) and motor (M-Intention) guidance and control of action. They form an intentional cascade [36; 37] in which higher intentions generate lower intentions. In this view the ability to feel "present" in a virtual reality system – a medium - basically does not differ from the ability to feel "present" in our body. When the subject is present during agency – he/she is able to successfully enact his/her intentions – he/she locates him/herself in the physical and cultural space in which the action occurs.

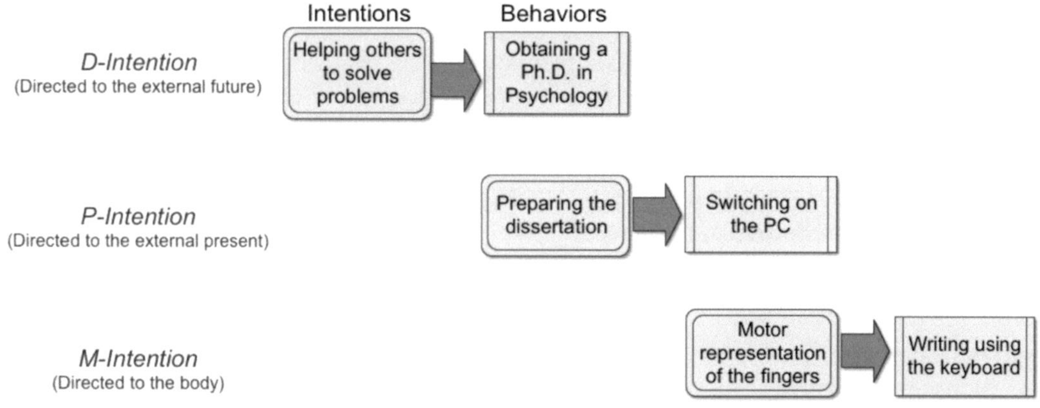

Figure 1. Intentional levels

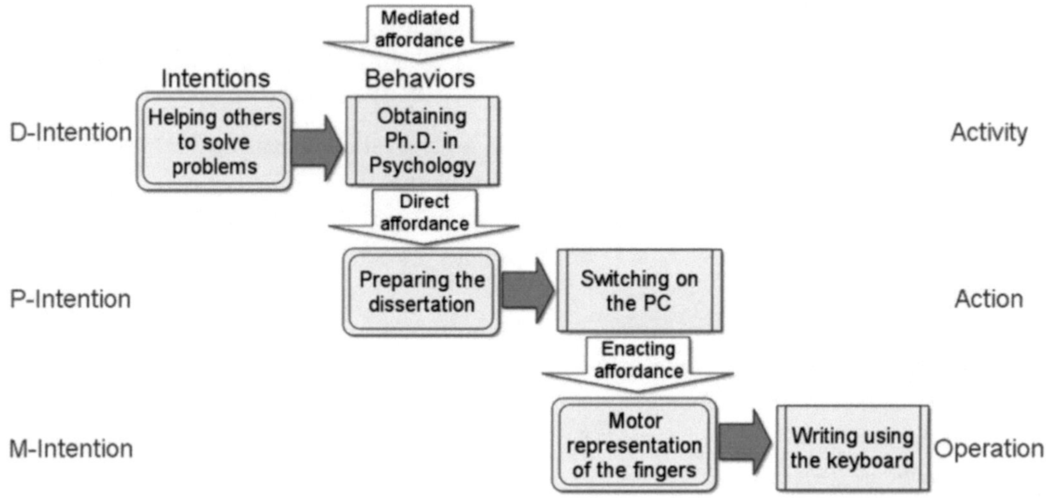

Figure 2. The intentional chain

More, it also suggest that even in the real world the feeling of presence will be different according to the ability of the subject to enact his/her intentions within an external environment. For instance, I'm in a restaurant for a formal dinner with some colleagues in a Korean restaurant, but I don't know how to use the chopsticks I have nearby my dish. In this situation I'm physically there, but the lack of knowledge puts me outside, at least partially, from the social and cultural space of the "formal Korean dinner". The result is a reduced presence and a limitation in my agency: I'm not able to enact my intention (pick up some rice) using the chopsticks, so I don't use them to avoid mistakes.

Finally, in this view presence can be described as a sophisticated but unconscious form of monitoring of action and experience: the self perceives the variations in the feeling of presence and tunes its activity accordingly. From a computational viewpoint, the experience of Presence is achieved through a forward-inverse model [40] (Figure 3):

- First, the agent produces the motor command for achieving a desired state given the current state of the system and the current state of the environment;
- Second, an efference copy of the motor command is fed to a forward dynamic model that generates a prediction of the consequences of performing this motor command;
- Third, the predicted state is compared with the actual sensory feedback. Errors derived from the difference between the desired state and the actual state can be used to update the model and improve performance.

The results of the comparison between the sensory prediction and the sensory consequences of the act (an intuitive process occurring at a sub-personal level) can then be utilized to determine both the agent of the action and to track any possible variation in its course. If no variations are perceived, the self is able to concentrate on the action and not on its monitoring. As suggested by *the simulation theorists* [41], the brain instantiates a sophisticated simulation, based on motor codes, of the outcome of an action and uses this to evaluate its course.

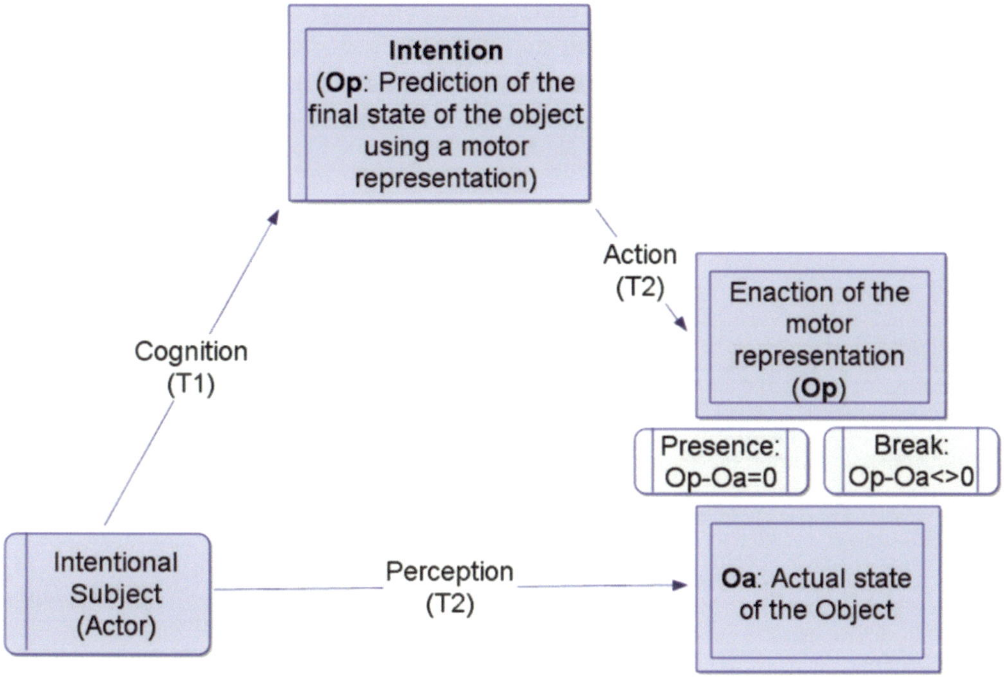

Figure 3. The feeling of presence

For this reason, the feeling of presence – *the prereflexive perception that the agent's intentions are successfully enacted* – is not separated by the experience of the subject but *is directly related to it*. It corresponds to what Heidegger [42] defined as "the interrupted moment of our habitual standard, comfortable *being-in-the-world*". A higher feeling of presence is experienced by the self as a better quality of action and experience [19]. In fact, the subject perceives consciously only *significant variations* in the feeling of presence: *breakdowns* and *optimal experiences* [43]. We will discuss more in detail this point in Paragraph 10.

6. The fourth feature of presence: it is divided in three layers

Even if presence is a unitary feeling, on the process side it can be divided into three different layers/subprocesses [44; 45], phylogenetically different, that correspond reasonably well (see Figure 4) to the three levels of intentions identified by Pacherie in her *Dynamic Theory of Intentions* [36]:

- *Proto Presence* (Self vs. non Self – M-Intentions);
- *Core Presence* (Self vs. present external world – P-Intentions);
- *Extended Presence* (Self vs. possible/future external world – D-Intentions).

We define "*Proto Presence*" as the process of internal/external separation *related to the level of perception-action coupling (Self vs. non-Self)*. The more the organism is able correctly to couple perceptions and movements, the more it differentiates itself from the external world, thus increasing its probability of surviving. Proto presence is based on proprioception and other ways of knowing bodily orientation in the world. In a virtual world this is sometimes known as "spatial presence" and requires the tracking of body parts and appropriate and rapid updating of displays, for example in response to head movements. Proto Presence allows the enaction of M-Intentions only.

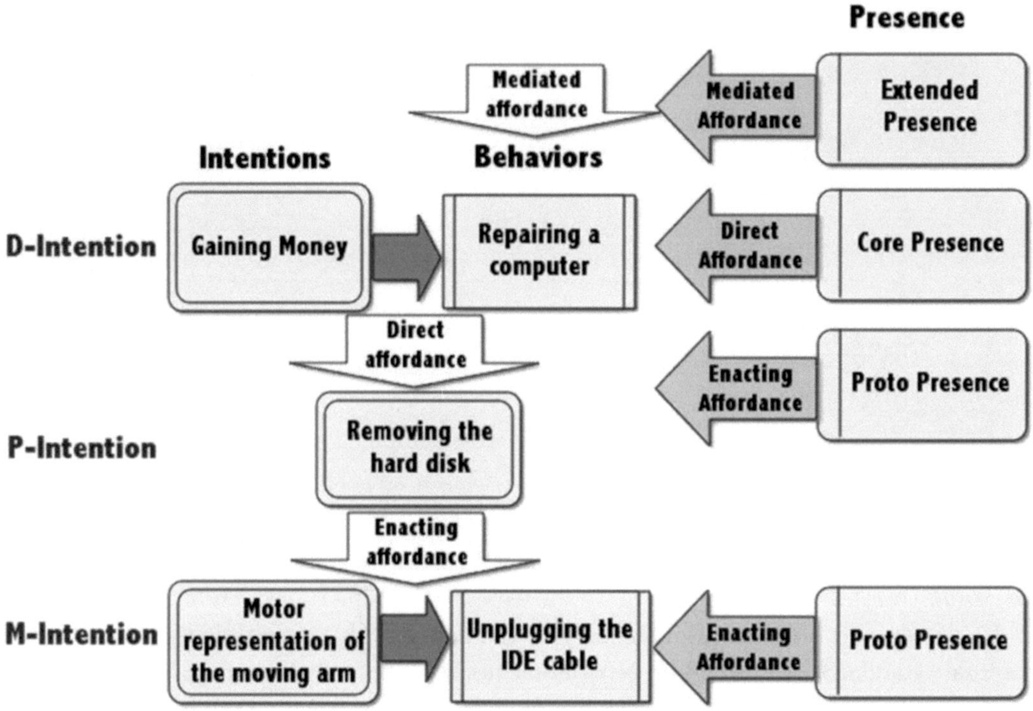

Figure 4. The layers of presence

"Core Presence" can be described as *the activity of selective attention made by the Self on perceptions (Self vs. present external world)*: the more the organism is able to focus on its sensorial experience by leaving in the background the remaining neural processes, the more it is able to identify events of the present moment and the direct affordances offered by the current external world, increasing its probability of surviving. Core Presence allows the enaction of M-Intentions and P-Intentions only. Core presence in media is based largely on vividness of perceptible displays. This is equivalent to "sensory presence" and requires good quality, preferably stereographic, graphics and other displays.

The role of "Extended Presence" is to *verify the relevance to the Self of possible/future events in the external world (Self vs. possible/future external world)*. The more the Self is able to forecast possible/future experiences, the more it will be able to identify relevant ones, increasing the possibility of surviving. Extended presence allows the enaction of M-Intentions, P-Intentions and D-Intentions. Following Sperber and Wilson's approach [46], an input is relevant when its processing yields a positive cognitive effect, a worthwhile difference to the Self's representation of the world. Extended Presence requires intellectually and/or emotionally significant content. So, reality judgment influences the level of extended presence - a real event is more relevant than a fictitious one.

As underlined by Dillon and colleagues [47], converging lines of evidence from different perspectives and methodologies support this three-layered view of Presence. In their analysis they identify three dimensions common to all the different perspectives, relating to a "spatial" dimension (M-intentions), a dimension relating to how consistent the media experience is with the real world, "naturalness" (P-intentions), and an "engagement" dimension (D-intentions). This view has two main consequences [11; 33].

On one side, the role of the different layers will be related to the complexity of the activity: the more complex is the activity, the more layers will be needed to produce a high level of Presence (Figure 4). At the lower level – motor intention (e.g., grasping a ball) – proto Presence is enough to induce a satisfying feeling of Presence. At the higher level – distal intention (e.g., improving stress management) – the media experience has to support all three layers (e.g., allowing movement, *proto presence*; allowing interaction with the environment, *core presence*; giving a sense to the experience, *extended presence*).

On the other side, subjects with different intentions will not experience the same level of Presence, even when immersed in the same virtual environment [13]: this means that understanding and supporting the intentions of the user will improve his/her Presence in a virtual world. More, maximal Presence is achieved when the environment is able to support the full intentional chain of the user.

7. Presence and clinical change

The use of virtual reality (VR) in clinical psychology has become more widespread [48]. The key characteristics of virtual environments for most clinical applications are the high level of control of the interaction with the tool, and the enriched experience provided to the patient [2]. Typically, in VR the patient learns to cope with problematic situations related to his/her problem. For this reason, the most common application of VR in this area is the treatment of anxiety disorders, i.e., fear of heights, fear of flying, and fear of public speaking [49; 50]. Indeed, VR exposure therapy (VRE) has been proposed as a new medium for exposure therapy [48] that is safer, less embarrassing, and less costly than reproducing the real world situations. The rationale is simple: in VR the patient is intentionally confronted with the feared stimuli while allowing the anxiety to attenuate. Avoiding a dreaded situation reinforces a phobia, and each successive exposure to it reduces the anxiety through the processes of habituation and extinction.

However, it seems likely that VR can be more than a tool to provide exposure and desensitisation [48]. As noted by Glantz and colleagues [51]: *"VR technology may create enough capabilities to profoundly influence the shape of therapy." (p.92).* Emerging applications of VR in psychotherapy include eating disorders and obesity (see Figure 5) [52-54], posttraumatic stress disorder [55], addictions [56], sexual disorders [57], and pain management [58]. But what is the potential role of presence in these treatments?

Figure 5. The use of VR in the treatment of Obesity: A phase of the therapy (*left*) and a screen-shot of the virtual environment (*right*)

To answer this question let's start from another question: How is it possible to achieve the desired change in a patient? This question has many possible answers according to the specific psychotherapeutic approach; however, in general, change occurs through an intense focus on a particular instance or experience [59]. By exploring this experience as thoroughly as possible, the patient can relive all of the significant elements associated with it (i.e., conceptual, emotional, motivational, and behavioral) and make them available for reorganization. Within this general model there exist many specific methods, including the insight-based approach of psychoanalysis, the schema-reorganization goals of cognitive therapy, the functional analysis of behavioral activation, the interpersonal relationship focus of interpersonal therapy, and the enhancement of experience awareness in experiential therapies.

What are the differences between them? According to Safran and Greenberg [60], behind the specific therapeutic approach there are two different models of change: bottom-up and top-down. Bottom-up processing begins with a specific emotional experience and leads eventually to change at the behavioral and conceptual level; top-down change usually involves exploring and challenging tacit rules and beliefs that guide the processing of emotional experience and behavioral planning. These two models of change are focused on the two different cognitive systems – intuition and reasoning – we discussed in Paragraph 4

Even if many therapeutic approaches are based on just one of the two change models, a therapist usually requires both [59]. Some patients seem to operate primarily by means of top-down information processing, which may then lead the way to corrective emotional experiences. For others, the appropriate access point is the intensification of their emotional experience and their awareness of both it and its related behaviors. Finally, different patients who initially engage the therapeutic work through top-down processing only may be able to make use of bottom-up emotional processing later in the therapy.

In this situation, the sense of presence provided by advanced technologies, VR in particular, offers a critical advantage [61]: used appropriately, it is possible to target a specific cognitive system without any significant change in the therapeutic approach. For instance, behavioral therapists may use a virtual environment for activating the fear structure in a phobic patient through confrontation with the feared stimuli; a cognitive therapist may use VR situations to assess situational memories or disrupt habitual patterns of selective attention; experiential therapists may use VR to isolate the patient from the external world and help him/her in practicing the right actions; psychodynamic therapists may use VEs as complex symbolic systems for evoking and releasing effects.

In fact, VR can be described as an *advanced imaginal system*: an experiential form of imagery that is as effective as reality in inducing emotional responses [62]. As underlined by Baños, Botella & Perpiña [63], the VR experience can help the course of therapy for "its capability of reducing the distinction between the computer's reality and the conventional reality." In fact, "VR can be used for experiencing different identities and… even other forms of self, as well" (p. 289). The possibility of structuring a large amount of realistic or imaginary controlled stimuli and, simultaneously, of monitoring the possible responses generated by the user of the technology offers a considerable increase in the likelihood of therapeutic effectiveness, as compared to traditional procedures [64].

More, As noted by Glantz and colleagues [51]:

"One reason it is so difficult to get people to update their assumptions is that change often requires a prior step – recognizing the distinction between an assumption and a perception. Until revealed to be fallacious, assumptions constitute the world; they seem like perceptions, and as long as they do, they are resistant to change." (p. 96).

Using the sense of presence induced by VR, it is easier for the therapist to develop realistic experiences demonstrating to the patient that what looks like a perception – e.g., the body image distortion – in fact is a result of his/her mind. Once this has been understood, individual maladaptive assumptions can then be challenged more easily.

However, as noted by Price and Anderson [65] presence is not enough to produce a clinical change. The two authors explored the relation between presence, anxiety, and treatment outcome in a clinical study that used a virtual airplane to treat individuals with fear of flying. The results support presence as a conduit that enabled phobic anxiety to be expressed during exposure to a virtual environment. Nevertheless, presence was not supported as contributing to treatment outcome: feeling present during exposure may be necessary but not sufficient to achieve benefit from VR therapy. These results echoed findings from Krinj and colleagues [66], who compared the efficacy of a highly immersive CAVE-like system and the less immersive but more affordable HMD technology. They reported more presence and more anxiety in the CAVE system, but no difference in treatment outcome.

To better understand the possible link between presence and clinical change in the next two paragraphs we will explore the connections between presence, emotions, optimal experiences and therapy.

8. Presence for clinical change: The role of emotions

One of the most important effects of presence for clinical practice is that a virtual experience may evoke the same reactions and emotions as a real experience. For instance, Slater and colleagues [67] used VR to reproduce the Stanley Milgram's 1960s experimental approach: the participants were invited to administer a series of word association memory tests to a female virtual human (avatar) representing the stranger; when the avatar gave an incorrect answer, the participants were instructed to administer an "electric shock" to her, increasing the voltage each time; the avatar then responded with increasing discomfort and protests, eventually demanding termination of the experiment.

Their results show that in spite of the fact that all participants knew for sure that neither the avatar nor the shocks were real, the participants who saw and heard the female virtual human tended to respond to the situation at the subjective, behavioral and physiological levels as if it was real. As noted by the researchers [67]:

"In the debriefing interviews many said that they were surprised by their own responses, and all said that it had produced negative feelings – for some this was a direct feeling, in others it was mediated through a 'what if it were real?' feeling. Others said that they continually had to reassure themselves that nothing was really happening, and it was only on that basis that they could continue giving the shocks."

Experimental manipulations of emotions and presence have been conducted. Bouchard and colleagues [68] immersed adults suffering from snake phobia to a virtual environment where anxiety was experimentally induced, or not, by manipulating the apprehension of the participants and keeping the content of the immersions identical. Using a single-item measure of presence, the results showed that presence was significantly higher when participants were anxious during the immersion than in the baseline or the non-anxious immersion.

Baños, Botella, Guerrero, Liaño. Alcañiz, & Rey [69] compared the sense of presence between virtual and imaginary environments. Participants were randomly assigned to one of the two conditions (imagined versus virtual spaces) and the subjective sense of presence was measured in three moments (beginning, middle, and end). Results shown that the participants in "imagery" spaces indicated a decrease of their sense of presence, whereas the opposite occurs in participants in "virtual" spaces.

Michaud et al. [70] experimentally manipulated presence in a sample of heights phobics who had to take an elevator and perform tasks on a scaffold outside of a 15-story building. When the immersion in the virtual environment was conducted in a high-presence setting, the level of anxiety was significantly higher than when the immersion was conducted in a low-presence setting.

Riva and colleagues [71] also analyzed the possible use of VR as an affective medium focusing on the relationship between presence and emotions. Their data showed a circular interaction between presence and emotions: on one side, the feeling of presence was greater in the "emotional" environments; on the other side, the emotional state was influenced by the level of presence. Taken together these results, in agreement with the model presented before, underline the existence of a bi-directional relationship between presence and emotions.

First, the higher the presence, the higher intensity of emotions the user experiences. Therefore, if the focus is on designing applications capable of eliciting emotions with the goal of reducing or modifying them (as in psychological therapy), the environments must be able to induce a high feeling of presence through a full support to the intentions of the user.

However, the opposite could also be claimed: the higher the intensity of the emotions and feelings, the higher the presence and reality judgment. From this point of view, the focus for psychological treatment would lie on designing relevant environments, providing intellectually and/or emotionally significant content for the specific sample involved in the treatment. For instance, a recent study by Gorini and colleagues [72] comparing a sample of 20 Mexican participants - 8 living in El Tepeyac, a small rural and isolated Mexican village characterized by a very primitive culture, and 12 high civilized inhabitants of Mexico City - clearly showed that VR exposure to a relaxing environment has different physiological and psychological effects according to the cultural and technological background of the users.

A study by Bouchard et al. [73] studied presence using a virtual environment designed to treat specific phobias (musophobia) with VR. Participants in both conditions were immersed in the same VE containing a rodent, yet in one condition they were deceived and led to believe that they were actually being immersed in real time in the physical room with the rodent. The deception used a blend of mixed videoconference-VR technologies, display of high-tech hardware relaying the videoconference and the VR computers, and false instructions stating that they were "currently live in the real room" or that they were "seeing a fake 3D copy of a room". Presence was significantly higher when participants were told they were seeing the "real" room that was being projected in the head-mounted display in real time [73]. This study confirms the possibility of manipulating presence without changing any objective properties of the VE.

9. Presence for clinical change: The role of optimal experiences

In Paragraph 6 we discussed a critical feature of presence: it *provides the self with a feedback about the status of its activity*. Specifically, the self perceives the variations in the feeling of presence (*breakdowns* and *optimal experience*) and tunes its activity accordingly [43].

Winograd and Flores [74] refer to presence disruptions as *breakdowns: a breakdown* occurs when, during our activity, an aspect of our environment that we usually take for granted becomes part of our consciousness. If this happens, we shift our attention from action to the object or environment to cope with it. To illustrate, imagine sitting outdoors engrossed in reading a book on a pleasant evening. As the sun sets and the light diminishes one continues reading, engrossed in the story until one becomes aware that the light is no longer suitable for reading. In such conditions, before any overt change in behavior, what we experience is a breakdown in reading and a shift of attention from the book to the light illuminating the book.

It is interesting to consider why we experience these breakdowns. Our hypothesis is that breakdowns are a sophisticated evolutionary tool used to control the quality of experience that ultimately enhances our chances of survival [13; 44]. As a breakdown occurs we experience a lower level of presence. This reduces the quality of experience, and leads us to confront environmental difficulties through an attentive shift.

On the other side we have optimal experiences. According to Csikszentmihalyi [75], individuals preferentially engage in opportunities for action associated with a positive, complex and rewarding state of consciousness, defined as "optimal experience", or "flow." There are some exceptional situations in real life in which the activity of the subject is characterized by a higher level of presence. In these situations the subject experiences a full sense of control and immersion. When this experience is associated to a positive emotional state, it can create a flow state.

An example of flow is the case where a professional athlete is playing exceptionally well (positive emotion) and achieves a state of mind where nothing else matters but the game (high level of presence). For Ghani and Deshpande [76] the two main characteristics of flow are (a) the total concentration in an activity and (b) the enjoyment which one derives from the activity. Moreover, these authors identified two other factors affecting the experience of flow: a sense of control over one's environment and the level of challenge relative to a certain skill level.

Following this vision, it is possible to design mediated situations that elicit optimal experiences by activating a high level of presence [77]. Optimal experiences promote individual development. As underlined by Massimini and Delle Fave, [78]:

"To replicate it, a person will search for increasingly complex challenges in the associated activities and will improve his or her skill, accordingly. This process has been defined as cultivation; it fosters the growth of complexity not only in the performance of flow activities but in individual behavior as a whole." (p. 28).

According to this vision, existing VR treatments should include positive peak experiences because they serve as triggers for a broader process of motivation and empowerment. Within this context, the *transformation of flow* can be defined as a person's ability to draw upon an optimal experience and use it to marshal new and unexpected psychological resources and sources of involvement. We hypothesize that it is possible to use VR to activate a transformation of flow to be used for clinical purposes [77]. The proposed approach is the following: first, identify an enriched environment that contains *functional* real-world demands; second, using the technology to enhance the level of presence of the subject in the environment and to induce an optimal experience; third, allowing cultivation, by linking this optimal experience to the actual experience of the subject.

To verify the link between advanced technologies and optimal experiences, the "V-STORE Project" investigated the quality of experience and the feeling of presence in a group of 10 patients with Frontal Lobe Syndrome involved in VR-based cognitive rehabilitation [79].

On one side, the project used the Experience Sampling Method for repeated on-line assessments of the external situation and the emotional, cognitive and motivational components of daily experience during one-week of these patients, including traditional cognitive rehabilitation and sessions of exposure to V-STORE VR environment.

On the other side, after the VR experience they used the ITC-Sense of Presence Inventory to evaluate the feeling of presence induced by the VR sessions. Findings highlighted the association of VR sessions with both positive affect and a high level of presence. In particular, during the VR sessions, the "spatial presence," the first scale of the ITC-Sense of Presence Inventory, was significantly correlated with the positive psychological feelings of "being free" ($r = 0.81$, $p < 0.01$) and "being relaxed" ($r = 0.67$, $p < 0.05$).

The transformation of flow may also exploit the plasticity of the brain producing some form of functional reorganization [80]. Recent experimental results from the work of Hunter Hoffman and his group in the treatment of chronic pain [81] also might be considered to foster this vision. Few experiences are more intense than the pain associated with severe burn injuries. In particular, daily wound care - the cleaning and removal of dead tissue to prevent infection - can be so painful that even the aggressive use of opioids (morphine-related analgesics) cannot control the pain.

However it is well known that distraction - for example, by having the patient listen to music - can help to reduce pain for some people. Hoffman and colleagues conducted a controlled study of the efficacy of VR as an advanced distraction by comparing it with a popular Nintendo video game. The results showed dramatic reductions in pain ratings during VR compared to the video game [82].

Further, using a functional magnetic resonance imaging scanner they measured pain-related brain activity for each participant during conditions of virtual reality and without virtual reality in an order randomized study [81]. The team studied five regions of the brain that are known to be associated with pain processing, the anterior cingulate cortex, primary and secondary somatosensory cortex, insula, and thalamus. They found that during VR the activity in all the regions showed significant reductions. In particular, they found direct modulation of pain responses within the brain during VR distraction. The degree of reduction in pain-related brain activity ranged from 50 percent to 97 percent.

10. Presence for clinical change: The neuroVR software

Although it is undisputable tha potential of VR – as presence inducing technology - for clinical and research applications, the majority of existing clinical virtual environments are still in the laboratory or investigation stage. In a review, Riva [48] identified four major issues that limit the use of VR in clinical practive:

- the lack of standardization in VR hardware and software, and the limited possibility of tailoring the virtual environments (VEs) to the specific requirements of the clinical or the experimental setting;
- the low availability of standardized protocols that can be shared by the community of researchers;
- the high costs (up to 200,000 US$) required for designing and testing a clinical VR application;
- most VEs in use today are not user-friendly; expensive technical support or continual maintenance are often required.

To help researchers to overcome these issues and to develop VR applications able to exploit the clinical potential of presence, Riva and colleagues presented at the Medicine Meets Virtual Reality conference in 2007 a free virtual reality platform based on open-source software [83]: NeuroVR (http://www.neurovr.org). This software allows non-expert users to adapt the content of different pre-designed virtual environments to the specific needs of the clinical or experimental setting. Following the feedbacks of the thousands of users who downloaded the first version, they developed in late 2011 a new version – NeuroVR 2 – that improves the possibility for the therapist to enhance the patient's feeling of familiarity and intimacy with the virtual scene, by using external sounds, photos or videos [84].

In NeuroVR 2, the user can choose the appropriate psychological stimuli/stressors from a database of objects (both 2D and 3D) and videos, and easily place them into the virtual environment. The edited scene can then be visualized in the Player using either immersive or non-immersive displays. Currently, the NeuroVR library includes 18 different virtual scenes (apartment, office, square, supermarket, park, classroom, etc.), covering some of the most studied clinical applications of VR: specific phobias, cognitive rehabilitation, panic disorders and eating disorders.

The VR suite leverages two major open-source projects in the VR field: Delta3D (http://www.delta3d.org) and OpenSceneGraph (http:// www.openscenegraph.org). Both are building components that integrates with ad-hoc code to handle the editing and the simulation. The NeuroVR2 Editor's GUI (see Figure 6) is now based on the QT cross-platform application and UI framework from Nokia (http://qt.nokia.com/) that grants an higher level of editing and customization over the editor functionalities, while the graphical rendering is done using OpenSceneGraph, an open source high performance 3D graphics toolkit (http://www.openscenegraph.org/projects/osg).

The new features include advanced action triggering based both on user behavior (proximity and collision) and on therapist choice (keyboard), realistic walk-style motion, advanced lighting techniques for enhanced image quality, and streaming of videos using alpha channel for transparency.

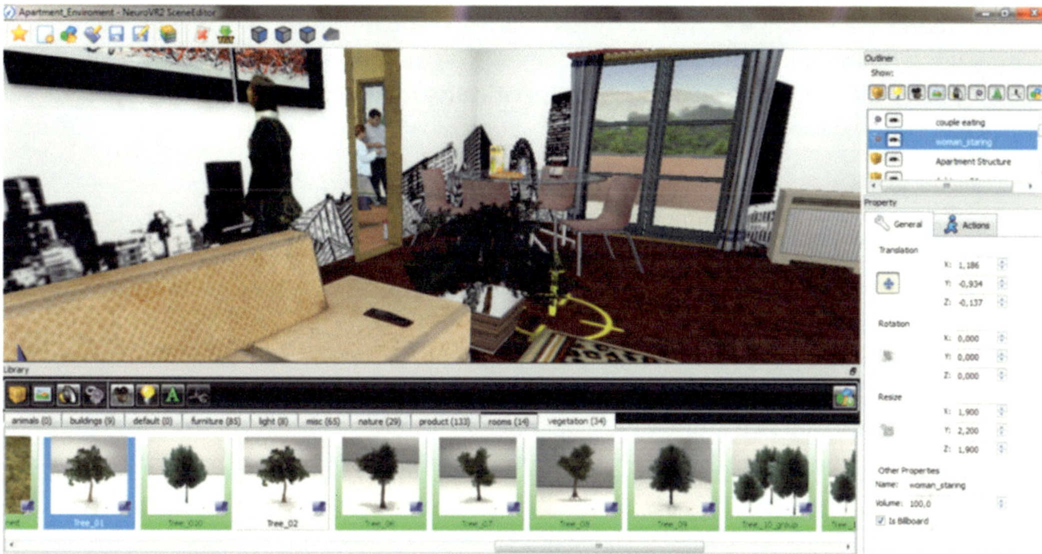

Figure 6. The NeuroVR 2 Editor

The NeuroVR 2 Player too has been largely rewritten to grant a more efficient workflow for the scenes playback and has a brand new startup interface written in QT. The whole suite is developed in C++ language, targeted for the Microsoft Windows platform but fully portable to other systems if needed.

11. Presence for clinical change: The interreality paradigm

Even if virtual reality is a very good presence-inducing technology, there is still room for improvement. Apparently, the main limitation of its actual use in clinical applications is the distance from reality [85]: the virtual experience is a distinct realm, separate from the emotions and behaviors experienced by the patient in the real world, In other words, the behavior of the patient in VR has no direct effects on the real life experience. More, the emotions and problems experienced by the patient in the real world are not directly addressed in the VR exposure [85].

To address this issue recently Fidopiastis and colleagues suggested the use of mixed reality (MR) [86]. The use of MR in clinical psychology is not new. Cristina Botella and her team used it for the treatment of small animal phobias [87]. The main advantage of this approach is that in MR virtual object are integrated into the real world: during the therapy the patient is seeing a real-world scene, and a series of computer-generated objects that, at that same moment, are super-imposed on the real physical environment. As noted by Botella and colleagues [87], this approach offers other advantages, too: it facilitates the experience of presence (the feeling of being there), and reality judgment (the fact of judging the experience as real) since the environment the patient is seeing is, in fact the "reality".

In this chapter we suggest that a further advancement might be offered by a new technological paradigm, Interreality: an hybrid, closed-loop empowering experience bridging physical and virtual worlds [88]. Specifically, the Interreality approach provides a twofold feedback activity:

- *behavior in the physical world influences the experience in the virtual world:*
  - For example, if the emotional regulation during the day was poor, some new experiences in the virtual world will be unlocked to address this issue.
  - For example, if the emotional regulation was okay, the virtual experience will focus on a different issue.
- *behavior in the virtual world influences the experience in the real world:*
  - For example, if I participate in the virtual support group I can send text messages during the day to the other participants.
  - For example, if my coping skills in the virtual world were poor, the decision support system will increase the chance of possible warnings in real life and will provide additional homework assignments.

On one side, the patient is continuously assessed in the virtual and real worlds by tracking the behavioral and emotional status in the context of challenging tasks (*customization of the therapy according to the characteristics of the patient*). On the other side, feedback is continuously provided to improve both the appraisal and the coping skills of the patient through a conditioned association between effective performance state and task execution behaviors (*improvement of self efficacy*).

Our claim is that bridging virtual experiences – fully controlled by the therapist, used to learn coping skills and emotional regulation - with real experiences – that allow both the identification of any critical problem and the assessment of what has been learned – using advanced technologies (virtual worlds, advanced sensors and PDA/mobile phones) is a feasible way to address the above limitations. This approach may offer the following innovations to current VR and/or MR protocols:

- *objective and quantitative assessment of symptoms using biosensors and behavioral analysis*: monitoring of the patient behavior and of his general and psychological status, early detection of symptoms of critical evolutions and timely activation of feedback in a closed-loop approach;
- *decision support for treatment planning through data fusion and detection algorithms:* monitoring of the response of the patient to the treatment, management of the treatment and support to the clinicians in their therapeutic decisions.
- *provision of warnings and motivating feedback to improve compliance and long-term outcome:* the sense of "presence" allowed by this approach affords the opportunity to deliver behavioral, emotional and physiological self-regulation training in an entertaining and motivating fashion.

For example, in the standard VR protocol used in the treatment of post-traumatic stress disorders [89] "*imagination and/or exposure evoke emotions and the meaning of the associated feelings can be changed through reflection and relaxation"*. We would suggest as an alternative that *"controlled experience evokes emotions that result in meaningful new feelings which can be reflected upon and eventually changed through reflection and relaxation"*.

More, if the typical VR exposure protocol focuses on directly modifying the content of dysfunctional thoughts through a rational and deliberate process, Interreality focuses on modifying the patient's relationship with his or her thinking through more contextualized experiential processes.

In conclusion we argue that the potential advantages offered to VR treatments by the Interreallity approach are:

- *a real-time feedback between physical and virtual worlds*: Interreality uses bio and activity sensors and devices (PDAs, mobile phones, etc) both to track in real time the behavior and the health status of the user and to provide targeted suggestions and guidelines.
- *an extended sense of community*: Interreality uses hybrid social interaction and dynamics of group sessions to provide each users with targeted – but also anonymous, if required - social support in both physical and virtual world.
- *an extended sense of presence*: Interreality uses advanced simulations (virtual experiences) to transform health guidelines and provisions in experience. In Interreality the patients do not receive abstract info but live meaningful experiences.

12. Conclusions

As explained previously, the feeling of presence induced by VR has helped this medium to find a significant space in clinical treatment. In particular, VR is playing an important role as a presence-enhanced supportive technique. Through presence, VR helps the patient to confront his/her problems in a meaningful yet controlled and safe setting. Furthermore, it opens the possibility of experiencing his/her life in a more satisfying way. In fact, VR therapists are using presence to provide meaningful experiences capable of inducing deep and permanent change in their patients. But what is presence? And how it can be used to improve the process of clinical change?

The International Society of Presence Research, defines "Presence" (a shortened version of the term "telepresence") as a "psychological state in which even though part or all of an individual's current experience is generated by and/or filtered through human-made technology, part or all of the individual's perception fails to accurately acknowledge the role of the technology in the experience". This approach describes the sense of presence as "Media Presence", a function of our experience of a given medium. The main outcome of this approach is the *"perceptual illusion of non-mediation"* [10] definition of presence. Following it, presence is produced by means of the disappearance of the medium from the conscious attention of the subject.

The main advantage of this approach is its predictive value: the level of presence is reduced by the experience of mediation during the action. The main limitation of this vision is what is not said. What is presence for? Is it a specific cognitive process? What is its role in our daily experience?

To address these questions a second group of researchers, including the authors of this chapter, considers presence as "Inner Presence", the feeling of being located in a perceived external world around the self [11-13]. In this view presence is broad psychological phenomenon, not necessarily linked to the experience of a medium, whose goal is the control of the individual and social activity.

In the chapter we used the following three research outcomes emerging from the recent work of cognitive sciences to build a cognitive theory of presence:

1. *Cognitive processes can be either rational or intuitive*: we showed that presence is an intuitive feeling tproduced by an experience-based metacognitive judgment;
2. *Skills become intuitive when our brain is able to simulate their outcome*: we suggested that presence monitors intuitively our activity processes using embodied simulations;
3. *Space is perceived in terms of the actions we could take towards them*: we argued that the feeling of presence in a real or virtual space is directly correlated to the outcome of the actions the subject can enact in it.

In sum, the feeling of presence can be described as the product of an intuitive experience-based metacognitive judgment related to the enaction of our intentions: We are present in an environment - real and/or synthetic - when we are able, inside it, to intuitively transform our intentions in actions.

From a clinical viewpoint presence transforms VR in an "empowering environment", a special, sheltered setting where patients can start to explore and act without feeling threatened [3]. Nothing the patient fears can "really" happen to them in VR. With such assurance, they can freely explore, experiment, feel, live, and experience feelings and/or thoughts. VR thus becomes a very useful intermediate step between the therapist's office and the real world. In other words, the key feature of VR for clinical goals is that it offers an effective support to the activity of the subject by activating a high sense of "presence", the feeling of being *inside* the virtual world.

However, as noted by Price and Anderson [65] presence is not enough to produce a clinical change: feeling present during VR exposure is necessary but not sufficient to achieve benefit from VR therapy. For this reason, in the last two paragraphs we explored the links between presence, emotions and optimal experiences.

First, the higher the presence, the higher is the intensity of emotions experienced by the user. Therefore, if the focus is on designing applications capable of eliciting emotions with the goal of reducing or modifying them (as in psychological therapy), the environments must be able to induce a high feeling of presence through a full support to the intentions of the user. However, the opposite could also be claimed: the higher the intensity of the emotions and feelings, the higher the presence and reality judgment. From this point of view, the focus for psychological treatment would lie on designing relevant environments, providing intellectually and/or emotionally significant content for the specific sample involved in the treatment.

Second, the higher the presence, the optimal is the experience for the user. Following this vision, it is possible to design mediated situations that elicit optimal experiences by activating a high level of presence. More, given the link between optimal experiences and individual development, VR treatments should promote positive peak experiences because they serve as triggers for a broader process of motivation and empowerment. The proposed approach is the following: first, develop a VR environment that contains *functional* real-world demands; second, use the technology to enhance the level of presence of the subject in the environment and to induce an optimal experience; third, allow cultivation, by linking this optimal experience to the actual experience of the subject.

To help therapists and researchers to test these ideas we provided two further suggestions.

On one side we introduced NeuroVR (http://www.neurovr.org). This software, that reached version 2, allows non-expert users to adapt the content different pre-designed virtual environments to the specific needs of the clinical or experimental setting. Using the software the user can choose the appropriate psychological stimuli/stressors from a database of objects (both 2D and 3D) and videos, and easily place them into the virtual environment. The edited scene can then be visualized in the Player using either immersive or non-immersive displays. Currently, the NeuroVR library includes 18 different virtual scenes (apartment, office, square, supermarket, park, classroom, etc.), covering some of the most studied clinical applications of VR: specific phobias, cognitive rehabilitation, panic disorders and eating disorders.

On the other side, even if virtual reality is a very good presence-inducing technology, there is still room for improvement. Apparently, the main limitation of its actual use in clinical applications is the distance from reality: the virtual experience is a distinct realm, separate from the emotions and behaviors experienced by the patient in the real world, In other words, the behavior of the patient in VR has no direct effects on the real life experience. More, the emotions and problems experienced by the patient in the real world are not directly addressed in the VR exposure. To overcome the above limitations, here we suggested a new paradigm for e-health – "*Interreality*" – that integrates assessment and treatment within a hybrid environment, bridging physical and virtual world.

The clinical use of Interreality is based on a closed-loop concept that involves the use of technology for assessing, adjusting and/or modulating the emotional regulation of the patient, his/her coping skills and appraisal of the environment (both virtual, under the control of a clinicians, and real, facing actual stimuli) based upon a comparison of that patient's behavioural and physiological responses with a training or performance criterion:

- the assessment is conducted continuously throughout the virtual and real experiences;

- the information is constantly used to improve both the emotional management and the coping skills of the patient.

In conclusion, we suggest that the feeling of presence, here described as an intuitive metacognitive judgment related to the enaction of our intentions - we are present in a real or virtual environment we are able, inside it, to intuitively transform our intentions in actions – is potentially very useful for improving the clinical practice. Our hope is that the present chapter and the ideas presented in it will stimulate a discussion within the clinical and research VR community about the potential, the advantages and the possible limitations that the use of presence inducing technologies – such as virtual reality, mixed reality and Interreality – may offer to clinical change.

Author details

Giuseppe Riva
Applied Technology for Neuro-Psychology Lab. – ATN-P Lab., Istituto Auxologico Italiano, Milan, Communication and Ergonomics of NEw Technologies Lab. – ICE NET Lab., Università Cattolica del Sacro Cuore, Milan, Italy,

Fabrizia Mantovani
Centre for Studies in Communication Sciences. – CESCOM, University of Milan-Bicocca, Milan, Italy

Acknowledgement

The ideas, concepts and tools described in this chapter were partially supported by the European funded project "Interstress" – Interreality in the management and treatment of stress-related disorders (FP7-247685 – http://www.interstress.eu).

13. References

[1] Riva G, Alcañiz M, Anolli L, Bacchetta M, Baños RM, Beltrame F, Botella C, Galimberti C, Gamberini L, Gaggioli A, Molinari E, Mantovani G, Nugues P, Optale G, Orsi G, Perpiña C, & Troiani R (2001) The VEPSY Updated project: Virtual reality in clinical psychology. CyberPsychology and Behavior. 4(4): 449-455.

[2] Schultheis MT, & Rizzo AA (2001) The Application of Virtual Reality Technology in Rehabilitation. Rehabilitation Psychology. 46(3): 296-311.

[3] Riva G, Molinari E, & Vincelli F (2002) Interaction and presence in the clinical relationship: virtual reality (VR) as communicative medium between patient and therapist. IEEE Transactions on Information Technology in Biomedicine. 6(3): 198-205.

[4] Botella C, Quero S, Banos RM, Perpina C, Garcia Palacios A, & Riva G (2004) Virtual reality and psychotherapy. Stud Health Technol Inform. 99: 37-54.

[5] Botella C, Riva G, Gaggioli A, Wiederhold BK, Alcaniz M, & Banos RM (2012) The present and future of positive technologies. Cyberpsychology, behavior and social networking. 15(2): 78-84.
[6] Burdea GC, & Coiffet P (2003) Virtual Reality Technology (2nd Ed.). New Brunswick, NJ: Wiley-IEEE Press.
[7] Sheridan TB (1992) Musing on telepresence and virtual presence. Presence, Teleoperators, and Virtual Environments. 1: 120-125.
[8] Schloerb D (1995) A Quantitative Measure of Telepresence. Presence: Teleoperators, and Virtual Environments. 4(1): 64-80.
[9] Sadowski WJ, & Stanney KM (2002) Measuring and managing presence in virtual environments. In K. M. Stanney (Ed.), Handbook of Virtual Environments Technology (Mahwah, NJ: Lawrence Erlbaum Associates.
[10] Lombard M, & Ditton T (1997) At the heart of it all: The concept of presence. Journal of Computer Mediated-Communication [On-line]. 3(2): Available: http://www.ascusc.org/jcmc/vol3/issue2/lombard.html.
[11] Waterworth JA, Waterworth EL, Mantovani F, & Riva G (2010) On Feeling (the) Present: An evolutionary account of the sense of presence in physical and electronically-mediated environments. Journal of Consciousness Studies. 17(1-2): 167-178.
[12] Revonsuo A (2006) Inner Presence, Consciousness as a Biological Phenomenon. Cambridge, MA: MIT Press.
[13] Riva G, Waterworth JA, Waterworth EL, & Mantovani F (2011) From intention to action: The role of presence. New Ideas in Psychology. 29(1): 24-37.
[14] Stanovich KE, & West RF (2000) Individual differences in reasoning: implications for the rationality debate? Behav Brain Sci. 23(5): 645-665; discussion 665-726.
[15] Epstein S (2008) Intuition From the Perspective of Cognitive-Experiential Self-Theory. In H. Plessner, C. Betsch & T. Betsch (Eds.), Intuition in judgment and decision making (New York: Lawrence Erlbaum Associates. pp. 23-37.
[16] Kihlstrom JF (1987) The cognitive unconscious. Science. 237(4821): 1445-1452.
[17] Koriat A (2007) Metacognition and consciousness. In P. D. Zelaso, M. Moscovitch & E. Thompson (Eds.), Cambridge Handbook of Consciousness (New York: Cambridge University Press. pp. 289-325.
[18] Price MC, & Norman E (2008) Intuitive decisions on the fringes of consciousness: Are they conscious and does it matter? Judgment and Decision Making. 3(1): 28-41.
[19] Zahoric P, & Jenison RL (1998) Presence as being-in-the-world. Presence, Teleoperators, and Virtual Environments. 7(1): 78-89.
[20] Slater M, Lotto B, Arnold MM, & Sanchez-Vives MV (2009) How we experience immersive virtual environments: the concept of presence and its measurement. Anuario de Psicología. 40(2): 193-210.
[21] Reber AS (1989) Implicit learning and tacit knowledge. Journal of Experimental Psychology: General. 118(3): 219-235.
[22] Rizzolatti G, Fadiga L, Gallese V, & Fogassi L (1996) Premotor cortex and the recognition of motor actions. Cognitive Brain Research. 3: 131-141.

[23] Gallese V (2005) Embodied simulation: From neurons to phenomenal experience. Phenomenology and the Cognitive Sciences(4): 23-48.
[24] Barsalou LW (2003) Situated simulation in the human conceptual system. Language and Cognitive Processes. 18: 513-562.
[25] Wilson M, & Knoblich G (2005) The case for motor involvement in perceiving conspecifics. Psychological Bulletin. 131(3): 460-473.
[26] Hommel B, Müsseler J, Aschersleben G, & Prinz W (2001) The Theory of Event Coding (TEC): A framework for perception and action planning. Behavioral and Brain Sciences. 24(5): 849-937.
[27] Fox I (1994) Our knowledge of the internal world. Philosophical Topics. 22: 59-106.
[28] Clark A (2008) Supersizing the mind: embodiment, action and cognitive extension. Oxford, UK: Oxford University Press.
[29] Metzinger T (2009) The ego tunnel: The science of the mind and the myth of the self. New York: Basic Books.
[30] Sirigu A, Grafman J, Bressler K, & Sunderland T (1991) Multiple representations contribute to body knowledge processing: evidence from a case of autotopagnosia. Brain. 114(1): 629-642.
[31] Slater M, Perez-Marcos D, Ehrsson HH, & Sanchez-Vives MV (2009) Inducing illusory ownership of a virtual body. Front Neurosci. 3(2): 214-220.
[32] Gamberini L, Seraglia B, & Priftis K (2008) Processing of peripersonal and extrapersonal space using tools: Evidence from visual line bisection in real and virtual environments. Neuropsychologia. 46(5): 1298-1304.
[33] Riva G (2009) Is presence a technology issue? Some insights from cognitive sciences Virtual Reality. 13(3): 59-69.
[34] Riva G (2008) From Virtual to Real Body: Virtual Reality as Embodied Technology. Journal of Cybertherapy and Rehabiliation. 1(1): 7-22.
[35] Russell JA (1996) Agency: Its role in mental development. Hove: Erlbaum.
[36] Pacherie E (2006) Toward a dynamic theory of intentions. In S. Pockett, W. P. Banks & S. Gallagher (Eds.), Does consciousness cause behavior? (Cambridge, MA: MIT Press. pp. 145-167.
[37] Pacherie E (2008) The phenomenology of action: A conceptual framework. Cognition. 107(1): 179-217.
[38] Kaptelinin V, & Nardi B (2006) Acting with Technology: Activity Theory and Interaction Design. Cambridge, MA: MIT Press.
[39] Riva G (2010) Dall'intenzione, all'azione, all'interazione: il ruolo di "presenza" e "presenza sociale". In F. Morganti, A. Carassa & G. Riva (Eds.), Intersoggettività e Interazione: Un dialogo fra scienze cognitive, scienze sociali e neuroscienze (Torino: Bollati Boringhieri. pp. 136-177.
[40] Blackemore SJ, & Decety J (2001) From the perception of action to the understanding of intention. Nature Reviews Neuroscience. 2: 561-567.
[41] Knoblich G, Thornton I, Grosjean M, & Shiffrar M (Eds.). (2005). Human Body Perception from the Inside Out. New York: Oxford University Press.
[42] Heidegger M (1959) Unterwegs zur Sprache. Neske: Pfullingen.

[43] Riva G (2006) Being-in-the-world-with: Presence meets Social and Cognitive Neuroscience. In G. Riva, M. T. Anguera, B. K. Wiederhold & F. Mantovani (Eds.), From Communication to Presence: Cognition, Emotions and Culture towards the Ultimate Communicative Experience. Festschrift in honor of Luigi Anolli (Amsterdam: IOS Press. Online:
http://www.emergingcommunication.com/volume8.html. pp. 47-80.

[44] Riva G, Waterworth JA, & Waterworth EL (2004) The Layers of Presence: a bio-cultural approach to understanding presence in natural and mediated environments. Cyberpsychology & Behavior. 7(4): 405-419.

[45] Riva G, & Waterworth JA (2003) Presence and the Self: A cognitive neuroscience approach. Presence-Connect. 3(1): Online:
http://presence.cs.ucl.ac.uk/presenceconnect/articles/Apr2003/jwworthApr72003114532/jwworthApr72003114532.html.

[46] Sperber D, & Wilson D (1995) Relevance: Communication and Cognition (2nd Edition). Oxford: Blackwell.

[47] Dillon C, Freeman J, & Keogh E (2003). Dimension of Presence and components of emotion. Paper presented at the Presence 2003, Aalborg, Denmark.

[48] Riva G (2005) Virtual reality in psychotherapy: review. CyberPsychology & Behavior. 8(3): 220-230; discussion 231-240.

[49] Wiederhold BK, & Rizzo A (2005) Virtual reality and applied psychophysiology. Applied Psychophysiology and Biofeedback. 30(3): 183-185.

[50] Emmelkamp PM (2005) Technological innovations in clinical assessment and psychotherapy. Psychotherapy & Psychosomatics. 74(6): 336-343.

[51] Glantz K, Durlach NI, Barnett RC, & Aviles WA (1997) Virtual reality (VR) and psychotherapy: Opportunities and challenges. Presence, Teleoperators, and Virtual Environments. 6(1): 87-105.

[52] Riva G, Bacchetta M, Cesa G, Conti S, Castelnuovo G, Mantovani F, & Molinari E (2006) Is severe obesity a form of addiction? Rationale, clinical approach, and controlled clinical trial. CyberPsychology and Behavior. 9(4): 457-479.

[53] Ferrer-Garcia M, & Gutierrez-Maldonado J (2012) The use of virtual reality in the study, assessment, and treatment of body image in eating disorders and nonclinical samples: A review of the literature. Body Image. 9(1): 1-11.

[54] Riva G, Manzoni M, Villani D, Gaggioli A, & Molinari E (2008) Why you really eat? Virtual reality in the treatment of obese emotional eaters. Stud Health Technol Inform. 132: 417-419.

[55] Reger GM, & Gahm GA (2008) Virtual reality exposure therapy for active duty soldiers. J Clin Psychol. 64(8): 940-946.

[56] Bordnick PS, Traylor A, Copp HL, Graap KM, Carter B, Ferrer M, & Walton AP (2008) Assessing reactivity to virtual reality alcohol based cues. Addict Behav. 33(6): 743-756.

[57] Optale G (2003) Male Sexual Dysfunctions and multimedia Immersion Therapy. CyberPsychology & Behavior. 6(3): 289-294.

[58] Hoffman HG (2004) Virtual-Reality Therapy: Patients can get relief from pain or overcome their phobias by immersing themselves in computer-generated worlds. Scientific American.

[59] Wolfe BE (2002) The Role of Lived Experience in Self- and Relational Observation: A Commentary on Horowitz (2002). Journal of Psychotherapy Integration. 12(2): 147-153.

[60] Safran JD, & Greenberg LS (1991) Emotion, psychotherapy, and change. New York: The Guilford Press.

[61] Riva G, & Gaggioli A (2008) Virtual clinical therapy. Lecture Notes in Computer Sciences. 4650: 90-107.

[62] Vincelli F, Molinari E, & Riva G (2001) Virtual reality as clinical tool: immersion and three-dimensionality in the relationship between patient and therapist. Studies in Health Technology and Informatics. 81: 551-553.

[63] Baños RM, Botella C, & Perpiña C (1999) Virtual Reality and Psychopathology. CyberPsychology & Behavior. 2(4): 283-292.

[64] Riva G, & Davide F (Eds.). (2001). Communications through Virtual Technologies: Identity, Community and Technology in the Communication Age. Amsterdam: Ios Press. Online: http://www.emergingcommunication.com/volume1.html.

[65] Price M, & Anderson P (2007) The role of presence in virtual reality exposure therapy. J Anxiety Disord. 21(5): 742-751.

[66] Krijn M, Emmelkamp PM, Olafsson RP, & Biemond R (2004) Virtual reality exposure therapy of anxiety disorders: a review. Clin Psychol Rev. 24(3): 259-281.

[67] Slater M, Antley A, Davison A, Swapp D, Guger C, Barker C, Pistrang N, & Sanchez-Vives MV (2006) A virtual reprise of the Stanley Milgram obedience experiments. PLoS One. 1: e39.

[68] Bouchard S, St-Jacques J, Robillard G, & Renaud L (2008) Anxiety Increases the Feeling of Presence in Virtual Reality. Presence: Teleoperators & Virtual Environments. 17(4): 376-391.

[69] Baños RM, Botella C, Guerrero B, Liaño V, Alcañiz M, & Rey B (2005) The Third Pole of the Sense of Presence: Comparing Virtual and Imagery Spaces. PsychNology Journal. 3(1): 90-100. On-line: http://www.psychnology.org/pnj103(101)_banos_botella_guerriero_liano_alcaniz_rey_abstract.htm.

[70] Michaud M, Bouchard S, Dumoulin S, Zhong XW, & Renaud P (2004) Manipulating presence and its impact on anxiety. Cyberpsychology & Behavior. 7(3): 297-298.

[71] Riva G, Mantovani F, Capideville CS, Preziosa A, Morganti F, Villani D, Gaggioli A, Botella C, & Alcaniz M (2007) Affective interactions using virtual reality: the link between presence and emotions. Cyberpsychology and Behavior. 10(1): 45-56.

[72] Gorini A, Mosso JL, Mosso D, Pineda E, Ruiz NL, Ramiez M, Morales JL, & Riva G (2009) Emotional response to virtual reality exposure across different cultures: the role of the attribution process. Cyberpsychol Behav. 12(6): 699-705.

[73] Bouchard S, Dumoulin S, Labonte-Chartrand G, Robillard G, & Renaud P (2006) Perceived realism has a significant impact on the feeling of presence. Cyberpsychology & Behavior. 9(6): 660.

[74] Winograd T, & Flores F (1986) Understanding Computers and Cognition: A New Foundation for Design. Norwood, NJ: Ablex Publishing Corporation.

[75] Csikszentmihalyi M (1975) Beyond Boredom and Anxiety. San Francisco: Jossey-Bass.

[76] Ghani JA, & Deshpande SP (1994) Task characteristics and the experience of optimal flow in Human-Computer Interaction. The Journal of Psychology. 128(4): 381-391.

[77] Riva G, Castelnuovo G, & Mantovani F (2006) Transformation of flow in rehabilitation: the role of advanced communication technologies. Behavior Research Methods. 38(2): 237-244.

[78] Massimini F, & Delle Fave A (2000) Individual development in a bio-cultural perspective. American Psychologist. 55(1): 24-33.

[79] Castelnuovo G, Lo Priore C, Liccione D, & Cioffi G (2003) Virtual Reality based tools for the rehabilitation of cognitive and executive functions: the V-STORE. PsychNology Journal. 1(3): 311-326. Online: http://www.psychnology.org/pnj311(313)_castelnuovo_lopriore_liccione_cioffi_abstract.htm.

[80] Johansson BB (2000) Brain plasticity and stroke rehabilitation. The Willis lecture. Stroke. 31(1): 223-230.

[81] Hoffman HG, Richards TL, Coda B, Bills AR, Blough D, Richards AL, & Sharar SR (2004) Modulation of thermal pain-related brain activity with virtual reality: evidence from fMRI. Neuroreport. 15(8): 1245-1248.

[82] Hoffman HG, Patterson DR, & Carrougher GJ (2000) Use of virtual reality for adjunctive treatment of adult burn pain during physical therapy: a controlled study. Clinical Journal of Pain. 16(3): 244-250.

[83] Riva G, Gaggioli A, Villani D, Preziosa A, Morganti F, Corsi R, Faletti G, & Vezzadini L (2007) NeuroVR: an open source virtual reality platform for clinical psychology and behavioral neurosciences. Studies in Health Technology and Informatics. 125: 394-399.

[84] Riva G, Gaggioli A, Grassi A, Raspelli S, Cipresso P, Pallavicini F, Vigna C, Gagliati A, Gasco S, & Donvito G (2011) NeuroVR 2 - A Free Virtual Reality Platform for the Assessment and Treatment in Behavioral Health Care. Stud Health Technol Inform. 163: 493-495.

[85] Repetto C, & Riva G (2011) From virtual reality to interreality in the treatment of anxiety disorders. Neuropsychiatry. 1(1): 31-43.

[86] Fidopiastis C, Hughes CE, & Smith E (2009) Mixed Reality for PTSD/TBI Assessment. Stud Health Technol Inform. 144: 216-220.

[87] Botella CM, Juan MC, Banos RM, Alcaniz M, Guillen V, & Rey B (2005) Mixing realities? An application of augmented reality for the treatment of cockroach phobia. Cyberpsychology & Behavior. 8(2): 162-171.

[88] Riva G (2009) Interreality: A New Paradigm for E-health. Stud Health Technol Inform. 144: 3-7.

[89] Riva G, Raspelli S, Algeri D, Pallavicini F, Gorini A, Wiederhold BK, & Gaggioli A (2010) Interreality in practice: bridging virtual and real worlds in the treatment of posttraumatic stress disorders. Cyberpsychol Behav Soc Netw. 13(1): 55-65.

Chapter 2

Virtual Realities in the Treatment of Mental Disorders: A Review of the Current State of Research

Christiane Eichenberg and Carolin Wolters

Additional information is available at the end of the chapter

http://dx.doi.org/10.5772/50094

1. Introduction

In the past decade, *virtual reality* (VR) technologies have been discussed as promising supplements in psychotherapy. Virtual realities enable users to interact in real time with computer-generated environments in three dimensions [1]. The fact that VR applications simulate real experiences and trigger anxiety, including physiological symptoms such as sweating or nausea, emphasizes their potential to replace conventional exposure therapy.

If users are to experience virtual environments as real, two conditions are required: immersion and presence. *Immersion* describes a state of consciousness in which the user's awareness of the physical self declines due to an increasing involvement in the virtual environment. A sensation of immersion can be achieved by creating realistic visual, auditory or tactile stimulation. Additionally, the usage of specific output devices (e.g. data-goggles and monitors) and input devices (e.g. data gloves, voice recognition and eye tracking software) may facilitate the user's perception of immersion. The feeling of being physically immersed can result in a sense of *presence*, that includes a perception of the environment as being real, shutting out real-life stimuli and performing involuntary, objectively meaningless body movements such as ducking to avoid an object displayed in VR. Moreover, persons seem to experience a strong sense of control in VR. A study [2] showed that persons who were told to have control over the movements of an elevator but actually did not, rated their perceived control as high as those who in fact had control over the elevator.

Another technology that has been developed in the past years is referred to as *Augmented Reality* (AR). AR describes the superimposition of virtual elements into the real world. Persons therefore see a visualization of the real world and virtual elements at the same time [3].

Advantages of AR in comparison to VR may include an enhanced feeling of presence and reality, since the environment is in fact real. Additionally, AR might be less expensive, because the real world environment can be used as a scheme. Thus, the setting does not need to be entirely developed.

Research on the usage of VR and AR technologies in psychotherapy has mainly focused on behavioral therapy and was proven to be effective particularly in the treatment of specific phobias [4]. According to well-established behavior therapy theories, clients have to be exposed to fear inducing situations in order to treat phobias, because avoidance of fearful stimuli might stabilize the assumption that they are dangerous. Corrective experiences would thus be prevented. Two kinds of exposure can be implemented in therapy. While in-vivo exposure involves the immediate exposure to a fear-enhancing situation or object in reality, in-sensu exposure describes the mere imagination of the exposure to fearful stimuli. In terms of a graduated exposure, stimuli that trigger low levels of anxiety are usually presented first, increasing up to the client's most extreme fear, which is called "flooding" (in-vivo exposure) or "implosion" (in-sensu exposure).

2. Benefits and costs of applying VR in psychotherapy

As already mentioned, exposure therapy supported by VR technologies exceeds imaginative exposure by adding a sense of presence. Moreover, including VR applications in psychotherapy offers a series of advantages. These include the possibility of adjusting virtual environments to each client's specific needs and controlling what is presented to the client. In addition, VR enables the therapist to expose the client to conditions that might be unsafe or only accessible at high cost in the outside world, and to improve confidentiality by avoiding spectators [5]. Furthermore, therapists seem to consider VR exposure to be less aversive than in-vivo therapy [6]. Presumably, the same applies to patients. For instance, García-Palacios et al. [7] showed that only 3% of 150 participants suffering from specific phobia refused VR exposure, while 27% refused in vivo therapy.

Nevertheless, the usage of VR entails considerable costs. First of all, despite recent findings, some groups might be reluctant to the use of VR technologies and might therefore be excluded from treatment. Furthermore, the handling of VR applications requires a certain amount of training for therapists. Besides, therapists are tied to the position of VR equipment, since it is usually too unhandy to transport [1]. Additionally, equipment acquisition is rather expensive, even though costs have sunk dramatically in the past ten years [5]. Finally, clients might experience dizziness and nausea while undergoing a VR application, a syndrome referred to as simulation sickness [4]. But even though the cited costs have to be taken into account, a recent study [1] indicated that therapists perceive the benefits of VR supported psychotherapy to be outweighing potential costs.

Self-evidently, those costs should only be accepted on condition that VR applications are able to effectively treat mental disorders. The present article aims to outline recent findings in order to examine the effectiveness of usage of VR technologies in psychotherapy.

3. Current state of research

Previous studies have mainly focused on the use of VR applications in the treatment of anxiety disorders and particularly specific phobias, such as fear of heights, fear of flying, fear of animals or social phobia. However, research has recently started to focus on the usage of VR in the treatment of other disorders as well, including eating disorders and sexual dysfunctions. In the following, an overview of the current state of research will be given. After briefly describing the search strategy, two meta-analyses that are concerned with the application of VR in the treatment of anxiety disorders will be presented. Subsequently, exemplary studies evaluating the effectiveness of VR-assisted psychotherapy of different specific disorders are summarized.

3.1. Method

In order to identify eligible studies, a search on the databases PsychInfo, PsychArticles and Pubmed was conducted. The search words *Virtual/ Augmented Reality, Exposure Therapy* and *effectiveness/ efficacy/ metaanalysis* were entered alone and in combination with *mental disorder* and derivatives of the different terms for disorders, particularly *acrophobia/ fear of heights/ aviophobia/ fear of flying/ arachnophobia/ fear of spiders/ social phobia/ fear of public speaking/ panic disorder/ posttraumatic stress disorder.* To ensure the currentness of the findings presented here, we focused on studies that were published within the past ten years, even though studies conducted before were not excluded if they contributed significantly to the current state of research.

3.2. Meta-analyses

Two current meta-analyses have been reported concerning the effectiveness of VR in the treatment of anxiety disorders. Parsons and Rizzo [8] analyzed *N*= 21 studies that used pre-post measurements but not necessarily a controlled study design. The authors found an average effect size of *d*= .95 (*SD*= .02) for the reduction of symptoms in VR-assisted therapy. The treatment of fear of flying (*d*= 1.5; *SD*= .05), and panic disorder with agoraphobia (*d*= 1.79; *SD*= .02) using VR applications accounted for the largest effect sizes concerning symptom reduction. They were followed by treatment of social phobia (*d*= .96; *SD*= .10), acrophobia (*d*= .93; *SD*= .06) and arachnophobia (*d*= .92; *SD*= .12), while the treatment of posttraumatic stress disorder (PTSD) by means of VR obtained the smallest effect size of *d*= .87 (*SD*= .01). In addition, a series of determining factors were assumed. These include the degree of immersion and presence, duration of disease and socio-demographic variables. However, due to a lack of data within the examined study, the authors were unable to make a valid statement about potential moderators.

Powers and Emmelkamp [9] examined *N*= 13 controlled studies, reverting to a more rigid design that excluded studies involving case reports, multiple components of treatment conditions, and an unequal amount of treatment sessions in the conditions compared. In general, effect sizes of VR exposure therapy were found to be large to very large, ranging from *d*= .85 to *d*= 1.67. A smaller effect size of *d*= .35 favored treatment with the aid of VR to in-vivo exposure and therefore demonstrates the superiority of VR in comparison to in-vivo treatment. Admittedly, studies considering therapy of specific phobias predominated.

Nevertheless, overall results prove that VR applications are highly effective in the treatment of anxiety disorders. However, difficulties common to the realization of meta-analyses, for instance a publication bias that favors publication of studies implying significant results, have to be taken into account. Moreover, small sample sizes as well as missing data about the point of time of follow-up ratings and therefore questionable lastingness of treatment effects, limit the meaningfulness of findings. Future research should include varied levels of immersion and ensure controlled study designs.

3.3. Exemplary studies of various syndromes

The meta-analyses presented here mainly focused on the effectiveness of VR as a supplement of behavior therapy for patients with anxiety disorders, some of the most frequently diagnosed psychological disorders. Nearly one out of five adults in the USA suffers from an anxiety disorder, whereat women are more often affected than men [10]. Therefore, the continuing development and evaluation of effective treatment methods seems crucial.

Anxiety disorders present the first syndrome category for which the use of modern media such as the Internet or VR technology as a setting for interventions was scientifically evaluated. They are usually assigned to the field of *behavior therapy*. Since anxiety disorders are frequently treated with the aid of exposure, they are suitable for VR settings. In contrast, *psychodynamic therapy* concentrates more on relationship aspects. However, there are conceptual considerations about how to integrate VR in psychodynamic therapy [11], and a few studies have already been conducted to examine the use of VR within the psychodynamic approach (e.g. [12]).

Anxiety disorders are classified differently within the two major diagnostic classification systems. While the Diagnostic and Statistical Manual of Mental Disorders (DSM-IV) [13] sorts them within a separate chapter, the International Classification of Diseases (ICD-10) [14] includes them in the chapter „Neurotic, Stress and Somatoform Disorders". The latter distinguishes between the subgroups of phobic disorders (agoraphobia, social anxiety disorder, specific phobias) and other anxiety disorders (panic disorder, generalized anxiety disorders). In both classification systems, posttraumatic stress disorder is discussed along with anxiety disorders. In the following, the effectiveness of VR-assisted treatment of various syndromes is presented.

3.3.1. Fear of heights

Acrophobia, classified as a specific phobia of the naturalistic type, describes an extreme fear of heights. It involves the avoidance of various height-related situations, such as stairs, terraces, high buildings, bridges, or elevators. The fear of heights is widely spread: In a survey of more than 8000 adults, 20% stated that they had already experienced an exaggerated fear of heights in the past, which did not meet the criteria for acrophobia [15]. In the aforesaid study, the prevalence of acrophobia amounted to 5.3% and therefore closely followed the prevalence of fear of animals. While women usually tend to develop specific phobias substantially more often than men [12], merely 55 to 70% of acrophobic persons are female.

The first successful application of VR in the treatment of acrophobia was presented in a case study of an acrophobic student who was successfully treated using graded VR exposure [16]. A more extensive study including a sample of 20 students furnished further evidence for the effectiveness of VR-assisted treatment [17]. However, due to study limitations such as the absence of a control group, the further conclusions can only be drawn under reserve.

The first clinical trial of the effectiveness of VR in the treating acrophobia was conducted by Emmelkamp and collaborators [18]. In a within group design, ten patients were treated with two sessions of VR, followed by two sessions of exposure in-vivo. Acrophobic symptoms were measured before treatment, after VR treatment and after in-vivo exposure. Results showed that after being treated by the means of VR, exposure to real situations did not lead to any significant improvement on the Acrophobia Questionnaire (AQ) or the Attitudes Towards Heights Questionnaire (ATHQ). Unexpectedly, the research design had created a ceiling effect, insofar as the VR treatment effects left little space for improvement during exposure in-vivo.

In a randomized controlled trial (RCT) conducted by the same research group, effectiveness of exposure by the means of VR and in vivo were compared [19]. The places used in the exposure in vivo were reproduced in a virtual environment. Exposure was affected in a real or virtual shopping mall in Amsterdam, a fire escape, and a roof garden. *N*= 33 acrophobic persons underwent three weekly sessions of one hour each. Anxiety levels were reported on the Subjective Units of Disturbance-Scale (SUDS). Results demonstrated that both kinds of treatment were equally effective and improvements were maintained at a six months follow-up.

Krijn et al. [20] examined the effectiveness of different VR systems. *N*= 37 acrophobic subjects were treated either with three VR sessions administered by a head-mounted display (HMD) or by a computer animated virtual environment (CAVE) or were assigned to the waitlist control group. Results showed no differences in effectiveness between the different VR systems. The higher degree of presence that was experienced in the CAVE condition did not affect outcome measures. In a following study [21] the same research group analyzed the role of cognitive self-statements in VR exposure therapy. In a crossover design, *N*= 26 acrophobic persons were randomly assigned to two sessions of VR treatment followed by two sessions of VR treatment plus self-statements or vice versa. Results indicated that VR-assisted treatment reduced symptoms of fear of heights as well as behavioral avoidance and improved attitudes towards heights. However, cognitive self-statements did not additionally enhance effectiveness of VR.

Another study series concentrating on treatment of acrophobia with the aid of VR was conducted by Coelho and collaborators [22, 23]. Initially, the authors compared effects of treatment in a VR (N= 10) and a real environment (N= 5). Both groups showed equally large improvements on the Behavioral Avoidance Test (BAT), the ATHQ, and the AQ, even though treatment time was substantially lower in the VR condition. A following study with eight persons suffering from fear of heights revealed that movement during VR exposure enhances anxiety. One of the virtual settings used in these studies can be seen in Figure 1.

However, VR-assisted treatment of acrophobia is not only effective and time efficient, but additionally represents a series of advantages. Anxiety inducing situations such as being on bridges or high buildings can be experienced without any great logistic efforts. Therefore, difficulties of accessing the actual place and potential disturbances by pedestrians can be avoided.

Figure 1. View from the real world (left) and the virtual reality system (right). Adapted from "Contrasting the Effectiveness and Efficiency of Virtual and Real Environments in the Treatment of Acrophobia" by C.M. Coelho, C.F. Silva, J.A. Santos, J. Tichon and G. Wallis, 2008, *PsychNology, 6*(2), p. 206. Copyright 2008 by PsychNology Journal. Adapted with permission.

3.4. Example case

Choi and collaborators [24] described the case of a 61-year old patient, who had been suffering from acrophobia for the past 40 years. He was not able to go up higher than the third floor of any apartment and therefore lived on the third floor on his 18-story building. In order to treat his acrophobia, the authors planned eight sessions of VR therapy that were supposed to take place three times a week and took about 30 minutes each.

The virtual environment was comprised of a steel tower which involved a lift within a steel frame structure that was open to all four sides. To enhance the sense of reality, sounds of wind and a moving lift were included, and the patient was isolated in a dark room in order to increase immersion. Prior to VR treatment, the patient received four sessions of relaxation training, including abdominal breathing and progressive muscle relaxation training, to be able to cope with body sensations during the VR sessions. Pretreatment assessment was completed and the patient accomplished a demo program to get used to VR.

In the first session, the patient stayed on the floor of the virtual lift to get accustomed to the environment. He was free to decide whether to go up on a higher floor or stay where he was. In this session, the patient went up to the fourth floor, experiencing dizziness and sweating and reporting 70 to 90 subjective units of disturbance (SUD). SUD was evaluated every two to five minutes. Whenever the patient stated to experience intense fear, he was instructed to relax. In the second and third session, the patient went up to the eighth floor, but still experienced high levels of SUD, breathlessness, and the sensation of falling down. After these sessions, the patient was already able to walk up to the eighth floor of his building for the first time in ten years. According to the patient, the virtual lift scared him more than going up his building in the real world. In the fourth session, the patient went up to 18th floor of the virtual tower, and then to the 25th floor, the top of the tower, in the fifth session. Even while looking down, the patient did not experience any particular symptoms and reported SUD scores below 30. After the sixth session, the patient claimed that he did not need VR anymore. The authors therefore changed treatment plans and assigned the patient to go up to the observatory of a mountain by cable car. Going up to and looking down from the observatory, the patient showed neither symptoms of anxiety nor avoidance. Subsequently, he suggested going up the highest building of Seoul. Looking outside from the elevator of this building, the patient expressed only little fear. Six months after the treatment, the patient stated that he did not have any fear of heights.

3.4.1. Fear of flying

Fear of flying, or *aviophobia,* is characterized by an intense fear of flying that often results in flight avoidance or experiencing substantial distress while flying. Acrophobia affects 10-20% of the general population and 20% of airline passengers consume alcohol or sedatives to deal with their fear of flying [25]. Most persons suffering from acrophobia fear a plane crash, while some fear being closed in and therefore often meet the DSM-IV criteria for claustrophobia. Further fears concern experiencing a panic attack and not being able to escape the situation or get medical attention, complying with the concept of panic disorder with agoraphobia, or a general fear of heights. Therefore, comorbidity with other anxiety disorders occurs very frequently.

The use of VR applications in the treatment of aviophobia could be advantageous to an exposure in-vivo because financial and logistical expenses are essentially lower. Furthermore, the privacy and confidentiality of a VR exposure in contrast to a regular flight should be emphasized.

The first RCT investigating the effectiveness of VR treatment of aviophobia was presented by Rothbaum et al. [26]. *N*= 49 participants were randomly assigned to VR exposure therapy, exposure in-vivo, or a waitlist control group. Both treatment groups received four sessions of anxiety management and four further sessions consisting of exposure to an airplane, either in reality or VR. The latter involved acoustic and visual simulations with the aid of a HMD, as well as vibration simulation. Exposure in vivo included preparation training at an airport as well as visualization of takeoff, flight and landing inside of an airplane. Both treatment groups showed significant symptom reduction on several standardized scales, while no improvements were observed for the control group. Effects remained stable after six and twelve months follow-up. However, flight situations differed between conditions, because an actual flight was not part of the exposure in-vivo. Moreover, both treatments were combined with anxiety management training with the result that treatment effects were not completely distinguishable. The findings were replicated in another sample of *N*= 75 aviophobic persons [27].

Another study compared VR exposure therapy with and without physiological feedback measures to self-visualization in *N*= 30 persons suffering from fear of flying [28]. Results showed significant improvements in flying behavior, physiological measures of anxiety, and standardized self-report measures of anxiety in the VR condition in contrast to imaginative exposure. Furthermore, the combination of VR treatment and biofeedback was more effective than VR treatment alone. The authors reported that after eight weeks of therapy, 20% of the patients in the imaginative condition, 80% of those in the VR condition and 100% of patients who received both VR treatment and biofeedback were able to fly again. In a follow-up study three years later, treatment effects were maintained [29].

Mühlberger et al. [30, 31] proved the effectiveness of VR-assisted treatment of aviophobia in a series of studies. *N*= 30 participants were randomly assigned to a VR treatment condition or a relaxation training group. While both treatment conditions resulted in significant symptom improvement, several outcome measures, including physiological fear responses, indicated larger effects of VR exposure therapy than self-visualization. In a following study, the research group demonstrated that one session of VR exposure therapy in combination with cognitive behavior therapy (CBT) achieved better results than CBT only and a control group (*N*= 45) [31]. Results remained stable at six months follow-up. Limitations of the study include the time spent with the therapist, which was much longer for the combined treatment than for CBT only. In addition, group assignment was not randomized. Nevertheless, the elucidated study demonstrates that VR-supported exposure can show persistent effects even after one single session. Furthermore, Mühlberger et al. [32] revealed that the completion of graduation flights might be important for long-term treatment effectiveness, but that the presence of a therapist is not necessarily required.

Comparing five sessions of VR exposure therapy to an attention placebo group, Maltby et al. [33] obtained mixed results. While the VR treatment condition was superior to the placebo condition on self-report instruments, BAT scores did not reveal any significant differences. Moreover, VR exposure was more effective on only one self-report measure at six months follow-up.

Furthermore, Krijn and collaborators [34] compared four sessions of VR exposure with four sessions of CBT and with five weeks of bibliotherapy, that involved reading a psycho-educative book about aviophobia. Results indicated that both VR treatment and CBT were effective and did not differ in symptom reduction. However, after undergoing an additional CBT program, including an exposure in-vivo, CBT group was superior to VR treatment group.

Finally, the efficacy of VR and computer-aided psychotherapy in the treatment of aviophobia was examined by Tortella-Feliu et al. [35]. *N*= 60 participants were randomly assigned to the following conditions: VR exposure, computer-aided exposure with a therapist's assistance, and self-administered computer-assisted exposure. Results demonstrated that all three conditions were equally effective in reducing flying phobia, even after one year. The findings indicate that therapist involvement might be reduced in VR and computer-aided treatment.

3.4.2. *Fear of spiders*

According to the ICD-10 [15], *arachnophobia* is categorized within the group of zoophobias and is characterized by a persistent fear of spiders, an immediate anxiety response to exposure to a spider, and avoidance of spiders. The category of "bugs, mice, snakes or bats", which includes spiders, accounts for about 40% of specific phobias [36]. Approximately 3.5 to 6.1% of the general population suffers from arachnophobia, whereof the majority is constituted by women. Even though most arachnophobic persons recognize that their fear is unreasonable, daily life can be restrained. For instance, persons suffering from fear of spiders might depend on the help of others when confronted with a spider, or be restricted in choosing an apartment.

VR applications seem to represent a potential treatment method for arachnophobia. Rinck et al. [37] examined spider fearful persons' attention and motor reactions to spiders on a VR. The authors demonstrated that spider fearfuls show increased state anxiety, spend more time looking at spiders, and exhibit behavioral avoidance of spiders.

A first single case report examining the effectiveness of treating arachnophobia with the aid of VR was conducted by Carlin et al. [38]. They used VR as well as mixed reality, which involved touching real objects that can be seen in VR, to treat a 37-year old female suffering severe fear of spiders. After twelve weekly sessions of one hour each, measures of anxiety, avoidance, and behavior towards real spiders improved significantly.

In 2002, a RCT was conducted that compared VR exposure therapy group and a waitlist group of altogether *N*= 23 participants [36]. The VR treatment group received four one-hour sessions on average. Effects were assessed by the Fear of Spiders Questionnaire (FSQ), a BAT, and severity ratings effected by clinicians. Results demonstrated that 83% of participants who received VR treatment showed clinically significant improvement compared with 0% in the waitlist control group.

In a following study with *N*= 36 participants, it was demonstrated that VR combined with touching an object that resembles a spider was more effective than only VR exposure [39]. *N*= 36 phobic students were randomly assigned to one of three conditions: No treatment, VR exposure without any tactile stimulation, and VR exposure including tactile stimulation.

After three sessions of VR exposure, the treatment groups showed less avoidance and lower levels of anxiety than the control group; and VR including tactile simulation was superior to VR without any tactile clues.

Michaliszyn et al. [40] found similar results comparing the effectiveness of VR treatment and in-vivo exposure to a waitlist condition. A total of N= 43 persons suffering arachnophobia were randomly assigned to the three conditions. Treatment groups received eight therapy sessions of one and a half hour each. Outcome measures included the Fear of Spiders Questionnaire, the Spider Phobia Beliefs Questionnaire (SBQ), a BAT, and the Structured Interview for DSM-IV (SKID). Both treatment groups showed clinically significant improvements in comparison to the waitlist control group, whereat in-vivo exposure was superior to VR treatment on the SBQ-F.

Furthermore, a study demonstrated that modified 3D computer games instead of actual VR software can be effective in the treatment of arachnophobia [41]. Modification of computer games could therefore represent a less expensive alternative to VR equipment.

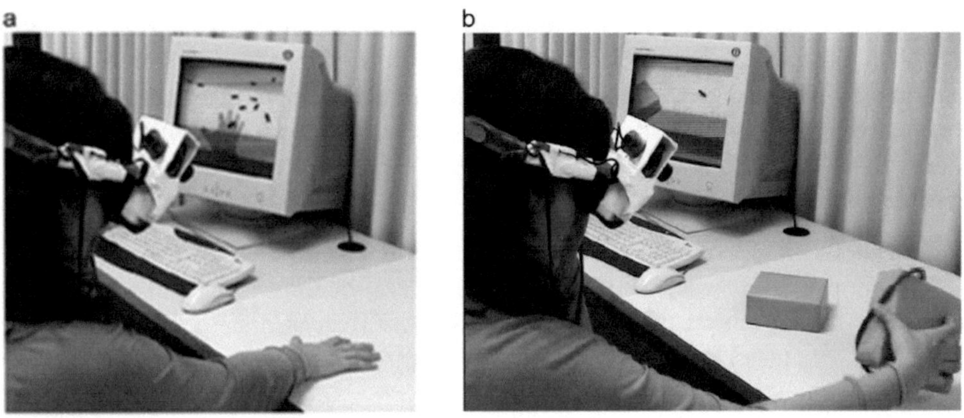

(a) (b)

Figure 2. (a) Participant putting her hand on the table and the cock- roaches crossing over it. (b) Participant searching for hidden cockroaches. Adapted from "A comparative study of the sense of presence and anxiety in an invisible marker versus a marker augmented reality system for the treatment of phobia towards small animals" by M.C. Juan and D. Joele, 2011, *International Journal of Human-Computer Studies 69*(6), p. 445. Copyright 2011 by Elsevier. Adapted with permission.

Research has also focused on the use of AR in treating phobia towards small animals. In doing so, virtual spiders or cockroaches are blended into the real world. In a first case study, a participant suffering from cockroach phobia showed significant decreases in fear and avoidance levels, being capable of approaching, interacting with, and killing real cockroaches following AR exposure and one month later [3]. In a following study evaluating the effectiveness of AR, nine participants with either spider or cockroach phobia were treated in a single session [42]. Firstly, progressively more virtual spiders or cockroaches were presented in the therapist's hand. Participants were asked to bring their hand closer to the animals. Subsequently, a box appeared which the participants had to pick up to see if there was an animal underneath. Finally, virtual animals had to be killed with an insecticide, flyswatter or dustpan and put into a box. After completion of the session, participants were asked to approach, interact and kill real spiders or cockroaches. All of the participants succeeded in doing so, showing considerable less avoidant behavior. A validation of the system used in these first studies demonstrated that all elements of the AR environment were highly fear inducing in *N*= 6 female participants with cockroach phobia. In addition, ratings of presence, reality and immersion obtained high scores [43]. In this AR system, visible markers were used to identify insecticide, flyswatter or dustpan approaching a virtual animal. To avoid this warning, a second version in which the markers were invisible was compared to the first one [44]. For an example of the AR setting, see Figure 2. Results indicated that the invisible marker-tracking system induced a similar or higher sense of presence and levels of anxiety and seems therefore superior to the visible marker-tracking system. In this context, it should me mentioned that we consider the killing of animals within the studies as ethically questionable.

3.4.3. Social phobia

Social phobia is defined as an unreasonable or excessive fear of social situations and the interaction with other people that automatically brings on feelings of self-consciousness, judgment, evaluation or inferiority [14]. Symptoms of social phobia include intense fear, blushing, sweating, a dry mouth, trembling, a racing heart and shortness of breath. There are two subtypes of social phobia: specific social phobia, that is limited to a small number of fear inducing situations, and generalized social phobia, that involves almost all social situations. Situations that may evoke fear include speaking in public, establishing contacts, protecting one's interests and being under scrutiny. Usually, persons suffering from social phobia are worried that their fear is being noticed by others. Social phobia is one of the most commonly observed mental disorders, showing a life-time prevalence of 13%, according to the USA National Comorbidity Survey [45].

Roy et al. [46] presented a clinical protocol to assess the effectiveness of VR treatment of social phobia, describing the study structure, assessment tools, and content of the therapy sessions. Four virtual environments were used to reproduce situations inducing high levels of anxiety in social phobics: Performance, intimacy, scrutiny, and assertiveness. In a preliminary study, the effectiveness of VR treatment was demonstrated in *N*= 10 persons suffering from social phobia in a between-subjects design. In a following study conducted by the same research group [47], the same virtual environments were used to examine the effectiveness of VR exposure in *N*= 36 social phobics. Participants were assigned to either VR treatment or cognitive-behavioral group therapy (CBGT). After twelve weeks of therapy, both treatment groups showed clinically and statistically significant improvement. In a more recent RCT, the effectiveness of VR treatment, a combination of CBT and VR exposure and a waitlist control condition were compared in *N*= 45 participants diagnosed with social phobia [48]. Results indicated a significant reduction of anxiety on all questionnaires for both treatment groups in contrast to the waitlist control group.

Furthermore, a few studies have focused specifically on the effectiveness of virtual environments in treating *public speaking anxiety*. Harris et al. [49] showed that four VR treatment sessions of 15 minutes each (see table 1) reduced self-reported anxiety as well as physiological reactions significantly in eight students suffering from public speaking anxiety in comparison to six students in a waitlist control group. In an open clinical trial, the effectiveness of four sessions of anxiety management and four subsequent therapy sessions was examined, using a virtual audience in *N*= 10 participants diagnosed with social phobia [50]. As a result, self-report measures indicated lower levels of public speaking anxiety at post treatment and three months follow-up. However, participation rates of giving a free speech to an actual audience did not differ before and after the treatment. A larger sample size of *N*= 88 persons with public speaking anxiety was used in a RCT that compared the effectiveness of CBT, VR and CBT combined, and a waitlist control condition [51]. Results demonstrated significant improvements of both treatment groups on self-rated anxiety during a behavioral task and four out of five anxiety measures in contrast to the control group. At one year follow-up, results remained stable.

Other studies have concentrated on specific aspects of treating social phobia with the aid of VR. For instance, Ter Heijden and Brinkmann [52] evaluated speech detection and recognition techniques in comparison to a human control condition in a VR surrounding. Interactions were observed in two phobic and 24 healthy persons. Results indicated that automatic speech techniques often did not show any significant differences compared to manual speech. Therapist workload of entering speech content might therefore be minimized in VR treatment.

Initial interview	Besides self-report instruments measuring social anxiety, a voice test sample was recorded while the participants answered a question and read a paragraph. The heart rate was measured during the speaking test and a brief relaxation exercise.
Session 1	Participants stood at a podium with a microphone, looking around a virtual empty auditorium to get accustomed to the environment. Subsequently, participants were asked to talk about their public speaking anxiety.
Session 2	Participants were asked to say the American Pledge of Allegiance. The auditorium was gradually filled with people, and applause was used to encourage participants. The pledge was repeated, with the virtual audience applauding at the end of the recitation.
Session 3	Participants were asked to deliver a 2-min speech with a small light on the clipboard. The room was gradually filled with audience, people were speaking to each other, laughing, asking the speaker to speak louder, and applauding at the end of the speech. Afterwards, the speech was repeated.
Session 4	Participants were asked to give the same or another speech. Manipulations of the scenario were made as in session 3.

Table 1. Procedure of VR treatment for public speaking anxiety [49]

Another study brought the aspect of presence within VR exposure into focus [53]. The relationship of three components of presence (spatial presence, involvement, and realness), fear ratings during VR, and treatment effectiveness were evaluated in *N*= 41 participants suffering from social phobia. The authors found an association between total presence as well as realness subscale scores and fear-ratings during treatment, while only scores on the involvement subscale were able to significantly predict treatment outcome.

VR environments may also facilitate research on specific aspects of social phobia. For example, Cornwell et al. [54] used a VR setting to examine physiological reaction of persons diagnosed with social anxiety disorder in social-evaluative threat situations. Participants were asked to deliver a short speech in front of a virtual audience. In this way, no actual audience has to be recruited in order to realize study designs of that kind.

3.4.4. Panic disorder

Around 5% of the US Americans suffer from *panic disorder* in their lifespan [55]. Panic disorder is diagnosed if a panic attack, including symptoms such as sweating, palpitations, trembling, nausea, derealization and depersonalization, results in consistent concern about having additional attacks, worries about its consequences or behavioral changes. Persons suffering from panic disorder frequently develop agoraphobic avoidance behaviors. Agoraphobia refers to anxiety about being in places or situations from which escape might be difficult or in which help may not be available in case of having an unexpected panic attack, such as being in a crowd, on a bridge, train or the like. Therefore, agoraphobia often has to be included in the treatment of panic disorder as well.

Vincelli et al. [56] presented a treatment protocol called Experiental Cognitive Therapy (ECT), which integrates VR in order to treat panic disorder and agoraphobia. Its effectiveness was demonstrated in *N*= 12 patients, who were randomly assigned to an ETC group and therefore undergoing VR exposure, a CBT group or a waitlist control group. Results indicated that eight sessions of ECT and twelve sessions of CBT equally reduced the number of panic attacks, the level of depression and state and trait anxiety.

In a following study examining the effectiveness of VR in the treatment of panic disorder, *N*= 40 participants received either four sessions of cognitive therapy including VR exposure, or twelve sessions of a panic control program [57]. Results indicated that both treatments were equally effective. However, findings did not remain stable at six months follow-up, where participants of the panic disorder program showed higher overall functioning. Botella et al. [58] used a more rigid study design in order to compare *N*= 37 persons receiving nine sessions of either CBT with VR exposure or CBT with in-vivo exposure and a waitlist control group. Both treatment groups obtained equal symptom reductions in comparison to the waiting list, and results were maintained at twelve months follow-up. A following study using a between-subjects design compared eleven sessions of CBT including exposure in-vivo with CBT and VR exposure in *N*= 28 participants diagnosed with panic disorder [59]. All participants additionally received antidepressant medication, and a BAT was applied to assess treatment effects. Results revealed that both treatments were equally effective, and results remained stable after three months. Findings were replicated in a RCT by the same research group in *N*= 27 participants with panic disorder and agoraphobia [60].

A later study evaluated effectiveness of interoceptive exposure in a virtual environment, simulating physical sensations through audible stimulation such as rapid heartbeat and panting, and visual stimulations such as blurry or tunnel vision [61]. Results indicated that both IE using VR and traditional IE significantly reduced symptoms of panic disorder, and that results were maintained or even improved at three months follow-up. Finally, Meyerbröker et al. [62] showed that varied levels of presence by using either a CAVE or a HMD did not influence effects of VR treatment of panic disorder.

3.4.5. Obsessive-compulsive disorder

Obsessive-compulsive disorder (OCD) is a debilitating mental disorder that is characterized by either obsessions, compulsions, or both. According to the DSM-IV, obsessions are defined as recurrent and persistent thoughts, impulses, or images that may cause anxiety, including the obsession of contamination, need for symmetry or aggression [13]. Compulsions refer to repetitive behaviors, such as hand washing, ordering, and checking, or mental acts such as praying, counting, or repeating words silently, that are performed to respond to an obsession or to rules in order to reduce distress. Lifetime prevalence rates are estimated about 2% worldwide.

While the benefits of computer-based assessment and treatment of OCD has already been demonstrated [63], only preliminary data concerning the use of VR in the treatment of OCD is available. A South Korean research group presented first results of VR exposure therapy of OCD [63]. *N*= 33 participants with OCD and *n*= *30* healthy controls navigated through a virtual environment, consisting of a training, distraction, and main task phase. Anxiety rates as well as decreased ratio of anxiety during the main task were significantly higher in participants with OCD than healthy controls. VR may therefore function as anxiety-provoking and a potential treatment tool for OCD. The same virtual environment was used to examine its potential efficacy in assessing OCD in *n*= 30 patients with OCD and *n*= 27 matched healthy controls [64]. Results indicated that OCD patients had significantly greater difficulties with compulsive checking than controls, and that task performance was positively correlated with self-reported symptoms as well as interviewer-rated measures of OCD. Another study by the same research group demonstrated that anxiety levels of *N*= 24 healthy participants decreased as a result of performing virtual arrangement tasks three times with three-day intervals [65]. However, the amount of anxiety reduction depended on the type of task, and only the Symmetry, Ordering and Arrangement Questionnaire (SOAQ) showed significant correlation with anxiety. Nevertheless, VR seems to be a potential device for the assessment and treatment of persons with symptoms of arranging compulsion.

3.4.6. Posttraumatic stress disorder

Posttraumatic stress disorder (PTSD) is a serious condition that persons experiencing a traumatic event may suffer from. According to the ICD-10, the traumatic event needs to be exceptionally threatening or catastrophic and would distress most people. Such disasters can be either manmade, as it is the case in war, torture, or sexual abuse, or they can be natural disasters, such as earthquakes, accidents, or life-threatening diseases. Criteria for PTSD include intrusions such as flashbacks and repeating dreams, avoidance of situations similar to the traumatic event, loss of memory about certain aspects of the event, and symptoms of hyperarousal. Experiencing psychological distress right up to PTSD is common among military members who are constantly confronted with threatening situations. Approximately 18% of warfighters returning from Iraq and 11% returning from Afghanistan were screened positive for PTSD.

Some authors suggest the application of new treatment approaches such as VR exposure, reasoning that conventional therapy approaches may be rejected by war veterans due to stigmatization and that in-vivo exposure is not possible. In fact, situations that caused the traumatization are difficult to frequent, but according to traumatherapy, this is neither necessary nor indicated [66]. On the contrary, exposure to virtual settings that are reconstructing traumatizing situations is ethically questionable, which is demonstrating by the following scenarios.

A case report describes the first application of VR for a Vietnam veteran suffering from PTSD [67]. As a result of VR treatment, he significantly improved on all PTSD measures and those gains remained stable at six months follow-up. A following open clinical trial demonstrated the effectiveness of VR in ten male Vietnam veterans diagnosed with PTSD [68]. They underwent eight to 16 sessions of VR exposure in two virtual environments: a virtual helicopter flying over Vietnam, and a clearing surrounded by jungle. Participants showed significant PTSD symptom reductions on the Clinician Administered PTSD Scale (CAPS) at six months follow-up, declaring symptom reductions ranging from 15 to 67% in an interview. Self-reported intrusion symptoms as measured by the Impact of Event Scale were significantly lower at three months follow-up in comparison to baseline, but not at six months follow-up. Another open trial of VR in the treatment of *N*= 21 Vietnam veterans was conducted by Ready and collaborators [69], simulating virtual environments in response to participants' memories. Participants showed significant symptom reductions at three and six-months follow-up, even though two participants experienced an increase in symptoms during VR exposure. A RCT using a small sample size was presented by the same research group [70]. Eleven Vietnam veterans were assigned to either ten sessions of VR exposure or present-centered therapy, utilizing a problem-solving approach. Results indicated no significant differences between treatment groups at posttreatment and six months follow-up.

Furthermore, several studies examined the use of a virtual environment to treat veterans returning from "Operation Iraqi Freedom" who suffered from PTSD. The "Virtual Iraq/ Afghanistan" environment was adapted from the Microsoft® X-box game "Full Spectrum Warrior". Scenarios include a Middle Eastern city and a Humvee driving down a desert highway. Auditory, visual, olfactory and vibrotactile stimulation such as gunfire, weather conditions, the smell of burnt rubber and the sensation of a moving car can be adjusted. Some examples of Virtual Iraq/Afghanistan are shown in Figure 3.

Several case reports were conducted [e.g. 71, 72]. The first clinical trial assessing the effectiveness of exposure using "Virtual Iraq" indicated clinically and statistically significant symptom reduction in *N*= 20 participants [73]. In addition, McLay and collaborators [74] presented a RCT comparing the efficacy of VR exposure therapy and usual CBT in a sample of *N*= 19 Iraq veterans diagnosed with PTSD. Within the VR condition, seven out of ten participants improved at least 30% on the CAPS, while only one out of nine within the CBT condition showed similar improvement. The effectiveness of "Virtual Iraq" was also supported by Reger et al. [75] in 24 Iraq and Afghanistan veterans. Moreover, a pilot study focusing on the combination of VR exposure and cognitive enhancing medication has shown promising results [76].

Another VR environment was created to treat Portuguese survivors of the 1961-1974 wars in Africa. Subsequently to a case study [77], Gamito and colleagues [78] examined the effectiveness of a VR war environment to imaginal exposure and a waiting list condition. Participants in the VR condition showed significant reduction of depressive and anxiety symptoms.

An increased incidence of PTSD was also detected among the survivors of the attacks of September 11, 2001. Consequently, Difede and Hoffman developed a virtual environment simulating jets crashing into the World Trade Center, people jumping to their deaths from the buildings, and towers collapsing. A study revealed that participants in a VR condition (*n*= 9) showed significantly greater improvement on CAPS scores than the waitlist control group (*n*= 8) [79]. Findings were replicated in a following study [80].

In a further study, a VR surrounding was developed in order to treat victims of a terrorist bus bombing in Israel. The potential effectiveness of "BusWorld" was demonstrated in a study examining 30 asymptomatic participants, who showed significantly higher mean subjective units of discomfort scores (SUDS) with increasingly distressful scenarios. Treatment of a 29-year-old victim of a bus bombing using VR resulted in significant reduction of PTSD symptoms as measured by the CAPS [81].

Another field of application in the treatment of PTSD by the means of VR exposure is made up by motor vehicle accident survivors. Saraiva et al. [82] presented a case study describing positive outcomes of VR exposure of a 42-year-old female in the aftermath of a serious vehicle accident. Findings were confirmed by Beck et al. [83], who demonstrated significant reductions of re-experiencing, avoidance, and emotional numbing in six persons reporting subsyndromal PTSD after completing ten sessions of VR treatment.

A new approach of treating PTSD was introduced by Fidopiastis et al. [84]: As aforementioned, AR, referred to as Mixed Realities (MR) by the authors, are supposed to blend virtual content into the real world, which means that computer-generated objects can be superimposed on the real-world environment. In a pilot study, first promising effects of MR in the assessment of PTSD by capturing the patient's interaction with the simulated environment were demonstrated. Riva et al. [85] further advanced the approach of MR by presenting the paradigm of Interreality, which is supposed to bridge the virtual and real world by using activity sensors, personal digital assistants or mobile phones.

However, even though treatment of persons suffering PTSD is crucial, study designs using VR seem questionable with regard to ethical concerns. Exposing war veterans or victims of terror attacks to simulated war scenarios is contra-indicated according to current research on trauma therapy. Certain phases of traumatherapy such as stabilization and development of a therapeutic relationship have to precede the processing of the traumatic experience [86]. In the studies presented here, none of these phases were considered so that VR exposure bore the risk of retraumatization. Besides, if the virtual setting does not in detail project the traumatic event, renewed traumatisation is risked. To date, long-term effects of exposing persons with PTSD to virtual environments are mostly unknown, because efficacy studies rarely collect follow-up data. Therefore, even though the use of VR technology seems feasible in the treatment of PTSD, ethical concerns and aspects with regard to therapy indication always need to be considered.

Figure 3. Virtual Iraq/ Afghanistan scenarios. Courtesy of Virtually Better Inc. and University of Southern California, Institute for Creative Technologies.

3.4.7. Other applications

Virtual environments have also been applied in the assessment, treatment and research of other mental disorders such as eating disorders, sexual dysfunctions, schizophrenia, attention deficit disorder, and addictions. In the following, a cursory overview of VR treatment approaches in those clinical pictures will be provided.

The first use of VR in treating *eating disorders* was accomplished by an Italian research group [87]. VR programs focused on the improvement of body image, body satisfaction and physical acceptance in obese patients and reduction of perfectionisms, body dissatisfaction and negative attitudes towards the body in anorectic patients. A few controlled studies demonstrated the effectiveness of VR in the treatment of eating disorders [88-90].

In contrast, research on the effectiveness of VR in the treatment of *sexual dysfunction* is still in an experimental stage. In one study, VR was integrated into psychodynamic psychotherapy for the treatment of erectile dysfunction and premature ejaculation in N= 160 men [12]. VR seemed to help to work through events and associations that were creating the sexual problems. Positive outcomes remained stable after one year follow-up [91].

Furthermore, VR has been used to assess attention impairments in order to diagnose *Attention Deficit Disorder* (ADD). A VR classroom scenario was created, in which children had to perform attention task while being distracted by classroom noises, activities occurring outside the window, or persons passing by [92]. In a clinical trial conducted by the same research group, it was demonstrated that the system could reliably distinguish between children with ADD and healthy controls.

Another virtual environment was developed to treat people with *schizophrenia* [93]. VR scenarios can be individually tailored to simulate patients' hallucinations, such as voices or walls appearing to close in, to teach patients to ignore hallucinations in real life. But even though VR treatment might be an effective adjunct in the treatment of schizophrenics, indications for VR exposure have to be carefully pondered.

The treatment of *addiction* by means of VR seems to be a promising area as well. Similar to exposure therapy in specific phobias, repeatedly showing cues of alcohol or tobacco should lead to extinction of craving. Virtual environments presenting virtual cigarettes [94] or bottles and glasses of alcohol [95] were able to significantly decrease craving.

4. Implications for research and therapy

Hereafter, the findings presented here will be discussed with regard to their implications for research and therapy.

4.1. Research

The studies examining the efficacy of VR treatment in psychotherapy that were conducted up to that point are various with respect to their designs, treatment methods, and results. A multitude of case reports and pilot studies with questionable generalizability were published to demonstrate that VR can be an effective tool in the treatment of mental disorders. To provide a clearer overview of the studies proving the effectiveness of VR in anxiety disorders, the study designs and results of all controlled trials were listed in table 2. All controlled trials that examined the VR or AR treatment of at least one group of participants suffering from an anxiety disorder and that used a standardized outcome measure of anxiety were included. As the table shows, most RTCs have been effected in the field of aviophobia. Particularly with reference to specific phobias, considerable systematic research has been conducted in the past years. However, while the effectiveness of VR and

Study	Clinical Sample	*N*	Design	No. sessions	Results
Fear of heights					
Coelho et al. (2008)	Acrophobia	15	Between-subjects	3	VR and in vivo exposure were equally effective, despite shorter treatment times of VR
Emmelkamp et al. (2001)	Acrophobia	10	Within-subjects	2	Exposure in vivo did not lead to any significant improvements after VR exposure
Emmelkamp et al. (2002)	Acrophobia	33	RCT	3	VR and in vivo exposure were equally effective; results stable after 6 months
Krijn et al. (2004)	Acrophobia	37	RCT	3	VR administered by HMD and CAVE were equally effective; results stable after 6 months
Krijn et al. (2007)	Acrophobia	26	RCT	4	Self-statements did not additionally enhance effectiveness of VR treatment
Fear of flying					
Krijn et al. (2007)	Aviophobia	59	RCT	4	VR treatment and CBT were equally effective and superior to bibliotherapy
Malty et al. (2002)	Aviophobia	45	RCT	5	VR treatment was superior to attention placebo group on self-report measures, but not avoidance test; results not stable after 6 months
Mühlberger et al. (2001)	Aviophobia	30	RCT	1	VR treatment and relaxation training were equally effective
Mühlberger et al. (2003)	Aviophobia	45	RCT	1	VR treatment in combination with CBT was more effective than CBT alone
Mühlberger et al. (2006)	Aviophobia	30	RCT	1	Presence of a therapist did not influence effectiveness of VR treatment
Rothbaum et al. (2000)	Aviophobia	49	RCT	8	VR and in vivo exposure were equally effective in comparison to a waitlist control group; results stable after 6 and 12 months
Rothbaum et al. (2006)	Aviophobia	75	RCT	8	VR and in vivo exposure were equally effective in comparison to a waitlist control group; results stable after 6 to 12 months
Tortella-Feliu et al. (2011)	Aviophobia	60	RCT	6 max.	VR exposure, computer-aided exposure with a therapist's assistance, and self-administered computer-assisted exposure were equally effective; results stable after 12 months
Wiederhold & Wiederhold (2003)	Aviophobia	30	RCT	8	VR exposure in combination with biofeedback was more effective than VR exposure alone
Fear of spiders/ cockroaches					
Bochard et al. (2006)	Arachnophobia	11	Within-subjects	5	Modified 3D computer games were effective in the treatment of arachnophobia
Garcia-Palacios et al. (2002)	Arachnophobia		Between-subjects	4 on average	83% of the VR exposure group showed clinically significant improvement, in comparison to 0% of the waitlist control group

Study	Clinical Sample	*N*	Design	No. sessions	Results
Hoffman et al. (2003)	Arachnophobia	36	RCT	3	VR treatment including tactile stimulation was more effective than VR without tactile stimulation; both treatment groups were superior to waitlist control group
Juan et al. (2005)	Arachnophobia, cockroach phobia	9	Open trial	1	AR treatment significantly reduced participants' fear and avoidance
Michaliszyn (2010)	Arachnophobia	43	RCT	8	VR and in vivo exposure groups showed clinically significant improvement in comparison to waitlist control group
Social Phobia					
Anderson et al. (2003)	Social Phobia	10	Within-subjects	8	The combination of VR and anxiety management resulted in reduction of public speaking anxiety on self-report; stable at 3 months follow-up
Harris et al. (2002)	Fear of public speaking	14	Between-subjects	4	VR treatment reduced self-reported anxiety and physiological reactions significantly in comparison to waitlist control group
Klinger et al. (2005)	Social Phobia	36	RCT	12	VR treatment and CBT showed equally significant improvements in anxiety and avoidance behavior
Price et al. (2011)	Social Phobia	41	RCT	8	Involvement score predicted therapy outcome
Robillard et al. (2010)	Social Phobia	45	RCT	16	VR treatment and combination of CBT and VR were both effective in comparison to waitlist control group
Roy et al. (2000)	Social Phobia	10	Between-subjects	12	VR treatment and CBT equally showed significant improvements in anxiety and avoidance behavior
Wallach et al. (2009)	Fear of public speaking	88	RCT	12	CBT as well as VR and CBT combined resulted in significant improvements of self-rated anxiety and 4 out of 5 anxiety measures in contrast to waitlist control group; results stable at 12 months follow-up
OCD					
Kim et al. (2008)	OCD	63	Matched between-subjects	1	Participants with OCD experienced significantly higher anxiety, but also showed a higher decreased ratio of anxiety than healthy controls
Panic disorder					
Botella et al. (2007)	Panic disorder	37	RCT	9	CBT including VR exposure and CBT including exposure in vivo resulted in equal symptom reductions in comparison to waitlist control group; results stable at 12 months follow-up

Study	Clinical Sample	*N*	Design	No. sessions	Results
Choi et al. (2005)	Panic disorder	40	RCT	12	CBT including VR exposure and a panic disorder program were equally effective; results did not stable at 6 months follow-up
Penate et al. (2008)	Panic disorder	27	Matched between subjects	11	CBT including VR exposure and CBT including exposure in vivo were equally effective in addition to antidepressive medication
Pérez-Ara et al. (2010)	Panic disorder		Between-subjects	8 max.	Interoceptive exposure using VR and traditional interoceptive therapy equally reduced symptoms; results stable at 3 months follow-up
Pitti et al. (2008)	Panic disorder	28	Matched between-subjects	11	CBT including VR exposure and CBT including exposure in vivo were equally effective in addition to antidepressive medication
Vincelli et al. (2003)	Panic disorder	12	RCT	9	VR treatment and CBT equally reduced the number of panic attacks, the level of depression and state and trait anxiety
PTSD					
Beck et al. (2007)	subsyndromal PTSD	6	Within-subjects	10	Motor vehicle accident survivors showed significant reductions of re-experiencing, avoidance, and emotional numbing after VR treatment
Difede et al. (2006)	PTSD	17	Between-subjects	14	Survivors of 9/11 undergoing VR exposure showed significantly greater improvement on CAPS scores than waitlist control group
Difede et al. (2007)	PTSD	21	Quasi-experimental	14 max.	Survivors of 9/11 undergoing VR exposure showed significantly greater improvement on CAPS scores than waitlist control group
Gamito et al. (2010)	PTSD	10	Between-subjects	12	Portuguese war veterans in the VR condition showed significant reduction of depressive and anxiety symptoms in comparison to waitlist control group
Ready et al. (2006)	PTSD	14	Open trial	20 max.	Vietnam veterans showed significant symptom reductions at 3 and 6 months follow-up; 2 participants experienced an increase in symptoms during VR exposure
Ready et al. (2010)	PTSD	11	RCT	10	No significant differences between VR and present-centered therapy at posttreatment and 6 months follow-up in Vietnam veterans

Study	Clinical Sample	*N*	Design	No. sessions	Results
Rizzo et al. (2009)	PTSD	20	Open trial	10	Participants of "Virtual Iraq" showed clinically and statistically significant symptom reductions
Reger et al. (2011)	PTSD	24	Open trial	3-12	Significant symptom reduction in Iraq or Afghanistan active duty soldiers
Rothbaum et al. (2001)	PTSD	10	Open trial	8-16	Vietnam veterans showed significant symptom reductions on the CAPS at 6 months follow-up; self-reported intrusion symptoms were significantly lower at 3 but not at 6 months follow-up

Table 2. Overview of VR treatment outcome studies

AR exposure in treating specific phobias seems to be proven, the application of VR in more complex disorders like panic disorder, obsessive-compulsive disorder, and PTSD needs to be further evaluated. Treatment protocols in this field of research are still in an experimental phase and lack controlled studies to prove their effectiveness. In addition, the majority of studies examining the effects of VR-based treatment combines different treatment approaches and therefore makes it difficult to analyze VR outcomes separately. Future research should also work out which groups of patients benefit most from VR and how environments can be adapted to patients' needs. Additionally, comparable outcome measures such as behavioral avoidance tests should be included in future studies. Finally, sample sizes are often too small to generalize study findings and longer-term catemneses are frequently missing.

Alongside the realization of further outcome studies, future research should focus on underlying cognitive and physiological processes of VR exposure. Moreover, the role of the therapist-patient-relationship should be further investigated. Although some studies indicate that the assistance of a therapist might be reduced (e.g. [35]), the consequences of a changing role of the therapist still need to be explored. For instance, the exposure of war veterans to frightening war scenarios might impair trust towards the therapist and therefore influence treatment outcome.

4.2. Therapy

A significant number of studies has furnished evidence for the effectiveness of using VR in psychotherapy. However, if therapists decide to include VR into treatment sessions, they should act in accordance with certain guidelines in order to abet positive outcomes and minimize negative treatment effects. To date, just a few treatment manuals have been published. For example, Rothbaum et al. [95] presented an abbreviated treatment manual for exposure therapy of acrophobia. VR was used to replace conventional exposure as a component of behavioral therapy. According to the manual, treatment sessions should include symptom assessment, breathing retraining, cognitive restructuring, hyperventilation exposure and VR exposure. The authors recommend arranging VR settings as follows:

1. Sitting on plane, engines off
2. Sitting on plane, engines on
3. Taxiing
4. Takeoff
5. Smooth flight
6. Landing
7. Thunderstorm and turbulent flight

Another treatment manual was developed by Spira et al. [96]. The authors describe in detail twelve steps to treat combat-related PTDS with the aid of meditation, biofeedback, and VR.Furthermore, Bouchard et al. [98] outlined a treatment manual for VR exposure therapy of specific phobias, can be used with different VR software. In approximately eight sessions, patients are supposed to overcome their fears and stop avoidance behaviors by participating in cognitive restructuring and graduated VR exposure. In addition, guidelines to enhance the sense of presence and minimize potential negative side effects of immersion are provided. However, even though first publications are promising, more evidence-based treatment manuals focusing on specific syndromes are required in order to advance VR usage in psychotherapy.

5. Conclusion

The current state of research presented in this article furnishes considerable evidence for the effectiveness of virtual and augmented environments in the treatment of several mental disorders. However, VR treatment is not yet part of ordinary mental health care. Possible explanations for that could be:

1. *Costs:* Acquisition of VR equipment is (still) expensive, and training is needed to apply VR tools.
2. *Reservations against technology:* A myriad of therapists have reservations regarding modern technologies and therefore do not consider using them. German studies demonstrated a relatively high readiness to make use of therapy that integrates text messages or e-mails, but not VR [99, 100].
3. *Limited indications:* VR exposure might be contraindicated in patients with PTSD or co-morbid mental disorders. Despite the potential benefits of using modern technologies in psychotherapy, indications with regard to the patient and specific disorder always need to be balanced. In some cases, VR treatment might not be as efficient as conventional therapy.

On the other hand, in the case of obvious indication of VR treatment, therapists should be open with respect to embedding VR technologies into therapy. Those who apply VR in therapy should be aware that VR tools always have to be part of a broader therapy plan and only complement, but cannot replace the skills of well-trained clinicians. Advantages of VR treatment include:

1. *Cost reduction:* With further technological advancements as well as increased amounts of research on the effectiveness and applicability of VR and AR, technical and financial costs of those tools will probably be reduced [1].

2. *Lower logistic efforts:* Using VR usually reduces logistic and therefore financial costs in the long term, because no real places have to be accessed in order to expose patients to fear inducing situations.
3. *Controllability of settings:* Virtual scenes can be easily controlled and adjusted to each patient`s needs.
4. *Therapy motivation:* Use of VR might increase therapy motivation, especially in younger patients. For these reasons, integration of virtual environments into day-to-day clinical practice will hopefully be extended in the future. According to a representative survey of the German population [101], 15.7% (*n*= 375) of the respondents would "maybe" make use of VR treatment in case of suffering from a phobia, and 7.4% (*n*= 177) estimated the use of VR as "rather" or "very probable". The results indicated that persons under the age of 35 who were already familiar with modern technologies were most likely to consider VR therapy. It can therefore be concluded that a certain group of the population would take advantage of an expanded offer of VR in psychotherapy.

Author details

Christiane Eichenberg and Carolin Wolters
University of Cologne, Germany

6. References

[1] Segal R, Bhatia M, Drapeau M (2011) Therapists' Perception of Benefits and Costs of Using Virtual Reality Treatments. *Cyberpsychology, Behavior, and Social Networking 14*(1-2): 29–34.

[2] Hobbs CN, Kreiner DN, Honeycutt MW, Hinds RM, Brockman CJ (2010) The Illusion of Control in a Virtual Reality Setting. *North American Journal of Psychology, 12*(3): 551-564.

[3] Botella CM, Juan MC, Banos RM, Alcaniz M, Guillén V, Rey B (2005) Mixing realities? An application of augmented reality for the treatment of cockroach phobia. *Cyberpsychology & Behavior, 8*(2): 162-171.

[4] Eichenberg C (2011) Application of „Virtual Realities" in Psychotherapy: Possibilities, Limitations and Effectiveness. In J.-J. Kim (ed.), *Virtual reality (pp 481-496).* Rijeka: InTech.

[5] Glantz K, Rizzo A, Graap K (2003) Virtual Reality for Psychotherapy: Current Reality and Future Possibilites. *Psychotherapy: Theory, Research, Practice, Training, 40*(1/2): 55-67.

[6] Garcia-Palacios A, Hoffman HG, See SK, Tsay A, Botella C (2001) Redefining therapeutic success with virtual reality exposure therapy. *CyberPsychology & Behavior 4*: 341–8.

[7] García-Palacios A, Botella C, Hoffman H, Fabregat S. (2007) Comparing Acceptance and Refusal Rates of Virtual Reality Exposure vs. In Vivo Exposure by Patients with Specific Phobias. *CyberPsychology & Behavior 10* (5): S. 722–724.

[8] Parsons TD, Rizzo A (2008) Affective Outcomes of Virtual Reality Exposure Therapy for Anxiety and Specific Phobias: A meta-analysis. *Journal of Behavior Therapy and Experimental Psychiatry, 39:* 250-261

[9] Powers MB, Emmelkamp PMG (2008) Virtual reality exposure therapy for anxiety disorders: A metaanalysis. *Journal of anxiety disorders 22* (3): 561-569

[10] Kessler RC, Chiu WT, Demler O, Merikangas KR, Walters EE (2005) Prevalence, severity, and comorbidity of 12-month DSM-IV disorders in the national comorbidity survey replication. *Archives of General Psychiatry, 62:* 617-627

[11] Eichenberg, C. (2007). Der Einsatz von „Virtuelle Realitäten" in der Psychotherapie: Ein Überblick zum Stand der Forschung. *Psychotherapeut, 52, 5,* 362-367.

[12] Optale G, Marin S, Pastore M, Nasta A, Pianon C. (2003) Male Sexual Dysfunctions and Multimedia Immersion Therapy (Follow-Up). *CyberPsychology & Behavior 6*(3): 289-294.

[13] American Psychiatric Association. (2000) *Diagnostic and statistical manual of mental disorders* (4th ed., text rev.). Washington, DC: Author.

[14] World Health Organization. (2008). *ICD-10: International statistical classification of diseases and related health problems (10th Rev. ed.).* New York, NY: Author.

[15] Curtis, GC, Magee, WJ, Eaton, WW, Wittchen, H-U & Kessler, RC (1998) Specific fears and phobias. Epidemiology and classification. *The British Journal of Psychiatry 173:* 212-217.

[16] Rothbaum BO, Hodges LF, Kooper R, Opdyke D, Williford J, North M (1995) Virtual reality graded exposure in the treatment of acrophobie: a case report. *Behavior Therapie 26*: 547-554.

[17] Rothbaum BO (1995) Effectiveness of computer-generated (virtual reality) graded exposure in the treatment of acrophobia. *The American Journal of Psychiatry, 152* (4): 626-628.

[18] Emmelkamp P, Bruynzel M, Drost L, van der Mast C (2001) Virtual realitiy treatment in acrophobia: a comparison with exposure in vivo. *Cyberpsychologie and Behavior 3*: 335-341.

[19] Emmelkamp P, Krijn M, Hulsbosch L, de Vries S, Schuemie MJ, van der Mast C (2002) Virtual reality treatment versus exposure in vivo: a comparative evaluation in acrophobia. *Behavior Research and Therapie 4:* 509-516.

[20] Krijn M, Emmelkamp PM, Biemond R, de Wilde de Ligny C, Schuemie MJ, van der Mast CA (2004) Treatment of acrophobia in virtual reality: the role of immersion and presence. *Behav Res Ther 42*(2): 229-39.

[21] Krijn M, Emmelkamp PMG, Olafsson RP, Schuemie MJ, van der Mast CA (2007) Do self-statements enhance the effectiveness of virtual reality exposure therapy? A comparative evaluation in acrophobia. *CyberPsychol Behav 10*: 362–370.

[22] Coelho CM, Santos JA, Silva C, Wallis G, Tichon J, & Hine, TJ (2008) The role of self-motion in Acrophobia Treatment. *CyberPsychology & Behavior, 11*(6): 723-725.

[23] Coelho, CM, Silva, CF, Santos JA, Tichon J, & Wallis G (2008) Contrasting the Effectiveness and Efficiency of Virtual and Real Environments in the Treatment of Acrophobia. *PsychNology, 6*(2): 203-216.

[24] Choi YH, Jang DP, Ku JH, Shin MB, Kim SI (2001) Short-term treatment of acrophobia with virtual reality therapy (VRT): A case report. *Cyberpsychology & Behavior, 4*(3): 349-354.

[25] Wiederhold BK, Gevirtz RN, Spira JL (2001) Virtual Reality Exposure Therapy vs. Imagery Desensitization Therapy in the Treatment of Flying Phobia. In G. Riva, C.

Galimberti (Eds), *Towards CyberPsychology: Mind, Cognitions and Society in the Internet Age* (Kap. 14). Amsterdam: IOS Press.

[26] Rothbaum, BO, Hodges L., Smith S, Lee JH (2000) A controlled study of virtual reality exposure therapy for fear of flying. *J Consult Clin Psychol, 68:* 1020–1026.

[27] Rothbaum BO, Zimand E, Hodges L, Lang D, Wilson J (2006) Virtual reality exposure therapy and standard (in vivo) exposure therapy in the treatment of fear of flying. *Behav Ther 37*: 80–90.

[28] Wiederhold B, Gevirtz R, Spira J (2001) Virtual reality exposure therapy vs. imagery desensitization therapy in the treatment of flying phobia, in G. Riva, C. Galimberti (eds). *Towards CyberPsychology: Mind, Cognition, and Society in the Internet Age.* Amsterdam: IOS Press. pp 254-272.

[29] Wiederhold BK, Wiederhold MD (2003) Three year follow-up for virtual reality exposure for fear of flying. *Cyberpsychol Behav 6*: 441-445.

[30] Mühlberger A, Herman MJ, Wiedemann G, Ellgring H, Pauli P (2001) Repeated exposure off light phobics to flights in virtual reality. *Behav Res Ther 39*: 1033-1050.

[31] Mühlberger A, Widemann G, Pauli P (2003) Efficacy of a one-session virtual reality exposure treatment for fear of flying. *Psychother Res 13*: 323-336.

[32] Mühlberger A, Weik A., Pauli P, Wiedemann G (2006) One-session virtual reality exposure treatment for fear of flying: 1-Year follow-up and graduation flight accompaniment effects. *Psychotherapy Research 16*(1): 26-40.

[33] Maltby N, Kirsch I, Mayers M, Allen GJ (2002) Virtual reality exposure therapy fort he treatment of fear of flying: a controlled investigation. *J Consult Clin Psychol 70*: 1112-1118.

[34] Krijn M, Emmelkamp PMG, Olafsson RP, et al. (2007) Fear of flying treatment methods: virtual reality exposure vs. cognitive behavioral therapy. *Aviat Space Environ Med 78*: 121–128.

[35] Tortella-Feliu M, Botella C, Labres J, et al. (2011) Virtual Reality Versus Computer-Aided Exposure Treatments for Fear of Flying. *Behavior Modification 35*(1): 3-30.

[36] Garcia-Palacios A, Hoffman H, Carlin A, Furness TA III, Botella C (2002) Virtual reality in the treatment of spider phobia: A controlled study. *Behaviour Research and Therapy, 40,* 983-993.

[37] Rinck M, Kwakkenbos L, Dotsch R, Wigboldus DHJ, Becker ES (2010) Attentional and behavioural responses of spider fearfuls to virtual spiders. *Cognition & Emotion 24*(7): 1199–1206.

[38] Carlin AS, Hoffmann HG, Weghorst S (1998) Virtual reality and tactile augmentation in the treatment of spider phobia: a case study. *Behaviour Research and Therapie 35*: 153-158.

[39] Hoffmann HG, García-Palacios A, Carlin A, Furness III TA (2003). Interfaces that heal: Coupling real and virtual objects to treat spider phobia. *International Journal of human-computer interaction, 16*(2): 283-300.

[40] Michaliszyn D, Marchand A, Bouchard S, Martel M-O, Poirier-Bisson, J (2010) A randomized controlled clinical trial of in virtuo and in vivo exposure for spider phobia. *Cyberpsychology, Behavior and Social Networking 13(6):* 689-695.

[41] Bouchard S, Côté S, St-Jaques J, Robillard G, Renaud P (2006) Effectiveness of virtual reality exposure in the treatment of arachnophobia using 3D games. *Technology and Health Care 14*: 19-17.

[42] Juan MC, Alcaniz M, Monserrat C, Botella C, Banos RM, Guerrero B (2005) Using augmented reality to treat phobias. *IEEE Computer Graphics and Application 05*: 31-37.
[43] Bretón-López J, Quero S, Botella C, García-Palacios A, Banos RM, Alcaniz M (2010) An augmented reality system validation for the treatment of cockroach phobia. *Cyberpsychology, Behavior and Social Networking 13*: 705-710.
[44] Juan MC, Joele D (2011) A comparative study of the sense of presence and anxiety in an invisible marker versus a marker augmented reality system for the treatment of phobia towards small animals. *International Journal of Human-Computer Studies 69*(6): 440–453.
[45] Kessler, RC, McGonagle KA, Zhao S, et al. (1994) Lifetime and 12-month prevalence of DSM-III-R psychiatric disorders in the United States. Results from the National Comorbidity Survey. *Archives of General Psychiatry 51*: 8–19.
[46] Roy S, Klinger E, Légeron P, Lauer F, Chemin I, Nugues P (2003) Definition of a VR-Based Protocol to Treat Social Phobia. *CyberPsychology & Behavior 6*: 411-420.
[47] Klinger E, Bouchard S, Légeron P, Roy S, Lauer F, Chemin I, Nugues P (2005) Virtual Reality Therapy Versus Cognitive Behavior Therapy for Social Phobia: A Preliminary Controlled Study. *Cyberpsychology and Behavior 1*: 76-88.
[48] Robillard G, Bouchard S, Dumoulin S, Guitard T, Klinger E (2010) Using virtual humans to alleviate social anxiety: Preliminary report from a comparative outcome study. Annual *Review of CyberTherapy and Telemedicine 8*: 46-48.
[49] Harris SR, Kemmerling RL, North MM (2002) Brief Virtual Reality for Public Speaking Anxiety. *CyberPsychology & Behavior 5*(6): 543-550
[50] Anderson, Page L.; Zimand, Elana; Hodges, Larry F.; Rothbaum, Barbara O. (2005): Cognitive behavioral therapy for public-speaking anxiety using virtual reality for exposure. In: *Depress. Anxiety* 22 (3), S. 156–158.
[51] Wallach HS, Safir MP, Bar-Zvi M (2009) Virtual Reality Cognitive Behavior Therapy for Public Speaking Anxiety: A Randomized Clinical Trial. *Behavior Modification 33*(3): 314–338.
[52] Ter Heijden N, Brinkman W-P (2011) Design and evaluation of a virtual reality exposure therapy system with free speech interaction. *Journal of CyberTherapy and Rehabilitation 41*(1): 41-56.
[53] Price M, Mehta N, Tone EB, Anderson PL (2011) Does engagement with exposure yield better outcomes? Components of presence as a predictor of treatment response for virtual reality exposure therapy for social phobia. *Journal of Anxiety Disorders 25*(6): 763–770.
[54] Cornwell BR, Heller R, Biggs A, Pine DS, Grillon C (2011) Becoming the Center of Attention on Social Anxiety Disorder: Startle Reactivity to a Virtual Audience during Speech Anticipation. *J Clin Psychiatry 72*: 942-948.
[55] Grant BF, Hasin DS, Stinson FS, et al. (2006) The epidemiology of DSM-IV panic disorder and agoraphobia in the United States: results from the National Epidemiologic Survey on Alcohol and Related Conditions. *J Clin Psychiatry 67: 363–74.*
[56] Vincelli F, Choi YH, Molinari E, et al. (2001) A VR-based multicomponent treatment for panic disorders with agoraphobia. *Studies in Health Technology and Informatics 81*: 544–550.

[57] Choi YH, Vincelli F, Riva G, Wiederhold BK, Lee JH, Park KH (2005) Effects of group experiental cognitive therapy for the treatment of panic disorder with agoraphobia. *CyberPsychol Behav 8*: 387-393.

[58] Botella C, García-Palacios A, Villa H, et al (2007) Virtual reality exposure in the treatment of panic disorder and agoraphobia: a controlled study. *Clin Psychol Psychother 14*: 164-175.

[59] Penate W, Pitti CT, Bethencourt JM, de la Fuente J, Gracia R (2008) The effects of a treatment based on the use of virtual reality exposure and cognitive-behavioral therapy applied to patients with agoraphobia. *Int J Clin Health Psychol 8*: 5-22.

[60] Pitti CT, Penate W, de la Fuente J, et al. (2008) Agoraphobia: combined treatment and virtual reality. Preliminary results. *Actas Esp Psiquiatra 36*: 94-101.

[61] Pérez-Ara MA, Quero S, Botella C, Banos R, Andreu-Mateu S, García-Palacios A, Bréton-Lopez J (2010) Virtual reality interoceptive exposure for the treatment of panic disorder and agoraphobia. *Annual Review of CyberTherapy and Telemedicine 8*: 61-64.

[62] Meyerbröker K, Morina N, Kerkhof G, Emmelkamp PMG (2011) Virtual Realitiy Exposure Treatment of Agoraphobia: a Comparison of Computer Automatic Virtual Environment and Head-Mounted Display. *Annual Review of Cybertherapy and Telemedicine 9:* 51-56.

[63] Kim K; Kim C-H, Cha KR, Park J, Han K, Kim YK, et al. (2008) Anxiety Provocation and Measurement Using Virtual Reality in Patients with Obsessive-Compulsive Disorder. *CyberPsychology & Behavior* 11 (6), S. 637–641.

[64] Kim, K, Kim SI, Cha KR, Park J, Rosenthal MZ, Kim J-J, et al. (2010) Development of a computer-based behavioral assessment of checking behavior in obsessive-compulsive disorder. *Comprehensive Psychiatry 51*(1): 86–93.

[65] Kim, K, Roh D, Kim, SI, Kim C-H (2012) Provoked arrangement symptoms in obsessive–compulsive disorder using a virtual environment: A preliminary report. *Computers in Biology and Medicine 42*(4): 422–427.

[66] Redemann L (2003). Die psychodynamisch imaginative Traumatherapie (PITT). *ZPPM 1*(2): 1-8.

[67] Rothbaum BO, Hodges L, Alarcon R, et al. (1999) Virtual reality exposure therapy for PTSD Vietnam Veterans: a case study. *J. Trauma. Stress 12:* 263–271.

[68] Rothbaum BO, Hodges L, Ready D, et al. (2001) Virtual reality exposure therapy for Vietnam veterans with posttraumatic stress disorder. *J Clin Psychiatry 62*: 617-622

[69] Ready, David J.; Pollack, Stacey; Rothbaum, Barbara Olasov; Alarcon, Renato D. (2006): Virtual Reality Exposure for Veterans with Posttraumatic Stress Disorder. In: *Journal of Aggression, Maltreatment & Trauma* 12 (1-2), S. 199–220.

[70] Ready DJ, Gerardi RJ, Backschneider AG, Mascaro N, Rothbaum BO (2010). Comparing Virtual Reality Exposure Therapy to Present-Centered Therapy with U.S. Vietnam Veterans with PTSD. Cyberpsychology, Behavior and Social Networking 13(1): 49-54.

[71] Gerardi M, Rothbaum BO, Ressler K, Heekin M, Rizzo A (2008) Virtual reality exposure therapy using a virtual Iraq: Case report. *J. Traum. Stress 21*(2): 209–213.

[72] Tworus R, Szymanska S, Illnicki S (2010). A Soldier Suffering from PTSD Treated by Controlled Stress Exposition Using Virtual Reality and Behavioral Training. *Cyberpsychology, Behavior, and Social Networking 13*(1): 103-107.

[73] Rizzo AA, Reger G, Difede J, et al. (2009) Development and clinical results from the virtual Iraq exposure therapy application for PTSD. *IEEE Explore: Virtual Rehabilitation* 2009.

[74] McLay RN, Wood DP, Webb-Murphy JA, Spira JL, Wiederhold MD, Pyne JM, Wiederhold BK (2011) A Randomized, Controlled Trial of Virtual Reality-Graded Exposure Therapy for Post-Traumatic Stress Disorder in Active Duty Service Members with Combat-Related Post-Traumatic Stress Disorder. *Cyberpsychology, Behavior, and Social Networking 14*(4): 223–229.

[75] Reger GM, Holloway KM, Candy C, Rothbaum BO, Difede JA, Rizzo AA, Gahm GA (2011) Effectiveness of virtual reality exposure therapy for active duty soldiers in a military mental health clinic. *J. Traum. Stress 24*(1): 93–96.

[76] Rothbaum BO, Rizzo A, Difede J (2010) Virtual reality exposure therapy for combat-related posttraumatic stress disorder. In: *Annals of the New York Academy of Sciences 1208*(1): 126–132.

[77] Gamito P, Oliveira J, Morais D, et al. (2007) War PTSD: a VR pretrial case study. Annual *Review of Cybertherapy & Telemedicine 5*: 191–198.

[78] Gamito P, Oliveira J, Rosa P, Morais D, Duarte N, Oliveira S, Saraiva T (2010) PTSD Elderly War Veterans: A Clinical Controlled Pilot Study. *Cyberpsychology, Behavior, and Social Networking 13*(1): 43-48.

[79] Difede J, Cukor J, Patt I, Giosan C, Hoffman H. (2006) The Application of Virtual Reality to the Treatment of PTSD Following the WTC Attack. *Ann. N.Y. Acad. Sci.1071:* 500-501.

[80] Difede J, Cukor K, Jayasinghe N, Patt I, Jedel S, Spielman L, Giosan C, Hoffman HG (2007) Virtual Reality Exposure Therapy for the Treatment of Posttraumatic Stress Disorder Following September 11, 2001. *J Clin Psychiatry 68*(11): 1639-1647.

[81] Freedman SA, Hoffman HG, García-Palacios A, Weiss PL, Avitzour S, Josman N (2010) Prolonged Exposure and Virtual Reality Enhanced Imaginal Exposure for PTSD following a Terrorist Bulldozer Attack: A Case Study. *Cyberpsychology, Behavior, and Social Networking 13*(1): 95-101.

[82] Saraiva T, Gamito P, Oliveira J, Morais D, Pombal M, Gamito L, Anastácio M (2007) The use of VR exposure in the treatment of motor vehicle PTSD: A case-report. *Annual Review of CyberTherapy and Telemedicine 5:* 199-205.

[83] Beck JG, Palyo SA, Winer EH, Schwagler BE, Ang EJ (2007) Virtual Reality Exposure Therapy for PTSD Symptoms After a Road Accident: An Uncontrolled Case Series. *Behavior Therapy 38* (1): 39–48.

[84] Fidopiastis C, Hughes CE, Smith E (2009) Mixed reality for PTSD/TBI assessment. *Annual Review of CyberTherapy and Telemedicine 7*: 216-220.

[85] Riva G, Raspelli S, Algeri D, Pallavicini F, Gorini A, Wiederhold BK, Gaggioli A (2010) Interreality in Practice: Bridging Virtual and Real Worlds in the Treatment of Posttraumatic Stress Disorders. *Cyberpsychology, Behavior, and Social Networking 13*(1): 55-65.

[86] Fischer G (2000) *Mehrdimensionale Psychodynamische Traumatherapie MPTT. Manual zur Behandlung psychotraumatischer Störungen.* Heidelberg: Asanger.

[87] Riva, G. (1998) Virtual reality in psychological assessment: The Body Image Virtual Reality Scale. *Cyber Psychology and Behavior, 1*: 37–44.

[88] Ferrer-García M, Gutiérrez-Maldonado J (2012) The use of virtual reality in the study, assessment, and treatment of body image in eating disorders and nonclinical samples. A review of the literature. *Body Image 9*(1): 1-11.

[89] Perpiná C, Botella C, Banos RM, Marco H, Alcaniz M, Quero S (1999) Body image and virtual reality in eating disorders: Is exposure to virtual reality more effective than the classical body image treatment? *CyberPsychology & Behavior, 2*: 149–159.

[90] Riva G, Bacchetta M, Baruffi M, Molinari E (2001) Virtual Reality–Based Multidimensional Therapy for the Treatment of Body Image Disturbances in Obesity: A Controlled Study. *Cyberpsychology & Behavior 4*: 511-526.

[91] Riva G, Bachetta M, Baruffi M, Molinari E (2002) Virtual-reality-based multidimensional therapy for the treatment of body image disturbances in binge eating disorders: A preliminary controlled study. *IEEE Transactions on Information Technology in Biomedicine: A Publication of the IEEE Engineering in Medicine and Biology Society, 6*: 224–234.

[92] Rizzo A, Buckwalter J, Bowerly Tvan der Zaag C, Humphrey L, Neumann U, Chua C, Kyriakakis C, van Rooyen A, & Sisemore D. (2000) The virtual classroom: A virtual reality environment for the assessment and rehabilitation of attention deficits. *CyberPsychology and Behavior 3*: 483–501.

[93] Nowak R. (2002, June 26) VR hallucinations used to treat schizophrenia. *New Scientist*. Retrieved from http://www.newscientist.com/news/news.jsp?id_ns99992459.

[94] Girard B, Turcotte V, Bouchard S, Girard B (2009) Crushing Virtual Cigarettes Reduces Tobacco Addiction and Treatment Discontinuation. *CyberPsychology & Behavior, 12*(5): 477-483.

[95] Lee J-H, Kwon H, Choi J, Yang B-H (2007) *Cue-exposure therapy to decrease alcohol craving in virtual environment*. *CyberPsychology & Behavior 10*(5): 617-623.

[96] Rothbaum BO, Hodges L, Smith S (1999) Virtual Reality Exposure Therapy Abbreviated Treatment Manual: Fear of Flying Application. *Cognitive and Behavioral Practice 6*: 234-244.

[97] Spira JL, Wiederhold BK, Pyne J, Wiederhold MD (2006). *Virtual Reality Treatment Manual. In Virtuo Physiologically-Faciliated Graded Exposure Therapy in the Treatment of Recently Developed Combat-related PTSD*. San Diego: The Virtual Reality Medical Center.

[98] Bouchard S, Robillard G, Larouche S, Loranger C (2012). *Description of a Treatment Manual for In Virtuo Exposure with Specific Phobia.* In: C. Eichenberg (ed.) (2012). Virtual Realities. Rijeka: InTech.

[99] Eichenberg, C. & Kienzle, K. (2011). Psychotherapeuten und Internet: Einstellung zu und Nutzung von therapeutischen Online-Angeboten im Behandlungsalltag. *Psychotherapeut.*

[100] Eichenberg, C. & Molitor, K. (2011). Stationäre Psychotherapie und Medien: Ergebnisse einer Befragungsstudie an Therapeuten und Patienten. *Psychotherapeut, 2,* 162-70. DOI: 10.1007/s00278-011-0833-4.

[101] Eichenberg C, Brähler E: Das Internet als Ratgeber bei psychischen Problemen: Eine bevölkerungsrepräsentative Befragung in Deutschland. *Psychotherapeut.* DOI 10.1007/s00278-012-0893-0.

Chapter 3

Games for Health: Have Fun with Virtual Reality!

Birgit U. Stetina, Anna Felnhofer,
Oswald D. Kothgassner and Mario Lehenbauer

Additional information is available at the end of the chapter

http://dx.doi.org/10.5772/50677

1. Introduction

> *"The beginning of wisdom is the statement 'I do not know.' The person who cannot make that statement is one who will never learn anything. And I have prided myself on my ability to learn"* (1).

Some would not expect a quote like this in a famous and heavily discussed game like "World of Warcraft" ®. And yet, it expresses our novel approach to use virtual realities (VR) in combination with the playful motivating character of digital games for clinical-psychological learning purposes. Games have a spontaneous quality, a potential for joy and flow, and are characterized by intrinsic motivation. According to our experience, playfulness in psychological intervention can dramatically increase insight processes in patients (e.g. modeling problematic situations with the help of plasticine, building blocks, paper and felt pens etc.). The progress of technology (especially virtual reality) offers possibilities for psychologists of integrating playful elements in virtual reality (serious games). In the present chapter we both, address the potential for psychology originating from the inclusion of technology and discuss problematic aspects of games as well as ethical concerns.

There are several definitions of a "game". In general, a game is a system within which players traditionally engage in an artificial conflict, trying to solve a specific problem. A game is defined by rules and measured by a quantifiable outcome (2). According to Jane McGonigal, a game has four key elements: (a) A specific goal that people are willing to work for, (b) rules that stimulate creativity, (c) a feedback system that lets individuals know how they are doing with respect to the goal and (d) voluntary acceptance of the goal, rules, and feedback systems (3). In our opinion, when it is possible to define the rules for a game, it is also possible to create tailor-made games to specifically enhance psychological well-being. The general attributes of playing a game involve spontaneity, intrinsic motivation, defined levels of active engagement and distinction from any other behavior with a make-believe quality (4). It is definitely a challenge to combine the motivating and spontaneous quality of a digital game with the characteristics of a psychological intervention.

Looking at psychological research about digital games over the past years, the main focus is clear. There are more studies about the negative side effects of digital games so far (e.g., addiction, social isolation, violence) than there are regarding the possible advantages of games. Given that in the 1980's the average game player was very young, early studies focused mainly on children and adolescents. Since this population grew older, digital games have become more popular and widespread. For example, the annual reports of the Electronic Entertainment Software Association (ESA) indicate that digital games have become very popular not only for adolescents, but for adults and elder people too. 72% of the American households play computer or video games; the average game player is 37 years old (34 years in 2008) and has been playing games for 12 years. Surprisingly, 29% of gamers were over the age of 50. In general, it is a 25.1 billion dollar industry (5). Another factor is the spread of videogames from playrooms to living rooms with the help of consoles and mobile devices.

At the same time, the percentage of people suffering from psychological disorders dramatically increased. Looking at the economic costs of social phobia, they appear substantial. In western societies, the concurrent estimated costs are 136 million euro per million people (6). These facts build the basis of our research purposes. When digital games are widespread and highly in use, and the need for psychological help is increasing, why not combine both? Psychological research indicates not only negative, but also a lot of positive aspects of playing digital games. There is definitely a need and a market for digital games which are combined with an evidence-based psychological protocol. Looking at peer reviewed publications in general, there is a sudden urge of "health games" research since 2008, attributed to advancements in gaming technology (e.g. Microsoft Kinect® or Nintendo Wii®) as well as the establishment of special conferences (e.g. "Games for Health Conference") (7-9). Unfortunately, so far only a few games for health have undergone a scientific evaluation to validate their effectiveness (10). In the following chapter we discuss positive effects of games for health as well as negative aspects of playing digital games in general. It is highly important to consider negative side effects, such as excessive gaming behavior as well as social isolation before integrating digital games into therapy. At the end of the chapter, we also address ethical aspects of research in gaming environments and integrating new media into psychological intervention.

2. The "fun part"

Exposition is a very powerful technique of confrontation with an anxiety or phobia related stimulus in cognitive behavioral therapy (CBT). This technique has already been used in the middle of the 20th century by behavior therapists and was afterwards adopted by cognitive pioneers such as Aaron Beck, describing the cognitive and behavioral rationale for example in one of his books 1979 (11). Early behavioral experiments showed that direct confrontation with phobic stimuli (without avoidance) leads to habituation and a substantial decrease in anxiety in most humans and animals. The development of different confrontational methods included gradient confrontation starting with stimuli that are perceived as less dangerous on a subjective scale and proceeding to the more dangerous ones well as non-gradient procedures or flooding, which starts directly with the feared object or situation or animal etc. More recent approaches even include concepts for so called one-session-treatment, as developed by Oest for specific phobias (12).

The traditional way of confrontation is to use imagination, confrontation in sensu, or to use real life situations, confrontation in vivo. Relatively early in media development the question of possible additional forms of confrontation has been asked, e.g. "confrontation in video". Already 2002 an article in the practice oriented German speaking journal "Psychotherapeutische Praxis" has been published that mentions the use of so-called computer simulations in psychological practice as equivalent method (13).

Virtual reality as an adjunct in psychological practice is not limited to confrontational techniques. Another relevant example is the inclusion of virtual environments in newer therapies of the so called "third wave". That term has been described by Steven Hayes (2004) as description for the mindfulness and acceptance-oriented cognitive behavioral therapies that have already been included in practical work in earlier years (14). All therapies in the third wave have that common goal to undermine destructive, or maladaptive using the traditional term, cognitions, emotions and behaviors. All mentioned therapies include experiments and a lot of training to accomplish the individual goal. As well as for confrontation, exercising in real life using mindfulness or similar concepts is quite complicated because a lot of unpredictable and uncontrollable situations happen in real life. Although of course relevant, because exactly that is the complicated aspect of life for many people, for working on a special skill or awareness is easier if disturbances are reduced.

Listening to the voices of nature around for example is a very powerful and meditative exercise in mindfulness training that can easily be ruined by an emergency helicopter landing near that spot chosen for the exercise. However, virtual environments offer 100% control including the option to simulate unpredictability in certain situations. In addition the virtual environment can easily be enhanced to increase motivation based on individual preferences of the participant.

That option to increase the motivation by using the technological and psychological possibilities is the "fun part" of working with virtual environments. The work itself is based on group work and a multiprofessional team. Only technological and psychological experts together, including aspects of treatment, design, development, programming and other relevant parts of every virtual environment, are able to develop a motivating game that is effective, includes enough options to generate individual settings based on clinical psychological diagnostic aspects and is motivating enough to actually be fun to play. What is known for children is the same for adolescents. Learning in combination with fun is more effective and guarantees long term changes. Fun in games is based on an interesting storyline, graphics and all other mentioned aspects and many others. Those basic requirements for fun already show that it is not possible that a single profession is able to create all that.

It is definitely a challenge to combine VR with the motivational character of games. Although games are an essential part of all known human cultures, the impact has grown over the past decades. Games for health offer a novel multidisciplinary working approach for psychologists, game designers and computer scientists, but they should solely concentrate on the player experience (15). We do not believe that psychologists can be game designers and vice versa, we need a strong interdisciplinary cooperation to focus on the player's needs. Psychologists can be advocates for players, providing fundamental knowledge about the impact of games, the side-effects and – last but not least – elements of therapy. Game designers provide creativity, inspiration and a motivational immersing player experience.

3. Side effects of playing digital games

Before speaking about side effects or problems that might occur with intensive or even dependent gaming behavior, an underlying methodological aspect has to be pointed out strongly. Although common knowledge, we tend to forget that the coexistence of two phenomena has nothing to do with causality. Even if we would find that gamers in general suffer significantly more often from psychological disorders than the general population, (besides the fact that there is no evidence that we could ever calculate a result in that direction) we would not be allowed to conclude that they are showing more psychological disorders because of gaming. Concluding the opposite; that people with psychological disorders are more likely to start gaming than the general population, is not acceptable as well.

In addition to the problem of causality an entire group of other methodological aspects that go along with gaming research has not been mentioned yet. Hence, a short discussion about other critical research aspects is included in the chapter about ethical aspects.

As mentioned above, most research studies concentrated on the negative side effects of playing digital games over the last years. In general, the perception of gamers in mass media is that they are highly addictive, that games can lead to social isolation or evoke depressive tendencies, and promote violence or cause aggressive behavior. As discussed the methodology of such findings has to be analyzed in depth before any conclusions can be drawn.

Yet, some studies indeed show negative aspects of using online games as mentioned above, such as problematic computer gaming behavior (16), social isolation (17), and aggressive tendencies of computer gamers (18). However, contrasting findings as in (19) exist as well, which suggest that the common stereotypes of online gamers are wrong.

3.1. Problematic computer gaming

Although widely and inflationary used, there is a huge lack of conceptualization in the concepts of "pathological computer gaming" or "computer game addiction". Several definitions, criteria and descriptive symptoms make it very difficult to characterize the maladaptive behavior of problematic computer gaming. An analysis of the studies over the past years reveals that some core criteria seem to be relevant over conceptual frameworks and might therefore have an especially high relevance for operationalization (a) withdrawal symptoms, (b) relapse/loss of control, (c) tolerance, (d) preoccupation, (e) negative consequences and (f) mood modification (19-21). Some authors suggested that the elements such as mood change, tolerance and cognitive (and not behavioral!) preoccupation should be explained better by using the term "highly engaged gamers", because the players show merely symptoms according to a high level of interest in the given activities. Therefore, it makes sense to distinguish between an "addictive" and an "engaged" behavior (20). There are significant differences between highly engaged players and "addicted" players, with highly engaged players often mistaken as addicted. "Addicted" players spend significantly more time playing MMORPGs, and there are more negative consequences in their lives (resulting in poorer quality of life), in comparison to highly engaged players. Therefore it makes sense to distinguish between these two types of gamers (20, 22).

Beyond these discussions about terms and definitions of problematic behavior in computer games, recent research as in (16) indicates that there are differences between several game types regarding problematic computer gaming behavior and psychopathology. Especially "Massively Multiplayer Online Role Playing Games" (MMORPGs) are one of the game types which are often associated with problematic behavior. Especially the social elements of MMORPGs lead to addictive behavior (23, 24). Other computer game types, such as real-time strategy games and ego-shooters, are less linked to problematic computer game behavior as a form of a behavioral addiction.

3.2. Social isolation of gamers

The general stereotype of a gamer is that he or she is socially isolated and sitting alone in front of a computer. Several studies succeeded in revealing the contrary. For example, some found MMORPG players not to be socially isolated, instead interacting in real time with other players to gain goals and achievements and in order to survive (25). MMORPGs require a certain level of social skills; unsocial behavior is not accepted in most guilds of MMORPGs such as World of Warcraft®. Yet, this observation is not limited to MMORPGs only. The majority of recent Ego-Shooters and Real-Time Strategy Games are also playable online with the possibility or exigency to interact with others (26). Studies show that playing computer games for reasons like relieving loneliness or escaping from everyday life problems leads to problematic behaviors mentioned as the first negative aspect (16, 27).

3.3. Violence in computer games

Another common stereotype, especially communicated by mass media, is that the average game player is aggressive and violent. Indeed, many popular computer games on the market include explicit violent and aggressive content. Usually, playing these games is associated with real life aggression. A meta-analytic review seems to support this opinion (28). Moreover, several current studies revealed that virtual violence may lead to desensitization in real life, partly confirming these assumptions using fMRI (29). These studies show that the act of killing virtually is connected with a decreasing potential in regions of the brain which are associated with empathy. Nevertheless, the impact of computer games with violent content on aggressive behavior should be discussed very critically. In fact, there is no clear evidence about the causality. We do not know whether games cause aggressive behavior, or players with aggressive behavior tend to buy games with violent elements. This would ignore the possibilities that people with higher trait aggression play violent games, or players discover this negative emotion of aggression in a controlled environment to explore their own self (30).

4. Positive effects of virtual reality and digital games

While some researchers focus on the negative effects of computer games such as the aforementioned "game addiction" or violent behaviour, we emphasize the positive aspects of VR and digital games. Literature suggests, that with the appropriate design and use, digital games have the potential to be very effective psychotherapeutic tools (31). Technology is changing the psychologist-patient relationship in every way, and helps to provide visual and auditory stimuli that are difficult to generate in real life settings. Psychologists are able to create new VR environments in which patients can engage during a therapy. To date, the effectiveness of VR in therapy is investigated in the treatment of anxiety disorders, phobias, addiction, depression and attention-deficit hyperactivity disorder (ADHD) as well as a tool for stress management (31).

Usually patients interact in VR with the help of peripheral devices, such as head-mounted displays and keyboards. The effects of virtual exposure to flight, heights and social settings are well studied (32, 33). They can be more cost and time effective for both therapists and patients, thereby improving the accessibility of therapy to individuals who may previously have been unable to afford treatment. Over the past decades, with the spread of new technologies (e.g. Internet, video game consoles, mobile devices) in households across various social classes, psychologists tried to integrate VR to provide low-threshold, efficacious and less time-intensive interventions (34). For example, VR combined with biofeedback (BFB) is a scientifically proven method that includes technological feedback to provide body relaxation. Decreased anxiety and physiological arousal have been demonstrated in patients who gained insights into their physical arousal immediately on a computer monitor and who consequently learned to control their bodily functions. In general, there is an estimated savings of \$540 – \$630 per client when new technology is integrated in therapy (34, 35). Furthermore, a variety of research has been conducted on technology-aided interventions such as online-based self-administered interventions, and promising results concerning anxiety and stress reduction have been reported (36). Self-administered treatment programs provide cognitive-behavioural therapy (CBT) with minimal therapist support, but effective outcomes (37, 38). These programs lack game plays and reward systems, they are not games per definition. However, these studies provide answers to the crucial basic research question, if technology-aided interventions are successful at all. VR has been used successfully for the treatment of phobias, using protocols based on CBT. CBT is the so called goldstandard when it comes to treatment of phobias (39). Usually, exposition in CBT treatment can be performed in vivo or in sensu. We propose a third approach "in virtu", demonstrating similar effects to in vivo or in sensu (33, 39-41). Garcia-Palacios and colleagues state that most patients suffering from phobias prefer VR over in vivo therapy (41). Accordingly, Piercey and colleagues find that the presentation of spiders in VR produces significantly increased skin conductance responses in arachnophobic patients, who perceive the virtual surrounding as "real" (39). In general, these results suggest that the use of technology in therapy can dramatically enhance insight processes (36).

Moreover, VR can be well combined with the motivating character of (serious) games. They offer plenty of varieties of social interaction with other users or non-playing characters/avatars. Almost any real life situation can be displayed in VR. Gaming in general is perceived by many gamers as a positive emotional experience, with positive effects on emotional health. Stress and psychophysiological arousal are important triggers and maintaining factors when it comes to the aetiology of many psychological disorders. Tailor-made games provide innovative ways for health improvements using stress management methods (4). The biggest impact of games with psychological intervention purposes is the potential to increase motivation through a mechanism that is usually activated in games within the context of play. There is a certain trend in the last years, most studies focus on the outcomes of exercise and rehab games and the games concentrate on physical activity and nutrition (7). So called "exergames" are interactive video games combined with exercise and movement. Exergames such as interactive dance videogames are very popular among adolescents; they can cause an increase in heart rate and physical expenditure. In their study of "Dance Dance Revolution", Maloney and colleagues found potential to use exergames to boost physical activity among overweight and obese adolescents. There was a significant improvement of physical activity in a treatment group (12 weeks of playing "Dance Dance Revolution"), compared to a decrease in a control group (42).

The use of videogame consoles as an analgesia treatment alternative has been studied at the Burn Center of New York Presbyterian Hospital (43). Most burn injury patients feel an intense amount of pain as the body repairs itself, together with extreme anxiety. These factors can lead to less commitment to medical exercise regimens, resulting in limitations in movement. Nurses often play videos or music to distract burn patients from pain. In the last years this treatment has been combined with VR environments such as "Snow-World", a virtual reality set in cold surroundings like the Arctic to distract burn recovery patients from pain (44). Video game consoles work on similar premises. Nintendo Wii ® is a popular console with motion sensors; it allows modulating therapeutic environments as well as the duration of the exercise. Although it is difficult because of the diversity of burn patients to devise standard treatment protocols, the authors conclude in general that the potential of digital games as analgesia is enormous. Burn injury patients can use a defined set of physical games, the gaming system records the progress and the movements of the player and sends the data automatically to a therapist, who is able to adjust the therapy according to the progress (43). Yohannan and colleagues state that burn patients can benefit therapeutically from videogames. In their case study a burn patient performed better after a game treatment over two weeks in terms of reaction time and maximum excursion than conventional therapy alone. The functional mobility increased faster, there was more motivation for the therapy when game-based interventions were included (45).

Some studies focus on games especially for older people. This is no surprise given the fact, that the average game player is 37 years old, and an estimated 29% of the gamers are over 50 years old (5). Older people can benefit from digital games as well; they have a positive impact on their health condition and improve their capability to carry on activities in their everyday life. The key factors of an interest in games for health are challenge, socialization, escape from a daily routine and fun. Focus groups stated that they would use specifically designed digital games on a daily basis (46). VR and digital games provide the opportunity to support the independence of older people. Especially in rural areas, where distance makes it inconvenient for older people to access adequate health services, technology can provide easier possibilities to measure the health status. In their narrative review, Marston and Smith identified several possibilities of using technology for older people. Commercially available game consoles, such as Sony Playstation with EyeToy ® or Microsoft Kinect ® make it easier to provide older people with exergames for rehabilitation purposes (47). Several studies report a positive outcome when it comes to fall prevention and stroke rehabilitation. Especially Nintendo Wii ® facilitates wrist rehabilitation by specifically designed programs (with the possibility to add joint therapies for knees or elbows) (48). In general, the results are promising. The use of games for health (both off-the-shelf or specially designed for rehabilitation purposes) can have a dramatically increasing effect of the physical abilities of older people (47).

Contrary to studies about exergames, there are only limited studies about games especially for psychological health. One example is "SPARX", an online role-playing game for the treatment of depression (http://sparx.org.nz), based on a CBT protocol. The player can choose an avatar and undertake a series of challenges in a virtual fantasy world. The main goal of the game is to restore a balance of "Gloomy Negative Automatic Thoughts". There are seven modules (e.g., psychoeducation about depression, activity scheduling and behavioral action, dealing with emotions, problem solving, cognitive restructuring and relapse prevention) over 5 weeks available. Intervention studies revealed promising effects. There were significant decreases of depression scores, compared to a waiting-list condition. "SPARX" seems to be an innovative and effective game-based treatment for students with symptoms of depression (49). Another study examined the effects of an online-based self-administered social skills training with playful elements, especially for shy students. This online-training consist of 14 text-based lessons, combined with playful elements such as "identifying negative automatic thoughts" with the help of a drag-and-drop game. Lehenbauer and colleagues revealed significant effects of this training, a highly significant increase of social skills as well as a highly significant decrease of social fears in the intervention group, with no results in a control group undergoing no intervention (8, 50).

5. Ethical challenges

In light of manifold possible applications of technology in therapy and research, there is a discussion about related ethical principles and their field of scope. Research in virtual environments should follow all "standard" ethical principles and has to take the special setting into consideration. Interestingly especially qualitative researchers have dealt with ethical aspects of researching gaming environments over the last years (51, 52).

We suppose that the same ethical principles, which pertain to conventional psychological treatment and therapy, should also be applied in technology-aided psychological treatment (53). The manifold risks psychologists are confronted with as well as the ethical dilemmas that may arise from the adoption of technology in the realms of treatment, may be significantly different from those encountered in conventional face-to-face counseling psychology. In online-based communication, psychologists are confronted with the same probability of encountering distressing information (e.g. a suicidal attempt of a client), yet online-therapists are a lot less likely to fully apprehend the severity of this attempt if not being confronted with the patient face-to-face (54). Furthermore, absolute certainty about a client's true identity is difficult to be achieved in digital settings. It seems that these facts make it more difficult to provide effective treatment (55).Last but not least, it is necessary to consider negative side effects of digital games, as pointed out above. We believe that we can't create digital games with a specific psychological protocol, even if evidence-based and studied before, and leave the patients alone. There has to be a certain level of guidance from a psychologist. It is necessary to consider these facts to engage in a discussion about implementing new, even more specific guidelines for technology-assisted therapy. Ethical guidelines regarding the use of VR and digital games are scarce at best. To date only few mental health organizations respectively psychologists' associations have issued specific regulations, most of which pertain to online-counseling or e-therapy (56).

To date, the American Psychological Association or the European Federation of Psychologists' Associations lack specific regulations. On the other hand, specialized organizations such as the International Society for Mental Health Online (ISMHO) or the American Mental Health Counselors Association (AMHCA) stress the importance of taking into account basic ethical principles like beneficence, nonmaleficence, autonomy and confidentiality when implementing any sort of technology-assisted training. Only then, ethically sound treatment of patients can be ensured. First of all, in terms of beneficence and nonmaleficence, it has to be ensured "that clients are intellectually, emotionally, and physically capable of using technology-assisted counseling services, and of understanding the potential risks and/or limitations of such services" (57). For example, some minor groups of people who never had any contact with technology (e.g., a personal computer) or elderly people may feel intimidated by VR or digital games and may not accept them as tools for psychological treatment. Compliance may be hampered for this group of clients. Additionally, there are some mental disorders for which specific technology based treatments are not recommended. Because of their immanent difficulties of separating reality from fiction, it is argued that VR can have negative side effects for some schizophrenic and psychotic patients (55). Patients in crisis situations as well as borderline patients, who are even in a greater need of face-to-face interaction than an average patient should not be treated exclusively via the Internet or using solely VR-applications.

We emphasize that special focus has to be put on the informed consent process when intending to use technology or conducting treatment via technological applications. Patients have the right to be fully informed about the potentials and the limitations of any technology such as the circumstances under which it can or cannot be applied to the patients' situation (58). Patients have to be provided with the information about what data is being collected about them when using the technology. They should furthermore be informed about the location it is stored at and about how long it is intended to remain there. To mitigate breaches of confidentiality, all measures have to be taken by the treating psychologist to ensure secure data storage no matter what kind of data is collected (e.g. physiological measures when using biofeedback). When providing Internet based services, psychologists should utilize secure web sites, firewalls and encrypting programs (56). In order to avoid any harm to patients, psychologists should possess adequate expertise concerning the technology they apply. In terms of serious games for health, this includes knowledge about maladaptive gaming behavior as well as any other negative side effects. Psychologists are to assure that their treatment method is in no violation of local laws or ethical regulations such as professional membership organizations (57).

Although not directly connected with ethical aspects regarding humans the possibilities of VR to reduce the suffering of animals "used" in treatment has to be mentioned and underlined (59). Confrontation in virtu seems to be an extremely powerful form of exposition that might not only be used as adjunct, but even as replacement for in vivo therapy. As described above there exist a lot of different forms of specific phobias and a relevant part of those pertain to animals. Exposition in vivo clearly means that animals have to be handled by the therapist and by the patient for a long enough time to let the anxiety level go down. Speaking of a session that might last up to a couple of hours the stress on the included animal(s) is obviously unbearable. That stress may reduce the quality of life of the animal and lead to behavioral disorders as well as physical illness or under the worst circumstances it might even end the animals' life. For example including wasps in order to treat "spheksophobia" leads to the death of many animals; they tend to kill each other in an area that's not large enough in relation to their natural environment. The question has to be asked if the suffering of animals is acceptable to treat a patients' phobia. Technology offers a very interesting alternative that shows a way without the discussion of ethics in human-animal-interaction or thoughts about animal experimentation. Working with animals in virtu includes all positive aspects of virtual environments that were mentioned before and does not harm any other living being while having a powerful effect on the patient. Rethinking the ethical problems of harming an animal or worse might be a burden for some professionals that can easily be reduced by employing technical adjuncts to therapy.

All in all, the integration of technology in psychological treatment has experienced a vast surge during the last couple of years. Ethical boards and psychological researchers are even more enforced to consider possible harm as well as potential that may arise from VR usage and transfer them into comprehensible, internationally applicable ethical guidelines.

6. Conclusion

Interestingly the focus of research on gaming has been relatively narrow in direction of problematic behaviour in combination with gaming. In relation the developmental side has been more or less neglected by some research groups. But especially experts on gaming behaviour are equipped with knowledge and skills that would enhance the development of interventions.

The development of interventions is mostly based on cognitive behavioural theories in psychology and psychotherapy. As these theories are based on well known learning research the fun aspect of games for health has been identified to be a core aspect for efficiency. To create fun in a virtual environment a game character of an intervention is the number one option. Good game design is difficult to achieve and not possible for psychologists. Multiprofessional teams and respectful interaction in those teams are needed to create an effective intervention with the mentioned fun aspect. In addition new fields are opening for computer scientists as well as for psychologists. Job opportunities are going to include less traditional positions. Psychologists might grow to be more developers and designers of interventions that are used with only limited help afterwards. Computer scientists might increase their focus on human-technique interaction and psychological aspects of technology. Both professions need to develop a sense of interdisciplinary work and different viewpoints of the same phenomenon.

Ethical considerations in the development of virtual environments have to be discussed with different professions and the intended target group. Of course the side effects need to be considered as well, and extensive differential diagnostics in clinical psychology are a prerequisite for participants as long as research was not able to show clear directions in causal relations.

A very relevant question that needs to be discussed for the future ist the availability of the developed and evaluated interventions. Who is going to be allowed to buy VR interventions? Based on open source ideas and the powerful empowerment aspect of knowledge free distribution is a very appealing way. A number of problematic thoughts are related with free distribution, amongst other aspects the following points have to be considered: (a) subjective value of the system: some theories suggest that free help is valued less or considered as not effective as help one has to pay for. (b) Use for other as the intended purposes. (c) Efforts of the developing team and their value system. (d) self help systems as an option or different modes (well known from self help guides with supplements for professionals – or the other way round).

We state that there is definitely more research needed to draw a clear picture of positive outcomes of games for health. Most studies in this area often included trials with small sample sizes, only a few include large scale trials. Some studies also lack explicit exclusion criteria as well as the establishment of control groups. Kharrazi and colleagues state that most studies regard a relatively short intervention period (1 ¼ months) as well as a lack in theoretical frameworks (7).

The third wave in behavioral psychotherapy looking at mindfulness, acceptance and commitment as well as compassion targets all needed prerequisites for personal change, enhancement of well-being, work-life-balance and many other concepts that are often present in the media. The fact that second life has a huge mindfulness group with people meeting for meditation and the growing number of mindfulness or (self-) compassion apps for iPhone, Android, iPad and other hardware shows that technology might as well be the cure to an enhanced life. A psychological and psychotherapeutical virtual environment allows us to change our viewpoint and monitor, develop and/or optimize our behavioral as well as cognitive patterns. Including assessment functions or psychophysiological measures the individual possibilities seem never ending for designing treatment protocols. Playing a game for health means change in the sense of self-actualization as well as in the sense of treating disorders with many additional possibilities real life can not offer.

Author details

Birgit U. Stetina
Department of Psychology, Webster University Vienna, Vienna, Austria

Anna Felnhofer, Oswald D. Kothgassner and Mario Lehenbauer
Working group "Clinical Psychology", Faculty of Psychology, Vienna, Austria

7. References

[1] DeCandido KRA. Cycle of Hatred. New York: Pocketstar.
[2] Griffiths MD, Davis MN, Chappell D. Breaking the stereotype: The case of online gaming CyberPsychology & Behavior. 2003;6(1):81-91.
[3] Ferguson B. The emergence of Games for Health. Games for Health Journal: Research, Development, and Clinical Applications. 2012;1(1):1-2.
[4] Kato P. Video games in healthcare: Closing the gap. Review of General Psychology. 2010;14(2):113–21.
[5] Association ES. Essential facts about the computer and video game industry. 2011 [June 1st, 2012]; Available from: http://www.theesa.com/facts/pdfs/ESA_EF_2011.pdf.
[6] Acarturk C, Smit F, de-Graaf R, van-Straten A, ten-Have M, Cuijpers P. Incidence of social phobia and identification of its risk indicators: a model for prevention. Acta Psychiatrica Scandinavica. 2009;119(1):62–70.
[7] Kharrazi H, Shirong A, Gharghabi F, Coleman W. A Scoping Review of Health Game Research: Past, Present and Future. Games for Health Journal: Research, Development, and Clinical Applications. 2012;2(1):153-64.

[8] Lehenbauer M, Stetina BU. Internet and Psychology: Use new media for preventing social fears. In: Sawyer B, editor. 8th Annual Games for Health Conference 2012; Boston2012.

[9] Games for Health Conference. 2012 [05.06.2012]; Available from: http://www.gamesforhealth.org/.

[10] Kato P. Evaluating Efficacy and Validating Games for Health. Games for Health Journal: Research, Development, and Clinical Applications. 2012;1(1):74-6.

[11] Beck A. Cognitive Therapy of Depression. New York: Guilford; 1979.

[12] Oest L-G. One-session treatment for specific phobias. Behaviour Research and Therapy. 1989;27(1):1-7.

[13] Drechsel G. Von der Expo in sensu über Expo in vivo zu Expo in video. Forum Psychotherapeutische Praxis. 2002;2(4):199-204.

[14] Hayes S. Acceptance and commitment therapy, relational frame theory, and the third wave of behavioral and cognitive therapies. Behavior Therapy. 2004;35(4).

[15] Fullerton T. Game Design Workshop. Burlington: Morgan Kaufman Publishers; 2008.

[16] Stetina BU, Kothgassner OD, Lehenbauer M, Kryspin-Exner I. Beyond the fascination of Online-Games: Probing addictive behavior and depression in the World of Online-Gaming. Computers in Human Behavior. 2011;27:473-9.

[17] Colewell J, Payne J. Negative correlates of computer game play in adolescents. British Journal of Psychology. 2000;91:295-310.

[18] Grüsser SM, Thalemann R, Griffiths MD. Excessive computer game playing: Evidence for addiction and aggression? CyberPsychology & Behavior. 2007;10:290-2.

[19] Griffiths MD, Davies MNO, Chappell D. Demographic factors and playing variables in online computer gaming. CyberPsychology & Behavior. 2004;7:479-87.

[20] Charlton JP, Danforth ID. Differentiating computerrelated addictions and high engagement. In: Morgan J, Brebbia CA, Sanchez J, Voiskounsky A, editors. Human perspectives in the Internet Society: Culture, Psychology, Gender. Southhampton: WIT press; 2004. p. 59-68.

[21] Meerkerk GJ, Van Den Eijnden RJ, Vermulst AA, Garretsen HJ. The Compulsive Internet Use Scale (CIUS). Some Psychometric Properties. CyberPsychology & Behavior. 2009;12(1):1-6.

[22] Charlton JP, Danforth ID. Distinguishing Addiction and High Engagement in the Context of Online Game Playing. Computers in Human Behavior. 2007;23(3):1531-48.

[23] Ng BD, Wiemer-Hastings P. Addiction to the internet and onlinegaming. CyberPsychology & Behavior. 2005;8:110-3.

[24] Kothgassner OD, Stetina BU, Lehenbauer M, Seif M, Kryspin-Exner I, editors. Behavior Beyond the World of Online Gaming. General Online Research; 2010; Pforzheim: Halem.

[25] Cole H, Griffiths MD. Social Interactions in Massively Multiplayer Online Role-Playing Gamers. CyberPsychology & Behavior. 2007;10(4):575-83.

[26] Jansz J, Tanis M. Appeal of Playing Online First Person Shooter Games. CyberPsychology & Behavior. 2007;10(1):133-6.

[27] Caplan SE, Williams D, Yee N. Problematic internet use and psychosocial well-being among MMO players. Computers in Human Behavior. 2009;25:1312-9.

[28] Anderson CA, Bushman BJ. Effects of violent video games on aggressive behavior, aggressive cognition, aggressive effect, physiological arousal and prosocial behavior: a meta-analytic review of the scientific literatur. Psychological science. 2001;12:353-9.

[29] Mathiak K, Weber R. Toward Brain Correlates of Natural Behavior:fMRI during Violent Video Games. Human Brain Mapping. 2006;27:984-56.

[30] Jansz J. The emotional appeal of violent video games for adolescent males. Communication Theory. 2005;15(3):219-41.

[31] Goh DH, Ang RP, Tan HC. Strategies for designing effective psychotherapeutic gaming interventions for children and adolescents. Computers in Human Behavior. 2008;24:2217-35.

[32] Garcia-Palacios A, Botella C, Hoffman H, Fabregat S. Comparing Acceptance and Refusal Rates of Virtual Reality Exposure vs. In Vivo Exposure by Patients with Specific Phobias. CyberPsychology & Behavior. 2007;10(5):722–4.

[33] Klinger E, Bouchard S, Légeron P, Roy S, Lauer F, Chemin I, et al. Virtual Reality Therapy Versus Cognitive Behavior Therapy for Social Phobia: A Preliminary Controlled Study. CyberPsychology & Behavior. 2005;8(1):76–88.

[34] Newman MG, Szkodny LE, Przeworski LA. A review of technology-assisted self-help and minimal contact therapies for drug and alcohol abuse and smoking addiction: Is human contact necessary for therapeutic efficacy? Clinical Psychology Review. 2011;31(1):178–86.

[35] Newman MG, Szkodny LE, Llera SJ, Przeworski A. A review of technology-assisted self-help and minimal contact therapies for anxiety and depression: Is human contact necessary for therapeutic efficacy? Clinical Psychology Review. 2011;31(1):89-103.

[36] Andersson G. Using the Internet to provide cognitive behaviour therapy. Behaviour Research and Therapy. 2009;47(3):175–80.

[37] Botella C, Gallego MJ, Garcia-Palacios A, Guillen V, Baños RM, Quero S, et al. An Internet-Based Self-Help Treatment for Fear of Public Speaking: A Controlled Trial. Cyberpsychology, Behavior, and Social Networking. 2010;13(4):407–21.

[38] Furmark T, Carlbring P, Hedman E, Sonnenstein A, Clevberger P, Bohman B, et al. Guided and unguided self-help for social anxiety disorder: randomised controlled trial. The British Journal of Psychiatry. 2009;195(5):440–7.

[39] Piercey CD, Charlton K, Callewaert C. Reducing Anxiety Using Self-Help Virtual Reality Cognitive Behavioral Therapy. Games for Health Journal: Research, Development, and Clinical Applications. 2012;1(2):124-8.

[40] Wiederhold BK, Wiederhold MD. A Review of Virtual Reality as a Psychotherapeutic Tool. CyberPsychology & Behavior. 1998;1(1):45–52.

[41] Garcia-Palacios A, Hoffman HG, Kwong-See S, Tsai A, Botella C. Redefining Therapeutic Success with Virtual Reality Exposure Therapy. CyberPsychology & Behavior. 2001;4(3):341–8.

[42] Maloney AE, Bethea TC, Kelsey KS, Marks JT, Paez S, Rosenberg AM, et al. A pilot of a video game (DDR) to promote physical activity and decrease sedentary screen time. Obesity. 2008;16(9):2074-80.
[43] Yohannan SK, Kwon R, Yurt RW. The Potential of Gaming in Rehabilitation of the Burn-Injured Patient. Games for Health Journal: Research, Development, and Clinical Applications. 2012;1(2):165-70.
[44] Sharar SR, Carrougher G, Nakamura D, Hoffman H, Blough D, Patterson D. Factors influencing the Efficacy of Virtual Reality Distraction Analgesia During Postburn Physical Therapy: Preliminary Results from 3 Ongoing Studies. Archives of Physical Medicine and Rehabilitation. 2007;88(12):43-9.
[45] Yohannan SK, Schwabe E, Sauro G, Kwon R, Polistena C, Gorga D, et al. Use of Nintendo Wii in Physical Therapy of an Adult with Lower Extremity Burns. Games for Health Journal: Research, Development, and Clinical Applications. 2012;1(1):62-9.
[46] Diaz-Orueta U, Facal D, Nap HH, Ranga M-M. What is the Key for Older People to Show Interest in Playing Learning Games? Initial Qualitative Findings from the LEAGE Project on a Multicultural European Sample. Games for Health Journal: Research, Development, and Clinical Applications. 2012;1(2):115-23.
[47] Marston HR, Smith ST. Interactive Videogame Technologies to Support Independence in the Elderly: A Narrative Review. Games for Health Journal: Research, Development, and Clinical Applications. 2012;1(2):139-52.
[48] Decker J, Li H, Losowyj D. Wiihabilitation: Rehabilitation of Wrist Flexion and Extension Using a Wiimote-Based Title System. 2009 [05.06.2012]; Available from: http://www.osd.rutgers.edu/gs/09papers/Wii.pdf.
[49] Merry SN, Stasiak K, Shepherd M, Frampton C, Fleming T, Lucassen MFG. The effectiveness of SPARX, a computerised self help intervention for adolescents seeking help for depression: randomised controlled non-inferiority trial. BMJ. 2012;344.
[50] Lehenbauer M, Stetina BU. An online social skills training: Results of a pilot study concerning new technology in psychology. 1st Annual Games for Health Conference Europe; Amsterdam2011.
[51] McKee HA, Porter JE. Playing a Good Game: Ethical Issues in Researching MMOGs and Virtual Worlds. International Journal of Internet Research Ethics. 2009;2(1).
[52] Buzinka M, Moore D. Ethical Standards in the Field of MMORPG Research. n.d. [June 12th, 2012]; Available from: http://www.buzinkay.net/texte/online-research-ethics.pdf.
[53] APA APA. Ethical Principles of Psychologists and Code of Conduct. [Website] 2010 [cited 2012 03.01.2012]; Available from: http://www.apa.org/ethics/code/index.aspx.
[54] Felnhofer A, Kothgassner O, Stetina BU. Cyberethics. Ethik im Kontext der Online Forschung. In: Felnhofer A, Kothgassner O, Kryspin-Exner I, editors. Ethik in der Psychologie. Wien: UTB facultas; 2011. p. 181-92.
[55] Alleman JR. Online counseling: The Internet and mental health treatment. Psychotherapy: Theory/Research/Practice/Training. 2002;39(2):199-209.
[56] Manhal-Baugus M, Risk A. E-health code of ethics. Journal of Medical Internet Research. 2000;2(2):e9.

[57] (AMHCA) AMHCA. Principles for AMHCA code of ethics. 2010; Available from: https://www.amhca.org/assets/news/AMHCA_Code_of_Ethics_2010_w_pagination_cxd_51110.pdf.

[58] Rippen H, Risk A. E-Health code of ethics. Journal of Medical Internet Research. 2000;2(2):e9.

[59] Stetina BU. Psychologische Überlegungen zu ethischen Aspekten der Mensch-Tier-Beziehung. In: Riether E, Weiss MN, editors. Tier-Mensch-Ethik. Berlin: Lit-Verlag; in press.

Chapter 4

Description of a Treatment Manual for *in virtuo* Exposure with Specific Phobia

Stéphane Bouchard, Geneviève Robillard,
Serge Larouche and Claudie Loranger

Additional information is available at the end of the chapter

http://dx.doi.org/10.5772/46417

1. Introduction

The DSM-IV-TR [1] defines a specific phobia as an intense and persisting fear that is excessive or irrational, usually triggered by the presence or the anticipation of a specific object or situation. The exposure to the fear-provoking object triggers an immediate and almost systematic anxious reaction that can take the form of a panic attack. The individual recognizes the excessive or irrational nature of his fear but avoids the situations that might put him in the presence of the fear-provoking stimulus or experiences these situations with a lot of anxiety. The avoidance and apprehension that are linked to the phobia impair the individual's daily functioning.

There are three important psychological components involved in the panic reaction of individuals suffering from phobias: thoughts, emotions and behaviour. Thought expresses how the patient interprets the stimulus. Thoughts are the first step of the anxious process, and two people could have different reaction in front of the same situation. In order for a situation to trigger anxiety, the person must consider the stimulus as dangerous, threatening or, in the case of some phobias, disgusting (e.g., worms). Second, the emotion is the panic itself. A panic attack is associated with many objective physiological reactions (e.g., cardiac palpitations, nausea, weak knees) which are traditional consequences of an increase in arousal, except in the case of blood-injuries subtype of phobias where a vasovagal response may lead to fainting [2]. Third, behaviour refers to the way people act following perceived threat. When danger is real, a simple and effective solution might require fighting or fleeing [3]. Avoiding the fear-provoking stimulus is a tempting solution, but in the case of phobias, it becomes a trap. If one avoids something that is not a real danger, how can one realise that danger is not real? In fact, avoidance is the key behaviour contributing to maintain anxiety and phobias. Avoidance can take many forms, from obvious behaviours such as refusing to visit someone who owns a dog to subtle behaviours such as carrying a rabbit foot for good luck. Avoidance is also referred to as safety seeking behaviour (i.e., a snake phobic taking a walk in a park could be proactively and continuously looking at the grass and scanning the surroundings) to highlight the fact that avoidance refers to the process of avoiding the occurrence of the feared consequence and aiming for safety, as opposed to the overt behaviour of moving away from the stimuli. In some instances, avoidance is also described as neutralization, when the phobic person tries to perform behaviours to reduce anxiety (e.g., asking for reassurance) or engage in mental processes in order to prevent to occurrence of the feared consequence. No matter how avoidance is called in the literature, the long term consequence is that avoidance prevents the phobic individual to develop mental representations of the stimuli that are associated with safety and with a sense of perceived control and self-efficacy, as opposed to perceived threat or disgust [3-9].

The recommended empirically supported treatment for phobias is CBT [10], and for specific phobia the key therapeutic technique is exposure. In the dictionary of behavioural interventions, Marshall [11] defines exposure as every procedure that confronts the individual to a stimulus generating an undesired behaviour or emotional response. There are many ways to face one's fears, from imaginary exposure to exposure in real situations (« *in vivo* ») [12,13]. To be effective, the treatment cannot blindly and solely rely on exposure; it must also include basic elements that will not be discussed in this chapter, such as a sound and individualized case formulation, a healthy and constructive therapeutic alliance, a competent therapist and an exposure plan that is well dosed and controls avoidance adequately (for more details, see [4,6-9,12-15]).

Traditional exposure techniques have some limits [15-16], such as potential breach in confidentiality during exposure in a public place, the lack of control over the stimuli in many exposure situations (e.g., traffic on a highway or behaviour of an animal), the financial costs associated with some stimuli (e.g., cost of flying tickets for the patient and the therapist in the treatment of flying phobia), the need to care for and feed reptiles or insects, the possibility of unexpected events (e.g., the elevator is out of order or jams), the impossibility to reproduce and graduate some stimuli at will (e.g., thunderstorms), etc. Fortunately, these limits stimulated researchers in the field of anxiety disorders to find new means of exposing patients. It led virtual reality to be used in several outcome studies that have shown that exposure conducted in virtual reality is an effective and empirically validated treatment of specific phobias [15-18].

Actually, more than 40 clinical trials have been conducted so far (for reviews see [15] to [18]). Although some studies correspond more to pilot or uncontrolled clinical trials, there has been many strong and convincing randomized clinical trials showing that exposure conducted in virtual reality is more effective than no treatment or a waiting list, more effective than some alternative treatments, and no less effective than *in vivo* exposure [17-18].

2. Exposure with virtual reality

According to Pratt et al. [19], virtual reality is defined as an application that allows a given user to navigate and interact in real time with a three-dimension and computer-generated environment. Conducting exposure in virtual reality allows a therapist to expose a person to fear-provoking stimuli (as with traditional means) in a situation that is computer-generated (see Figure 1 for example of various stimuli).

Figure 1. Screenshots of virtual environments used to treat phobia of cat, dogs, snakes and germs.

Many technologies are available to immerse patients in VR, and the currently most popular and affordable one relies on head-mounted-display (HMD) technology, where small monitor screens are mounted in a pair of glasses (see the patient on the left in Figure 2) and paired with a motion tracker, allowing to immersed the patient (usually referred to as the user) in a virtual environment where he or she can be gradually exposed to his or her fear. Semantically, it is important not to describe conducting exposure in virtual reality as "virtual exposure" or "virtual therapy" since it is the stimulus that is virtual, not the exposure, and the therapy is quite real. Instead of using the lengthy and precise expression "exposure conducted in virtual reality", authors are increasingly using the expression "*in virtuo*", a term coined by Tisseau [20]. *In virtuo* was created through an analogy with adverbial phrases from Latin such as *in vivo* (an expression commonly used to describe exposure conducted in real life situations, as opposed to require the patient to imagine the feared stimuli), *in vitro* (meaning in glass) and *in silico* (meaning in silicon). When learning to use *in virtuo* exposure, therapists must become familiar with two concepts, presence and virtual reality induced negative side effects.

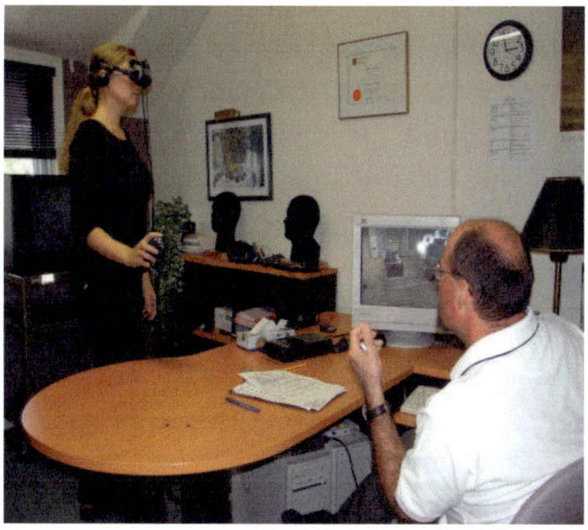

Figure 2. Illustration of an in virtuo exposure session using a HMD.

The notion of presence is considered as very important in virtual reality [16]. Presence is often defined as the perception of "being there" inside the virtual environment in which the individual is immersed, even if the person is physically in another environment [21]. Researchers usually agree on this simple definition, although most of them add somewhat different nuances to it [22]. Bouchard et al. [23] proposed a more elaborated view based on the notion of presence resting on a continuum, from a basic but complex and automatic perceptual illusion caused by multisensory integration, up to a sense of meaning and being there in the virtual environment. The gradient on the continuum begins at the most cognitively implicit and automatic level referred to as proto-presence, it evolves into core presence and, at the highest end of the continuum, the perceptual illusion enters a range referred to as extended presence. This terminology was proposed to match Damasio's [24] three neurological layers of consciousness. It builds on propositions by Riva and Waterworth [25-27], although it focuses on defining presence as essentially a strong perceptual illusion leading to a sense of meaning about the events occurring in the virtual environment. The subjective feeling of being "there" therefore refers to presence experienced at the highest end of the continuum and automatic behaviours such avoiding to collide with a virtual wall refers to core presence.

Some conditions seem associated to the development of the feeling of presence and increasing its intensity on the continuum from proto-presence to extended presence. Sadowski and Stanney [22] summarized seven factors contributing to the feeling of presence, and additional empirical evidences were provided by Youngblut [28]. These factors are summarized in Table 1.

The relationship between the feeling of presence and the treatment outcome of *in virtuo* exposure for phobias is still unclear. However, because developing new internal representations and associations between the feared stimuli and the lack of threat is the well known key ingredient in the process of change underlying exposure for phobias [4,8,12], it is unlikely that presence in itself could be an essential mediator of treatment outcome. Why would feeling more or less present in the virtual environment cure phobias? It is more likely that feeling present allows, at least to some extent, the virtual stimuli to elicit anxiety, exposure to occur, and traditional treatment mechanisms to take place. The feeling of presence during *in virtuo* exposure may therefore be more directly related to the anxiety reaction triggered by the virtual stimuli, play the role of a moderator of treatment outcome, and involve in a bi-directional relationship where presence influences anxiety and vice-versa.

Clinical experience suggests that the relationship between anxiety and presence is probably not linear. First, people who are predisposed to relate emotionally to the virtual stimuli, such as phobics in the case of phobogenic stimuli, react with stronger emotions and report more core and extended presence than "normal controls" [29-30]. Second, clinicians and experimenters often mention that users who do not feel present at all do not feel anxious when exposed to anxiety provoking stimuli. Third, there seems to be an unknown trigger point where some level of presence is sufficient to lead to a strong sense of anxiety and more presence contribute only moderately to further increase in anxiety, as illustrated in Figure 3. These clinical observations need to be substantiated by empirical evidences, but in the meantime they may guide therapists in understanding why a minimal amount of presence

	Description	Explanation
System-related factors	The immersion should foster interactions and replicate well the physical reality by stimulating the senses as it would be the case in the physical reality and help forgetting the interface between the user and the system. Examples include tracking head movements, offering a larger field of view, stimulating multiple senses, using stereoscopic displays, providing multimodal interactions and using ergonomically good sensors and effectors.	A poorly designed system might cause a degradation of the immersive experience. The immersion does not require perfect realism, but the VE itself must be well developed to allow natural multisensory integration that is credible and is meaningfully interpreted within the perceived internal state of the user.
Ease of interaction	The immersion offers seamless interactions that allow the user to be oriented in, to navigate in and to interact with the VE*.	Unrealistic or asynchronous interactions might prevent a meaningful integration of information coming from the various senses, including proprioception, and deviate attention away from the experience in the VE and towards the interface and the synthetic nature of the experience.
User-initiated control	The immersion allows an immediate response, the correspondence of the patient's actions and the natural aspect of the control mode.	An implicit sense of agency, or user's natural impression that he or she is the cause of the actions happening in the VE is important, and therefore controls should be seamless and concordant between the user's actions and the effectors reactions.
Objective realism	The stimuli allow continuity, consistency, connectedness and meaningfulness of the presented stimuli. The stimuli can be visual, auditory, olfactory, etc.	Poorly designed or replicated stimuli may impair how stimuli are detected, appraised, interpreted as well as the meaning emerging from the immersive experience. Objective realism must be interpreted in the global context of the immersion instead of essentially based on how perfectly each stimulus is replicated.
Social factors	The immersion should provide opportunities to interact and communicate with virtual humans, or other users involved in the immersion, and allows the virtual humans, or other users, to acknowledge the user's existence in the VE.	If the virtual humans or avatars of other users in the VE do not acknowledge the existence of the user, it can impair the perception that the patient « exists » in the VE.
Duration of immersion	The immersion should provide sufficient time allows the user to be familiarized with the task and the VE, as well as for sensory adaptation.	Avoids unnecessarily prolonged immersion that could be associated with unwanted negative side effects induced by the immersion.
Internal factors	Identifies the individuals' characteristics that help to increase presence.	Individual differences influencing how a user will process the information afforded by the immersion.

* VE = Virtual environment

Table 1. Seven factors influencing the feeling of presence, adapted from Sadowski and Stanney [22].

is required, and why the treatment is not more effective when delivered with immersions in very expensive technologies such as immersive rooms (see the illustration of the immersive room in Gatineau in Figure 4) [31,32] or hydrolic platforms [33] than the much more affordable HMD technology.

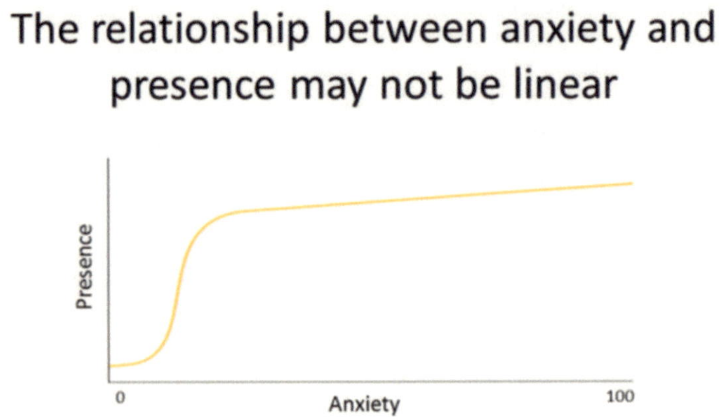

Figure 3. Illustration of the potential relationship between anxiety and presence.

Figure 4. A room-size immersive system where the user stands up in a cube with stereoscopic images retro-projected on its wall to create a very strong sense of presence.

Another important concept to address before describing the treatment manual is the potential negative side effects that can be induced by the immersion in VR. These unwanted side effects are often called "cybersickness" [34], although they do not refer to an actual sickness or illness. The term was coined following the use of common expressions such as sea sickness, motion sickness and simulator sickness. Unwanted side effects are often reported in the literature, but interpreting the data is difficult because the occurrence of unwanted side effects is influenced by several factors, including what the user does during the immersion, the physical fitness of the user and the use of older less powerful technologies. Early studies [35] report that between 50% to 100% of users feel some dizziness, 20% to 60% feel some abdominal symptoms, at least 60% experience some side effects during their first immersion, and about 5% feel severe symptoms. Negative unwanted side effects are routinely measured using a well established instrument, the Simulator Sickness Questionnaire (SSQ) [36]. More recent report confirmed some of these incidence rates. For example, Sharples et al. [37] found increase in SSQ scores post-immersion in almost 70% of users of HMD or large projection screens technologies, compared to 37% of users immersed using traditional computer monitors. Two reports [38-39] have been published on data gathered with adults immersed in situations similar to *in virtuo* exposure with phobics. Bouchard et al. [38] examined SSQ scores post *in virtuo* exposure therapy sessions and found the majority (94%) of users reported having at least one slight symptom listed on the SSQ. But this observation is hard to interpret since these symptoms might have been present before the immersion or may be symptoms of anxiety. Nevertheless, 20 % of their sample reported high scores on the SSQ, which is consistent with other studies revealing the occurrence of side effects in users. None of the participants in Bouchard et al. [38] had to stop the immersion dues to negative side effects and for 92% of the users the intensity of symptoms was rated as "slight". Another study in the same article replicated what was found in experimental settings, showing that therapy sessions with *in virtuo* exposure requiring more movements from the user, such as walking to different locations in the virtual environment, induced more unwanted side effects than immersion requiring only to sit and look around, such as *in virtuo* exposure for fear of flying. Finally, they [38] followed-up some participants 24 hours post-immersion with the SSQ. Their data indicated that side effects were not an issue after the therapy session. Interestingly, the SSQ scores were higher before than after the immersion, suggesting that apprehension and other factors may inflate SSQ scores. In sum, this publication shows that negative unwanted side effects do occur, should be monitored by therapist, but should not be a source of concern for the vast majority of patients.

Bouchard et al. [39] targeted more specifically the assessment of unwanted side effect of a VR immersion using the SSQ and the potential overlap with anxiety symptoms. First, results of their factorial analysis conducted with more than 500 users questioned the scoring method of the SSQ and led to propose: (a) to systematically report the total score without following the weighting procedure suggested for the SSQ items, and (b) a 2-factor solution consisting of nausea and oculomotor symptoms. In another experiment reported in the same paper, scores on the SSQ correlated with a self-report measure of anxiety after participants were subjected to standardized stressor that did not involve any immersion at all. In a third experiment, they found a few items of the SSQ that were more strongly associated with anxiety than cybersickness. This study [39] did not provide a definitive answer on the confound between anxiety induced by *in virtuo* exposure and symptoms of unwanted negative side effects rated on the SSQ. But is raises concern that post-immersion scores on the SSQ may be inflated by anxiety experienced during *in virtuo* exposure. Together, the studies by Bouchard's team also show the importance of administering the SSQ before the immersion, as participant's apprehension toward both the immersion and exposure may artificially inflate scores on the SSQ, and of interpreting the scores with caution.

To understand the causes of the unwanted negative side effects, it is possible to analyse two global types of factors: those related to the hardware itself, and those related to interacting with the virtual environment. Wearing a HMD that is too heavy, bulky or too tightly strapped around the head may induce some neck strain and headache. Holding a heavy computer mouse or other interface in one had for a long time could also be unpleasant during long *in virtuo* exposure sessions, especially with children. These problems are becoming less of a problem as technology improves and therapists are paying attention to the user's comfort during the immersion. Tension in ocular muscles could also occur during lengthy immersions. Just as watching television from a very short distance or focusing on the computer monitor while working for a long period of time create eye-strain, keeping the eyes in a fixed position to look at tiny images displayed from a few centimetres away can induce ocular discomfort. Using stereoscopy without correction for interpupillary distance is also a potential source of negative side effects. A simple solution to reduce these problems is available for therapists and based on simple ergonomic principles: limit the duration of immersions to allow the eyes focusing at a different point than the displays in the HMD. The rule of thumb is to take a pause in the immersion after about 20 minutes or so. It can be a nice occasion for the therapist to discuss what happened during the *in virtuo* exposure, before continuing the immersion if time allows it. The duration of the immersion should also be gauged based on what is happening in the exposure session in order not to interrupt the psychotherapeutic change process.

A second set of factors inducing unwanted negative side effects relates to motion, user's behaviour and the task required by the therapist. In the physical reality and under normal conditions, physical movements are perceived by the visual, vestibular and proprioceptive systems and all converging information from theses senses should be consistent and match with each other. In addition to using visual cues, postural imbalance is assessed and maintained by the organism based on the head's orientation as detected by the vestibular system and by proprioception. While immersed in VR, quite as much as in other situations inducing motion sickness, a mismatch between these senses can occur. For example, a mismatch between the visual and vestibular system can occur if the motion tracking system is not fast enough to accurately track head rotations, if the computer is not powerful enough to update and send to the HMD a matching version of the virtual environment, or if the content of the virtual environment was not optimized when it was created and requires too much computing power. A lag between actual motion and delivering its corresponding effect to the user might induce nausea and disorientation. A mismatch with the proprioceptive system can also contribute to some symptoms, probably to a lesser extent. Other effects could also induce cybersickness, such as vection and rotations or movements perceived in the peripheral areas of the visual field [35,37,40]. The simplest explanation for the occurrence of nausea symptoms induced by immersion is referred to as the sensory conflict theory [41] and, as suggested by Treisman [42] and Money [43] based on an evolutionary perspective, could be a reaction from the organism to get rid of potential poisonous or intoxicating substances. These explanations are not without criticisms (see [44]) and a lengthy description of this topic would digress for the objective of the current chapter. Lawson et al. [35] added the possibility that some side effects might also be linked to the Sopite Syndrome, which is an excessive drowsiness, difficulty concentrating and apathy induced by motion.

Based on factor analyses of symptoms experienced mostly in military training simulators, Kennedy et al. [36] proposed to organize the temporary unwanted negative side effects found in VR in three groups: (1) *ocular* (e.g., blurred vision, headache), (2) *disorientation* (e.g., vertigo, dizziness) and (3) *nausea* (e.g., nausea, vomiting). Following factor analyses of symptoms after immersion of users recruited from the general population or among people diagnosed with anxiety disorders [39, 45], the best factor structure of the SSQ in these samples appears to be two factors describing nausea and oculomotor symptoms. The list of symptoms measured by the SSQ is reported on Table 2 and organized according to the factor structure found in a sample of adults suffering from anxiety and phobic disorders.

Nausea symptoms	Oculomotor symptoms
• General discomfort • Increased salivation • Sweating • Nausea • Dizzy (eyes open) • Dizzy (eyes closed) • Vertigo • Stomach awareness • Burping	• Fatigue • Headache • Eyestrain • Difficulty focusing • Difficulty concentrating • Fullness of head • Blurred vision

Table 2. Nausea and oculomotor symptoms of the SSQ in anxious patients [39]

What the therapists ask users to do during the immersion may also affect the induction of side effects. For example, asking users to rotate frequently on themselves during an *in virtuo* exposure session may easily lead to disorientation and nausea. On the opposite, guiding users to move at a normal or slower pace would limit the occurrence of unwanted side effects. Other factors fall under the control of the therapist. For example, a few researchers found that some users report a linear increase in symptoms during long immersions [46]. Kennedy et al. [46] also reported habituation and a decrease in side effects with repeated immersions. As a rule of thumb, it is suggest that if there is no reason to stop the immersion prematurely, the unwanted negative side effects should be within acceptable levels for immersions lasting between 10 to 60 minutes. Other user's characteristics that may need to be taken into account are suffering from migraine headache [47], being prone to motion sickness [60] and age. Stanney et al. [40] suggested that susceptibility is greater in children aged between two and twelve years old, although this was not supported in a least one sample of children [47]. In order to reduce the incidence and intensity of unwanted negative side effects induce by immersions, clinicians can take some basic precautions (see [37,40] for a more detailed list). Following the preventive strategies listed in Table 3 contributes to reduce patients discomfort while conducting *in virtuo* exposure, at least based on past experience in experimental and clinical research centers.

List of useful strategies therapists should consider using with their patients
• Inform users about cybersickness before the first immersion.
• Inform users about behaviours that can exacerbate side effects.
• Allow users to control their movements in the virtual environment.
• Monitor the user to detect unwanted side effects, provide reassurance and do not confound cybersickness and signs of anxiety. Look for red flags, such as excessive sweating, postural imbalance, burping, restricted movements for a significant duration, nausea.
• Assess symptoms with the SSQ if necessary, and administer the instrument before and after the immersion.
• Terminate the immersion if side effects are disturbing.
• Use adequate hardware and software.
• Limit the duration of the immersion.
• Exclude users who are highly susceptible to motion sickness, or have conditions that would preclude the use of VR, or otherwise conduct the immersion with caution.
• Assess the presence of side effects after the therapy session.
• Do not let the user leave the clinic unless there is no side effect. Waiting 15 minutes post-immersion before allowing the user to leave is a routine procedure in our clinics and lab.

Table 3. Simple preventive strategies to control "cybersickness"

3. The program's philosophy, objectives and procedures

The program developed at the Université du Québec en Outaouais to treat specific phobias in older children, adolescents and adults has been used with success at the university clinic, at the Pierre-Janet Hospital, in other research centers and in private practice [15,49-51]. It was designed for people receiving a primary diagnosis of specific phobia [1]. It was not indented to be used without significant modifications for people suffering from a primary disorder that is not specific phobia (e.g., other more complex anxiety disorders, depression, pathological gambling), for people suffering from a phobia accompanied with a more severe comorbid disorder that is more urgent to address, and for people under the effect psychotropic substances that can alter consciousness during the immersion (e.g., alcohol, drugs) or suffering from conditions that may significantly increased the incidence of unwanted negative side effects (e.g., Ménière disease). We strongly discourage people suffering from a specific phobia to use virtual environments and apply this program by themselves without the use of a professional trained in cognitive-behaviour therapy. Virtual environments do not treat phobias; they only represent a tool that is used in the context of a more elaborated treatment. The program has been created to be applied by professionals who have received training in mental health and in the CBT of anxiety disorders. In addition, as exposure in a virtual environment and its associated equipment add a small layer of complexity, it is also suggested that professionals who are interested in this type of intervention familiarise themselves with the VR equipment, learn how to use the available environment and remain aware of the probability of side effects. The program was prepared in such a way it can be used with a variety of virtual environments and is therefore not written as a user's guide to any specific software.

This program is based on cognitive-behavioural therapy (CBT). Literature shows that CBT is an efficient treatment mode for anxiety disorders. The goal of the program is to eliminate the symptoms of anxiety in phobic individual through exposure to fear-provoking stimuli in a virtual environment. The program was developed for the clinical trials cited above, where it was administered over five to seven weekly 60-minute sessions, and has been slightly modified since then. When not used in a standardized research protocol, the number of sessions should be tailored to the specific need of each patient. The distribution of sessions' frequency and contents is thus left to the discretion of each therapist. The exposure's pace and phobia severity must also be taken in consideration after a thorough assessment and case conceptualization [3,4,6,8,12,14]. In research trials, using homework exposure between therapy sessions was restricted to circumscribe the potentially active therapeutic ingredients. Such restriction is not required in standard clinical practice. Recent developments in the use of portable VR systems, for example based on iPad and iPhone technologies, are currently being tested and will soon allow patients to bring virtual environments at home to complete more *in virtuo* exercices. Affordable haptic solutions are also being implemented to allow patients perform behaviours where the sense of touch is relevant in the treatment, such as crushing and killing a spider (see Figure 5).

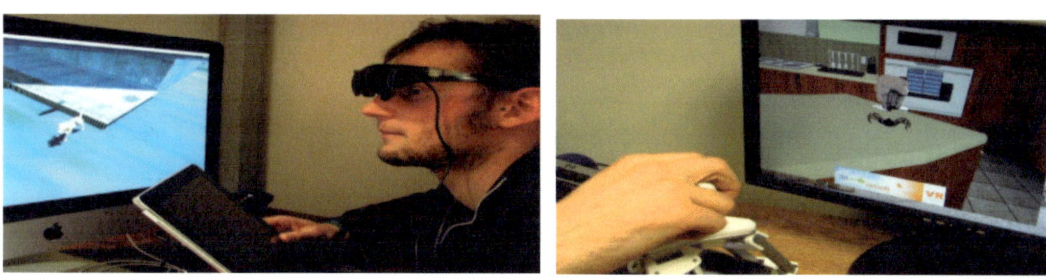

Figure 5. A user immersed in an environment for dog phobia using an iPad (left) and a haptic device allowing crushing and killing virtual spiders during the immersion (right).

Overall, the program is structured around a first introductory session, a few core sessions devoted essentially to *in virtuo* exposure, and the addition of a relapse prevention module to the last session.

3.1. First session

Goals for Session 1:

- Describe CBT principles that will guide the treatment;
- Establish a behavioural contract and the structure of the sessions;
- Present a models of factors maintaining anxiety and specific phobias;
- Warn against the trap of avoidance;
- Identify anxiety-provoking thoughts;

- Familiarize the patient with the VR system and demystify unwanted negative side effects of immersions;
- Brief cognitive restructuring;
- Establish a hierarchy for the exposure;
- First immersion in VR using a neutral environment that is devoid of the phobic stimuli.

The therapist begins Session 1 by introducing himself briefly. The general clinical picture and patient's problem are assessed to perform a nuanced diagnosis. The information gathered should allow for a good and individualized case formulation (also referred to as functional analysis or case conceptualisation [4,6,8,13,14]).

The therapist explains some of the CBT principles to the patient, such as:

- It is based on a model that puts an individual's thoughts, emotions and behaviours in interaction;
- The patient will develop new skills through an active collaboration with the therapist to achieve autonomy in facing his or her own difficulties and feel empowerment;
- The therapist will use the Socratic method to help the patient become aware of his thoughts and actions;
- CBT is brief, structured and focused on current maintaining factors and how one can solve current problems.

It is possible that the expected number of sessions scheduled for the treatment might not be sufficient to completely eradicate the phobia. But efforts must be invested to mobilize the patient to change within the expected time frame, knowing that some adjustment may ultimately be possible. The patient will have to do exposure exercises at home and in "natural" situations in order to maintain the treatment gains. CBT is therefore an emotionally challenging endeavour requiring motivation and efforts.

The therapist explains that CBT, when applied to specific phobias, can is divided in three stages:

- Case formulation and overview of key concepts (first session);
- Exposure (sessions 2 to 6);
- Relapse prevention (session 7).

Following the presentation of this information, the therapist must ask the patients to rephrase in their own words what they understand is required in terms of implication, homework, and the time that will have to be devoted to therapy. Progress will heavily rely on these factors. The therapist has to answers questions (if any) and must agree with the patient on a therapeutic contract that could include, among other things, the number of sessions, their length and a schedule that is convenient to both.

The therapist will provide a model of anxiety, emotions and specific phobias that allows patient to understand what are the maintaining factors involved in the problem and which dysfunctional factors must be target in the treatment. This preliminary step is essential to foster treatment adherence. The following paragraph is an example of the information that could be transmitted to the patient. It is important to point out that therapists must not recite this information automatically but rather understand it and explain it with their own style. Concrete examples and metaphors are useful tools to consider.

Example of information to provide about anxiety in general

"Anxiety is an unpleasant emotion triggered by the perception of a threat. Anxiety is also described as an alarm reaction produced by our body in order to protect us against danger. This emotional state affects both the body and the mind. When we are anxious, several physical symptoms manifest themselves such as, for example, muscular tension, perspiration, sweaty palms, faster heart rate. On a psychological level, anxiety is characterized by a state of tension, worry and apprehension. Anxiety is a normal and healthy reaction: it allows our body to be ready to react quickly in front of a potential danger (for example, avoiding, by running, to be hit by a car when we cross the street). However, it becomes a problem when it is triggered when no real danger is present. Indeed, the sensations that we feel when anxious or having a panic attack quickly are unpleasant. When anxiety interferes with our daily life (for example, a person who moves to another country because his snake phobia is too strong), it becomes necessary to learn how to change this automatic association with perceived threat. I sometimes hear people suffering from phobias saying they would have die from a « heart attack » if they had not fled from a specific situation or circumstance. Actually, anxiety generally follows a curve like in this Figure 6 if you do not run away flee from the source of your fear. It will progressively go down after a while. Let's discuss this for a while. What do you think about this? Do you have examples from your own experience that we can examine?"

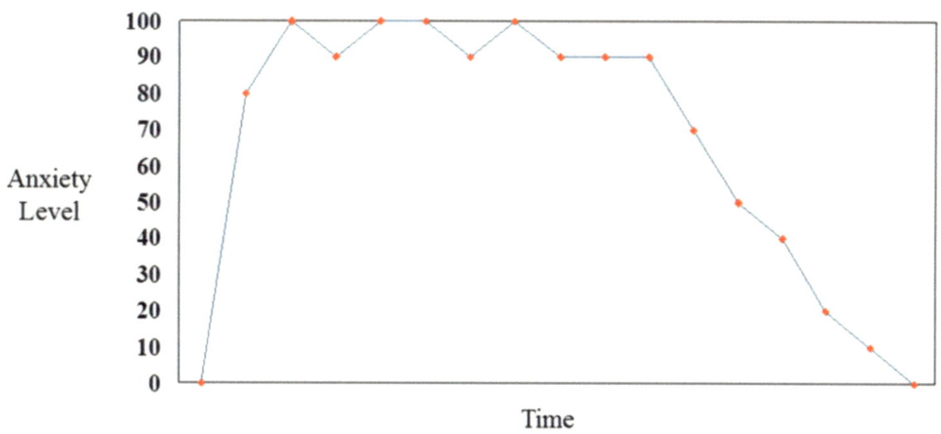

Figure 6. What happens to anxiety over time when there is no avoidance.

After discussing with the patient that anxiety is a normal reaction and it becomes a problem when it interferes with normal functioning, you can introduce the concept of avoidance. "When a person suffering from a phobia sees what she is afraid of, a snake for example, her anxiety will rise rapidly, sometimes up to 90% or 100%. Who would blame that person to avoid the situation? It is normal not wanting to be confronted with that situation and prevent the feared consequences to occur. But can you see how this avoidance could lead to a trap? Can you tell me how avoidance can prevent learning, and feeling with confidence, the disconfirming evidences about the real consequences of facing this stimuli. What about your impressions on the likelihood that such catastrophic consequence will happen? And what about your confidence that you can actually cope with the situation?" Is it possible that if you don't engage in avoidance, anxiety will cease to increase, and remain stable and later decline? Let's take an example from your own situation and examine this."

The four take home messages for the patient are that: (a) avoidance may appear as a logical solution, (b) avoidance maintains perceived threat, (c) stopping avoidance allows gathering disconfirming evidences, and (d) gathering evidence through empirical and emotional experiences allows us to "feel" convinced.

The therapist must also ensure the patient understands the role of appraisal, or the notion it is not the situation that brings the emotion (i.e., anxiety) but rather the *interpretation* that we make of the situation, how it is *perceived*. Using an example or the classical metaphor of the branches hitting the window can be more effective to convey this idea than a long lecture. The branches metaphor can be summarized as follow:

"As I was sleeping one night, my partner woke me up in panic saying there is a burglar in the house. I was about to go back to sleep when I heard it myself. Bing, bang, coming from the kitchen. As I was walking to the kitchen, anxious and worrying about all the bad things a burglar can do to us, I shouted that I had called the police, then hold my breath and hid by the kitchen door. As I leaned to see what was happening in the kitchen... I discovered that branches from a tree were blown by the wind and hitting the kitchen's window. There was no burglar, actually it was only branches and the wind ! You (sir / madam), do you think it was ok to feel anxious in that situation?" As the patients contemplate the reasons to be anxious in that situation, make sure he or she can see both sides of the situation: it was natural to be anxious but at the same time there was no danger. Let the patient state that it is because "you didn't know" and then discuss the fact "What you are saying is, it's not as much the events that matter, but how we interpret them".

As a therapist, it is often useful to understand what psychological mechanisms are involved behind perceived threat (or disgust). A heuristic formula made popular by Thorpe and Salkovskis [52] highlights the important target that can be used for cognitive restructuring with anxious patients. It reads as follow:

$$\text{Perceived threat} = \frac{\text{negative consequences (or disgust) X high probabilities X short imminence}}{\text{Not knowing what to do (low perceived self-efficacy)}}$$

When working on dysfunctional beliefs with anxious patients, the therapist can target exaggerated consequences, and / or over estimation of probabilities of occurrence of the feared consequences, and / or the imminence of the feared consequence, and / or the perceived self-efficacy that one can cope with the situation. Considering perceived self-efficacy, or perceived control, on the denominator is quite useful as it can buffer the impact of actually negative, realistic and imminent events. It also explains how experience makes people less frightened by some situations. Finally, it guides the therapists in setting the exposure exercises, as *in virtuo* exposure must disconfirm how negative, likely or imminent the feared consequences are, and how well the patient can cope with the situation [53].

The first 60-minute session is charged with material to learn, which is why the above anxiety equation is rarely explained to the patients and metaphor or examples are regularly used instead of theoretical lectures. The next information that needs to be addressed before beginning exposure is the description of a model of specific phobia. It is more fruitful to describe this model by starting with the patient describing a typical phobic episode and sketching the model on a white board or a piece of paper, than lecturing the patient based on an impersonal and already drawn theoretical model. The therapist could explain the maintaining factors involved in phobias using a simplified model like the one in Figure 7. In order to facilitate patient's understanding, the therapist can use the following information:

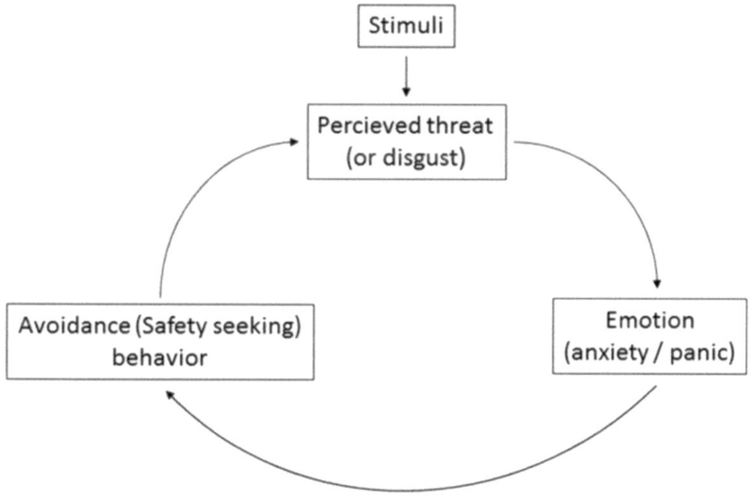

Figure 7. A simplified descriptive model of phobias.

- The triggering or fear-provoking stimuli usually are the phobogenic stimuli themselves (e.g., driving a car, highways, heights, cats), but often times the case formulation will lead to more precise cues (e.g., during take-off in the case of fear of flying, or the claws in the case of a phobia of cats).
- Perceived threat, or disgust, can be acquired in many ways [5], but what matters is not how phobias were acquired but how they are maintained by avoidance.
- At a cognitive level, when the phobic person is confronted with the fear-provoking situation or object, some inner speech occurs. This speech is sometimes so automatic that it may be difficult to notice. In the case of heights phobia, for example, the patient can think that the cliffs are dangerous, if he walks closer to the edge it is sure he will fall and it is sure he will die. The physiological reactions that inevitably follow (trembling, heart pounding, etc.) and the impression that he would not know what to do if the feared situation occurred all confirm that danger is near.

- The triggering stimuli can take many forms, from the actual feared animal to a cartoonish representation, and avoidance can be as subtle as seeking reassurance from the therapist, positive thinking, or carrying lucky charms.

It is important to have the patient share is or her experience and see how it fits, or not, with the model. Review with examples from the patient how anxiety is maintained by avoidance. Let the patient recall examples illustrating how avoiding situations maintains the dysfunctional beliefs. It is true that avoiding provides a short term relief from anxiety, only until next time he or she is confronted with the stimuli. In addition, a state of general apprehension might remain as the patient is trying to prevent future encounter with the phobogenic stimuli. It is important to work with the patient to detect the avoidance strategies that are used, as it will be essential during exposure to monitor them and progressively get rid of them.

Cognitive restructuring may be used [6], based on the therapist's judgement. In many cases, the patients' either already know their fear is irrational or are extremely well informed. Our experience with the current program is to limit cognitive restructuring to a minimum, mostly because it is not as efficient as exposure. However, an introduction to cognitive restructuring remains in the program because it was found to help patients put in context what they will achieve through exposure, to accept exposure more readily and to provide tools to generalize what they have learned to other areas in their life. It seems easier for the patient to be convinced both logically and « in his guts » about the validity of the alternative functional thoughts if he or she has previously identified and attempted to restructure the dysfunctional beliefs.

The first part of cognitive restructuring is identifying the dysfunctional anxiety-provoking beliefs [6]. For example, the individual can be asked what thoughts are evoked when he is confronted to the phobic stimulus. He can think that if he goes on the plane, there are high chances that it will crash. This is a good example of a dysfunctional belief: even if it is true that such an eventuality could happen, it is important to keep in mind that probabilities are very small (for statistics, see [9]). This new information may not suffice to convince the patient that the fear is irrational, but it can help de-dramatize the situation and, therefore, reduce anxiety levels. Other questions that can help identify dysfunctional anxiety-provoking beliefs: *What thought runs in my mind when I am anxious?? What is the worst thing that could happen to me in that situation? What makes this situation more frightening than others? What could make the situation more difficult or easier? What image comes to my mind?*

The second part of cognitive restructuring is modifying the dysfunctional beliefs [6]. Following the identification of dysfunctional anxiety-provoking beliefs, the patient learns to become efficient in reframing these thoughts, with the help and support of the therapist. Note that cognitive restructuring is not the equivalent of positive thinking; it is a search for a realistic appraisal based on empirical facts. The role of the therapist is not to argue with the patient, but to collaborate with him in developing alternative and more functional beliefs. Here are a few examples of questions that can help the therapist with cognitive restructuring: *What proofs do you have for and against these thoughts? Is there another way to see the situation (another explanation)? How would someone else see the situation? Are you I fixing yourself irrational or unattainable goals? Are you reasoning with your emotions instead of your logic? Are your overestimating the degree of control over the events? Are you underestimating the things you can do to cope with the problem? Are you "catastrophizing"? Is your judgement based on impressions or facts? To what extent are you thinking in terms of certitude instead of probabilities? "So what if it happens!!!", is the consequence you fear so bad after all?* Let's reinstate that this techniques aims to at least build some confidence in alternative interpretations based on more functional beliefs. With specific phobias, it may be difficult for patients to have a strong confidence in the functional beliefs, or they might say they are convinced logically but not emotionally. At that stage, it is time to test the alternative beliefs with behavioural experiments [5] or exposure. When using the treatment protocol in private practice, it has been our experience to reach this point with most patients within the first therapy session or early in the second session.

The therapist must introduce exposure and provide directions and procedures [4,8,54]. In short, exposure can be introduced either as a way to build confidence in the alternative functional beliefs, as a tool to overcome fears by progressively facing them, or as a method to teach the emotional part of the brain (the limbic system) that there is no danger. The stimuli or situation that triggers a small or manageable amount of anxiety will be used first, based on what is available in the virtual environment. For fear of flying, it could be at the gate waiting for boarding or seated inside the airplane, for fear of snakes it could be with a garden snake lying still on the grass and for fear of heights it could be on the fourth floor of a virtual building. The therapist must therefore be aware of the different opportunities afforded by the available virtual environments. For example, a VR environment developed for the treatment pathological gambling may include a virtual bar where the patient can be exposed to social interactions with virtual humans, even if the environment was not designed for that purpose.

After informing the patient about exposure, the therapist should also how exposure will proceed. Firstly, the patient is informed that teamwork is an important treatment component and the therapist, like the patient, possesses equal responsibilities in treatment results. Exposure requires hard work and can only be efficient if the patient accepts his or her role in the team. The patient is also invited to provide suggestions on the exposure plan. It favours empowerment, facilitates treatment adherence and contributes to a sound working alliance. Some patients might apprehend to be exposed to extreme levels of anxiety and would not be able to cope mentally or physically. Subtle avoidance strategies should also be eliminated or otherwise tolerated until they can be stopped, such as taking a benzodiazepine pill before the therapy session or rescheduling the session because "today is a bad day".

The last part of the session could allow for a first contact with the virtual reality system so patients could familiarize themselves with VR before beginning exposure on the next session. Cybersickness should be discussed to see if the patient is prone to motion sickness or suffers from condition that would counter indicate the use of virtual reality in the following session. Patients are often curious about virtual reality and fearful apprehension may be reduced by an immersion in a virtual environment that does not contain any cues that would be phobogenic to the patient. In research settings, a first immersion is useful to reduce the effect of novelty on the first contact with the virtual phobogenic stimuli. With children, it is essential to assess patient's expectations toward the virtual stimuli [51]. Children are expecting virtual stimuli used for exposure to be larger, scarier and more dangerous than adults do [55], probably in part because they do not know what to expect from therapists and have a more active imagination.

3.2. Virtual reality exposure sessions (usually Sessions 2 to 6)

Goals for Sessions devoted to *in virtuo* exposure:

- Overcoming fears;
- Stopping avoidance behaviours;
- Developing new associations between the stimuli and the absence of threat;
- Decreasing perceived threat;
- Increasing perceived self-efficacy.

The core of the treatment protocol is exposing the patient to fear-provoking situations in a controlled virtual environment. It is important to underline the difference between therapeutic exposure and ordinary day to day confrontations with the feared stimuli. In therapy, exposure is planned, gradual, controlled and set-up to minimize avoidance, whereas in naturally occurring situations the events are unplanned, far from gradual, uncontrolled and faced with various forms of avoidance, safety seeking behaviours and neutralizations.

Elaboration of the exposure hierarchy is an important step and must be performed with the goals of disconfirming the dysfunctional mental associations and building confidence in new associations with the lack of threat. Therefore, the hierarchy does not have to be linear, evenly paced or with items systematically set on every step of the scale. What matters is to be able to adjust the difficulty level so the patient can successfully engage in emotionally charged situations that will modify the dysfunctional mental representations of the phobic stimuli. The rationale for a progressive hierarchy is to ensure that patients are: (a) not exposing themselves to feared situations that are too great for what they can cope with, (b) building self-efficacy, (c) not engaging in disruptive avoidance behaviours, (e) conclude the exposure with success, and (f) do not set-up for a pace that is too slow. Alternative can be implemented [3,4,8,12,13,16], such as flooding, guided mastery, combining biofeedback, or using antiphobic behaviours to challenge the notion that one must be careful in phobic situations (e.g., jumping off the virtual cliff to show how much control patients have over their actions).

Before initiating the actual exposure session, the therapist should have assessed the patient's understanding of the treatment rationale and the role of avoidance. At this stage, misconception should have been addressed. Although the therapist should begin exposure rapidly in the session to prevent increases in apprehension, the patient should be instructed to focus on what is happening right now in the virtual environment, instead of distracting himself or herself from the feared stimuli.

When the patient is ready, the patient can put on the HMD and begin exposure in virtual reality, starting with the first element in the pre-established hierarchy. The duration of the immersion should be planned to last between 20 to 30 minutes in order to reduce unwanted negative side effects. However, many immersions can be cumulated in a session if they are spaced by small pauses. Between each immersion, the therapist can discuss what happened during exposure. Since the duration recommended for the immersion is based on a very conservative approach to prevent eyestrain and cybersickness, the therapist should not rely solely and rigidly on the proposed time frame and gauged when it is appropriate to stop based on the psychotherapeutic processes occurring in the session (e.g., it may not be appropriate to stop if anxiety is at its peak value).

Cognitive restructuring, breathing retraining, relaxation and any other strategy that can become reassuring or a distracter must be eliminated during exposure [3,4,8,12,54]. During the immersion, the therapist interactions with the patient have to stay minimal and limited to: (a) asking for verbal ratings (from 1 to 100%) of anxiety, presence and cybersickness, (b) checking and removing avoidance behaviours, (c) guiding the patient to the next step in the hierarchy, and (d) providing a narrative or instructions that could increase presence (e.g., "Look at that spider, it is so tiny and quick it can probably climb on you hairs) or increase the value of the corrective experience (e.g., "Pay attention to how many people are actually laughing at you during your conference"). Therapists should not hesitate to be creative and use any cue in the virtual environment that can trigger more anxiety during exposure. Exploiting patient's fearful imagination is a very strong asset of *in virtuo* exposure. For example, in a virtual environment designed for claustrophobia where it is possible to lock the user in a wooden closet and a room with brick walls, a patient once said being immersed in the room in brick walls was much more frightening because it would be impossible to break the door. Since the objective way out was the same in both rooms - removing the HMD to stop the immersion - the therapist noticed the strong sense of presence during that session and later suggested he had lost the key to unlock the brick door and must leave the patient in the room for a few more minutes. It allowed exposing the patient to a much higher step on the hierarchy with exactly the same stimuli.

To foster presence during the immersion, it is important that the therapist attempts to maximize the feeling of presence by helping the patient feel as if the synthetic environment is real. For example, the therapist can prepare the immersion with a narrative providing a context for the exposure (e.g., "We will be leaving Ottawa by plane for a trip to Montreal. Our flight has been delayed for unknown reasons, but I guess since it is raining outside it may be due to poor weather conditions").Guiding the patient in the virtual environment should be done in ways that suggest the experience is real and not computer generated (e.g., ask the patient "Can you walk closer to the spider" and not "Can you push the mouse button to move closer to the virtual spider", or ask "Please turn off the engine before you exit the car" even if it is pointless in a virtual car). During the immersion, it is useful to observe the patient for signs of avoidance (e.g., always looking in a specific direction) and look for cues in the interaction between the patient and the environment to improve the exposure session (e.g., a patient may be invited to turn her back from a bed under which a spider just disappeared and wait until it returns).

When a patient accomplishes an exposure task in a rigid or unnatural way, the therapist should provide suggestions and guidance in order to improve the quality of his performance (e.g., talking in front of an audience, petting a dog) and eliminate safety seeking behaviour (e.g., always looking through the airplane window). Intermediate objective are sometimes needed. It is a technique that consists in asking during a difficult exposure session to concentrate efforts in reaching an intermediate objective if the planned objective turns out to be too difficult. For example, if a patient suffering from heights phobia cannot walk closer than a few feet away from the edge of a cliff, the therapist could suggest beginning by approaching the edge of the cliff from a different path where a railing is preventing him to fall.

Ideally, exposure should last until anxiety comes back to an acceptable baseline level (for example 20%) or the alternative functional belief is sufficiently strong to counter the association with perceived threat. The patient should remain in the exposure situation a few more minutes to make sure anxiety is gone and safety is well established. Exposure must be repeated to generalize across contexts and across therapy sessions. Afterwards, the patient can progress to situations that trigger more anxiety, either by visiting a new virtual environment or by confronting the fear-provoking stimuli more intensely. The therapist must assist the patient during this progression while again paying close attention to potential avoidance behaviours.

Reviewing post-immersion what happened during the exposure exercise is essential. Now it is time to apply other CBT strategies such as cognitive restructuring without risking distracting the patient or facilitating avoidance. Post processing of the emotional experience occurring during exposure serves several purposes. It allows the therapist to: (a) make sure the patient did not engage in avoidance strategies, (b) confirm if the exercise's was set at an adequate level of difficulty and implement corrections if required, (c) help the patient process the emotions that were experiences and the conclusions that should be reached, and (d) plan the next step to be conducted in therapy or as homework. After the session, the patient should remain in the waiting room for a while to make sure no there is no lasting unwanted side effects. It is an ideal moment to administer questionnaires, if needed.

3.3. Final ession for exposure, relapse prevention and closure (usually Session 7)

Goals for the final Session:

- Making final touch-up through exposure, if needed;
- Planning further exposure at home, if needed;
- Fostering patient's appropriation of treatment successes;
- Implementing relapse prevention strategies.

There is usually enough time in the last therapy session to conduct some *in virtuo* exposure if the therapist needs to tie some loose ends and push to reach the top of the exposure hierarchy. The therapist should allocated the last 15 to 20 minutes of this final session to review the progress, boost empowerment and appropriation of treatment success, plan homework or further in vivo exposure if needed, detect high risk situations, discuss the eventuality of a relapse, and set-up a relapse prevention plan [24]. The therapist should reinforce the patient's perceived self-efficacy, both toward facing phobogenic stimuli and using a difficult emotion regulation technique. Patient should be able to trust their abilities to detect avoidance strategies, implement an exposure plan and deal with fear. All they need now is time to strengthen and generalize their gains.

As with *in vivo* exposure, the therapist must now encourage patients to keep exposing themselves to stimuli that were feared before the treatment. Some patients may have not reached their objectives entirely. Any improvement during therapy must first be highlighted. The next exposure steps should be written down, taking into account roadblocks that might have occurred during the treatment. For those who are free from their phobia, creating opportunities to consolidate their newly acquired set of mental representations is important. For example, a patient treated for flying phobia should schedule a flight in the near future, and keep flying once in a while afterwards. It does not mean the treatment is incomplete. But safety seeking behaviours might have been over learned and avoiding contacts with the feared stimuli was one of these behaviours. For example, a patient treated for spider phobia must make a habit of taking the lead in getting rid of spiders when one is found in the family home.

At the final therapy session, it is important to address relapse. Just like an insurance policy that people pay with the hope of never having to claim it, relapse must be discussed in order to prevent a lapse to become a relapse. The program focuses on three ingredients of relapse prevention: (a) patients must be aware that experiencing anxiety is normal and does not mean the occurrence of a relapse, (b) managing well the emotional reaction triggered by a lapse significantly reduces the risk of a relapse, and (c) patients must prepare a cognitive-behavioural plan to deal with lapse and relapse. It must be clear for patients that feeling anxious under certain circumstances is normal. In the case of a lapse, or the reoccurrence of phobic fear, it is unproductive to react in a catastrophic manner, as if everything accomplished in therapy was useless and lost. As with any problem, it is more fruitful to interpret the situation as an opportunity to find out was went wrong and correct it. Actually, the intensity of the guilt and perceived loss of control that a patient feels when phobic fear is returning are associated with the crystallisation of a momentary lapse into a relapse of the phobic disorder. The best attitude is to identify what did not work out, look at what was learned during therapy and make the necessary corrections. The therapist can give the example of falling from a bicycle to illustrate this principle: *"I will tell you a story about something that happened to me a few years ago in the Spring. I was going down a hill and had to stop. As I braked forcefully using both the front and back brakes, I was thrown off balance and fell on the street. As I stood up, I was very angry and I lost control for a few moments and give several kicks to the bicycle. Do you think it was a good idea to lose control, feel bad and kick my bicycle?* As the patient is telling you that these emotions and behaviour were unproductive, you can add *"What you are saying is I must consider my fall as an accident or a glitch; falling down from my bicycle does not mean that I don't know how to ride one, but rather that I should be more vigilant in the future and use more the back brake and less the front brake. A glitch is a source of information, not a matter for overacting. So, let's apply what we discussed about my fall from the bicycle to a potential return of phobic fear in your life. You must not blame yourself for what happened, overreact and lose control. Instead, you must look at the way the incident happened and analyse it: have you been taken by surprise? If yes, it is understandable that you have reacted the way you did. Have you been avoiding? A good way to get rid of the problem is to assess the situation, recall what worked for you in therapy and do a few exposure exercises on your own."*

As a practical relapse prevention exercise, the therapist should ask the patient to write a letter to himself or herself. The letter must include three sections: (a) information about the differences between a lapse and a relapse, factors and situations at risk for a relapse, and that over reacting aggressively, with guilt and a feeling of losing control is not fruitful; (b) a summary of what was learned in therapy and recommendations the patient would like to formulate; and (c) a list of specific actions and homework to do in order to get rid of the phobia, including suggestions for exposure. Once written, the Letter to Myself is to be sealed and kept by the patient at home with his or her personal documents. In the case of a lapse, or if the patient calls a few years post-therapy, the first thing to do for the patient is to read the letter and engage in the recommended homework.

4. Conclusion

The program described in this chapter builds on decades of experimental and clinical research on exposure and CBT for specific phobias [4-8, 11-14]. Conducting exposure with an immersion in virtual reality is not much different than traditional CBT, except for the importance of the feeling of presence and monitoring cybersickness. The key ingredients for an effective *in virtuo* exposure treatment are the same as *in vivo* exposure, with a dedication to remove avoidance in all its forms. The program was intended to be delivered in seven active sessions, plus a few preliminary sessions for diagnosis and case formulation. The treatment is described without reference to specific software, which should facilitate its use in various settings. Outcome studies and randomized outcome trial have been conducted with success using this protocol [15,49-51] or other quite similar protocols developed for a specific disorder or software (for reviews[15-18]).

CBT being the only empirically validated and effective treatment for specific phobias [10], it is not surprising that VR applications have been developed for CBT. The consequence is that very few researchers have yet tempted to apply VR technologies in the context of other schools of thoughts. A literature search revealed a few papers from the field of psychodynamic therapy or reflections from a psychodynamic perspective. The research of Optale and collaborators [56] addressed the treatment of erectile dysfunction and premature ejaculation in 160 males. In their experiment, the virtual environment depicted different pathways through a forest in order to symbolically bring male patients back to their childhood, adolescence and later teens. The authors suggest the immersion accelerates a psychodynamic process that eludes cognitive defenses and directly stimulates the subconscious. Hence, the obstacles that lead to sexual dysfunction would be brought to light and modified under the therapist's guidance much faster than with conventional psychodynamic psychotherapy. Optale reported that positive results were found with his technique. Another approach to refine the understanding of the possibilities of VR in psychodynamic therapy is to use VR within a cognitive-behavioural program but analyse the results from a psychodynamic point of view. In the research of Wiederhold and collaborators [57], a case report of a patient treated with *in virtuo* exposure for fear of flying is discussed based on psychoanalytic principles. During her treatment, the patient was said to "progressively learned to abandon her defensive dissociation mechanism while in immersion". The authors suggest this was possible because the transference that the patient developed occurred primarily with the virtual environment, rather than with the therapist, so it can serve as a transformational object.

Virtual reality, conceived in a much broader perspective that include everything related to cyberspace, from Internet profiles, video games and television, also stir interest in some psychoanalysts. One of their preoccupations is how new technologies will influence the therapeutic process. At the 46th Congress of the International Psychoanalytical Association, Moreno proposed that by simulating an object, virtual reality may occlude the space between the represented and the representation, thus reducing some frustrations [84]. For instance, children who play video games can now live in the illusion of becoming others. In other words, virtual reality could reduce the difference between fantasy and reality. Similarly, if the representation of an object may subsume the effects of its presentation, the question of the necessity of both parties being physically present for the analysis arises. Thus, according to Moreno, clinicians need to address the difference between the traditional analytic process and the analytic process in treatments of anxiety disorders delivered through e-mail, telephone or Skype™. With the growing popularity of social networks like Facebook, some psychoanalysts also think there is a need to redefine the traditional views of privacy and anonymity in therapeutic settings [58]. Research about psychodynamic therapy and VR is still limited to just a few papers and creative reflections. Neubeck and Neubeck [59] proposed a hypothetical description of psychodynamic possibilities in VR. For instance, they suggested that VR could be used to create symbolic reproductions, to allow patients to associate with diffusely-built fictitious worlds, and to create pictures that can help them cope with hidden memories. However, these ideas still need to be implemented before we know how VR could complement and support psychodynamic treatment.

Another matter of interest in psychologists from other approaches than CBT, and also among CBT therapists, is the impact of technology on therapeutic alliance. Indeed, little is known about the role of therapeutic alliance in technology-assisted interventions. Therapeutic alliance between the patient and the therapist is often defined as the strength of: (a) the collaborative agreement on the general goals of the treatment, (b) the collaborative agreement on the specific tacks required in the treatment, and (c) the quality of the affective bond between therapist and patient. In 2008, Meyerbröker and Emmelkamp [60] published a study on the role of the therapeutic alliance during *in virtuo* exposure for fear of flying and acrophobia. Their results indicated that the quality of the therapeutic alliance according to the patient was positively related to anxiety reduction following exposure, but only for the fear of flying participants. The authors suggested that one explanation for the inconsistent results between both types of phobias could be the nature of the questionnaire used to measure the therapeutic alliance.

In summary, VR does not have to be used only by CBT therapists. Other therapeutic approaches are starting to explore what can be done with immersions in synthetic environment. In the meantime, real-time interactions with virtual stimuli allow CBT therapist to create emotionally curative experiences that can be used to treat phobias. In the near future, applications will be widely available using augmented reality, portable technologies, or live biofeedback interaction between the user and the virtual environment. But using technology should not become an aim in itself; it remains a tool to assist well trained professionals in the delivery of mental health services.

Author details

Stéphane Bouchard*, Geneviève Robillard and Claudie Loranger
Université du Québec en Outaouais, Gatineau, Québec, Canada

Serge Larouche
Centre hospitalier Pierre-Janet, Gatineau, Québec, Canada

Acknowledgement

The preparation of this chapter was made possible thanks to financial support from the Canada Research Chairs and Canadian Foundation for Innovation programs. The authors wish to acknowledge the contribution of Belle Paquin and Sophie Côté to the preparation of an earlier version of the treatment program. All pictures and screenshots, courtesy of the Cyberpsychology Lab of UQO.

5. References

[1] American Psychiatric Association (2000) Diagnostic and Statistical Manual of Mental Disorders – Text Revision (4e ed.). Washington: APA. 943 p.

[2] Martin M Antony, Mark A Watling (2006). Overcoming Medical Phobias. Oakland: New Harbinger Publications, Inc. 164 p.

[3] Martin M Antony, Randy E McCabe (2005) Overcoming Animal and Insect Phobias. Oakland (CA): New Harbinger Publications, Inc. 149 p.

[4] Jonathan S Abramowitz, Brett J Deacon, Stephen PH Whiteside (2011). Exposure Therapy for Anxiety: Principles and Practice. New York: The Guilford Press. 398 p.

[5] David H Barlow (2002). Anxiety and its Disorders (2nd Ed.). New York: Guilford Press. 704 p.

[6] David A Clark, Aaron T Beck (2010). Cognitive Therapy of Anxiety Disorders: Science and Practice. New York: The Guilford Press, 628 p.

* Corresponding Author

[7] Graham CL Davey (1997) Phobias: A Handbook of Theory, Research and Treatment. Brighton: John Wiley & Sons. 451 p.
[8] Michael W Otto, Stefan G Hofmann (2011) Avoiding Treatment Failures in the Anxiety Disorders. New York: Springer. 405 p.
[9] Martin M Anthony, Richard P Swinson (2000) Phobic Disorders and Panic in Adults: A Guide to Assessment and Treatment. Washington: APA. 422 p.
[10] Peter E Nathan, Jack M Gorman (2007) A Guide to Treatments that Work (3rd ed.). New York: Oxford.784 p.
[11] Marshall, W.L (1985) Exposure. In: Bellack AS & Hersen M, editors. Dictionary of Behaviour Therapy Techniques. New York: Pergamon Press. pp.121-124
[12] David CS Richard, Dean L Lauterbach (2007) Handbook of Exposure Treatment. Burlington: Elsevier, inc. 437 p.
[13] William T O'Donohue, Jane E Fisher (2009). General Principles and Empirically Supported Techniques of Cognitive Behaviour Therapy. New Jersey: John Wiley & Sons, Inc. 743 p.
[14] Jacqueline B Persons (2008) The Case Formulation Approach to Cognitive-Behaviour Therapy. New York: The Guilford Press. 273 p.
[15] Bouchard S, Côté S, Richard DS (2006) Virtual Reality Applications of Exposure. In: Richard DS and Lauterbach D, editors. Handbook of exposure. New York: Academic Press. pp. 347-388.
[16] Brenda K Wiederhold, Mark D Wiederhold (2005) Virtual Reality Therapy for Anxiety Disorders. Advances in Evaluation and Treatment. Washington, DC: American Psychological Association Press.
[17] Côté S, Bouchard S (2008) Virtual Reality Exposure for Phobias: A Critical Review. Journal of CyberTherapy and Rehabilitation. 1(1): 75-91.
[18] Meyerbröker K, Emmelkamp, PMG (2010) Virtual Reality Exposure Therapy in Anxiety Disorders: A Systematic Review of Process and Outcome Studies. Depression and Anxiety. 27(10): 933-944.
[19] Pratt DR, Zyda M, Kelleher K (1995) Virtual reality: In the Mind of the Beholder. IEEE Computer. 28(7): 17-19.
[20] Tisseau J (2008) In Vivo, In Vitro, In Silico, In Virtuo. 1st Workshop on SMA in Biology at Meso or Macroscopic Scales, Paris, July 2, 2008.
[21] Draper JV, Kaber DB, Usher JM (1998) Telepresence. Human Factors. 40(3): 354-375.
[22] Sadowski W, & Stanney KM (2002) Presence in Virtual Environments. In: Stanney KM, editor. Handbook of virtual environments : Design, implementation and applications. Mahwah : IEA. 791-806.
[23] Bouchard S, Dumoulin S, Monthuy-Blanc J, Labonté-Chartrand G, Robillard G, Renaud P (2011, submitted). Perceived Realism Contributes to the Presence Experienced in a Virtual Environment. Interacting with Computers.
[24] Antonio Damasio (1999) The Feeling of What Happens: Body and. Emotion in the Making of Consciousness. New York: Harcourt Brace. 386 p.

[25] Riva G, Waterworth JA, Waterworth EL (2004) The Layers of Presence: A Bio-cultural Approach to Understanding Presence in Natural and Mediated Environments. CyberPsychology & Behaviour. 7(4): 402-416.

[26] Riva G (2009) Is presence a technology issue? Some Insights from Cognitive Sciences. Virtual Reality. 13: 159-169.

[27] Waterworth JA, Waterworth EL, Mantovani F, Riva G (2010) On Feeling (the) Present: an Evolutionary Account of the Sense of Presence in Physical and Electronically-Mediated Environments. Journal of Consciousness Studies. 17(1-2): 167-188.

[28] Youngblut C (2007) What a Decade of Experiments Reveals About Factors that Influence the Sense of Presence: Latest Findings. Virginia: Institute for Defense Analyses. 214 p.

[29] Robillard G, Bouchard S, Fournier T, Renaud P (2003) Anxiety and Presence During VR Immersion: A Comparative Study of the Reactions of Phobic and Non-Phobic Participants in Therapeutic Virtual Environments Derived from Computer Games. CyberPsychology and Behaviour. 6(5): 467-476.

[30] Loranger C, Bouchard S, Boulanger J, Robillard G. (2011) Validation of Two Virtual Environments for the Prevention and Treatment of Pathological Gambling. Oral presentation at the 16th Annual Cybertherapy & Cyberpsychology Conference, Gatineau, June 20-22.

[31] Krijn M, Emmelkamp PMG, Biemond R, de Wilde de Ligny C, Schuemie MJ, van der Marst CAPG (2004) Treatment of Acrophobia in Virtual Reality : The Role of Immersion and Presence. Behaviour Research and Therapy. 42: 229-239.

[32] Bullinger A, Angehrn I, Wiederhold BK, Nueller-Spahn F, Mager, R (2005) Treating Acrophobia in a Virtual Environment. In: Wiederhold BK, Riva G, Bullinger MD, editors. *Annual Review of CyberTherapy and Telemedecine*. San Diego, CA: Interactive Media Institute. pp. 93-100.

[33] Mülhberger A, Herrmann MJ, Wiedemann G, Ellrring H, Pauli P (2001) Repeated Exposure of Flight Phobics to Flights in Virtual Reality. Behaviour Research and Therapy. 39:1033-1050.

[34] McCauley ME, Sharkey TJ (1992) Cybersickness: Perception of Self-Motion in Virtual Environments. Presence. 1(3): 311-318.

[35] Lawson BD, Graeber DA, Mead AM (2002) Signs and Symptoms of Human Syndromes Associated with Synthetic Experience. In: Stanney KM, editor. Handbook of Virtual Environments: Design, Implementation, and Applications. Mahwah: IEA. pp. 589-618.

[36] Kennedy RS, Lane NE, Berbaum KS, Lilienthal MG (1993) Simulator Sickness Questionnaire: An Enhanced Method for Quantifying Simulator Sickness. International Journal of Aviation Psychology. 3: 203-220.

[37] Sharples S, Cobb S, Moody A, Wilson JR (2008) Virtual Reality Induced Symptoms and Effects (VRISE): Comparison of Head Mounted Display (HMD), Desktop and Projection Display Systems. Displays. 29: 58–69.

[38] Bouchard S, St-Jacques J, Renaud P, Wiederhold BK (2009) Side Effects of Immersions in Virtual Reality for People Suffering from Anxiety Disorders. Journal of Cybertherapy and Rehabilitation. 2(2): 127-137.

[39] Bouchard S, Robillard G, Renaud P, Bernier F (2011) Exploring New Dimensions in the Assessment of Virtual Reality Induced Side-Effects. Journal of Computer and Information Technology. 1(3): 20-32.

[40] Stanney KM, Kennedy RS, & Kingdon K (2002) Virtual Environment Usage Protocols. In: Stanney KM, editor. Handbook of Virtual Environments: Design, Implementation, and Applications. Mahwah: IEA. pp. 721-730.

[41] James T Reason, J J Brand (1975) Motion Sickness. London: Academic Press.

[42] Treisman M (1977) Motion Sickness: An Evolutionary Hypothesis. Science. 197: 493-495.

[43] Money KE, Lackner J, Cheung R (1996) The Autonomic Nervous System and Motion Sickness. In: Yates BJ, Miller AD, editors. Vestibular Autonomic Regulation. Boca Raton, FL: CRC Press. pp. 147-173,

[44] Stoffregen TA, Riccio GE (1991) An Ecological Critique of the Sensory Conflict Theory of Motion Sickness. Ecological Psychology, 3: 151-194

[45] Bouchard S, Robillard G, Renaud P (2007) Revising the Factor Structure of the Simulator Sickness Questionnaire. Oral presentation at the 12th Annual CyberTherapy Conference 2007, Washington (DC), June 12-14.

[46] Kennedy RS, Stanney KM, Dunlap WP (2000) Duration and Exposure to Virtual Environments: Sickness Curves During and Across Sessions. Presence: Teleoperators and Virtual Environments, 9, 466-475.

[47] Nichols S, Ramsey AD, Cobb S, Neale H, D'Cruz M, Wilson JR (2000) Incidence of Virtual Reality Induced Symptoms and Effects (VRISE) in desktop and projection screen display systems, HSE report 274/2000.

[48] St-Jacques J, Bouchard S (2005) Clinical Applications of Virtual Reality and Cybersickness. In B.K. Wiederhold, G. Riva & A. H. Bullinger, (Eds). Annual Review of CyberTherapy and Telemedicine, Volume 3, pp. 296-297.

[49] Côté S, Bouchard S (2005) Documenting the Efficacy of Virtual Reality Exposure with Psychophysiological and Information Processing Measures. Applied Psychophysiology and Biofeedback. 30(3): 217-232.

[50] Michaliszyn D, Marchand A, Bouchard S, Martel MO, Poirier-Bisson J (2010). A Randomized, Controlled Clinical Trial of In Virtuo and In Vivo Exposure for Spider Phobia. Cyberpsychology, Behaviour and Social Networking. 13(6): 689-695.

[51] St-Jacques J, Bouchard S, Bélanger C (2010) Is Virtual Reality Effective to Motivate and Raise Interest in Phobic Children Towards Therapy? Journal of Clinical Psychiatry, 71(7): 924-931.

[52] Thorpe SJ, Salkovskis PM (1997) The Effect of One-Session Treatment for Spider Phobia on Attentional Bias and Beliefs. British Journal of Clinical Psychology. 36: 225–241.

[53] Côté S, Bouchard S (2009) Cognitive Mechanisms Underlying Virtual Reality Exposure. Cyberpsychology & Behaviour. 12(2): 121-129.

[54] Bouchard S, Mendlowitz SL, Coles ME, Franklin M (2004) Considerations in the Use of Exposure with Children. Cognitive and Behavioural Practice. 11(2): 56-65.

[55] Silva C, Bouchard S, Bélanger C (2011) Youths are More Apprehensive and Frightened than Adults by a Virtual EnvironmentUsed to Treat Arachnophobia. Journal of Cybertherapy and Rehabilitation. 4(2): 200-201.

[56] Optale G, Marin S, Pastore M, Nasta A, Pianon C (2003) Male Sexual Dysfunctions and Multimedia Immersion Therapy (Follow-Up). CyberPsychology & Behaviour. 6(3), 289-294.

[57] Wiederhold BK, Gavshon L, Wiederhold MD (2010) A Psychodynamic View of Virtual Reality Exposure Therapy. Journal of CyberTherapy & Rehabilitation. 3(4): 395-403.

[58] Cairo I, Fischbein SV (2010) Psychoanalysis and Virtual Reality. The International Journal of Psychoanalysis. 91(4): 985-988.

[59] Neubeck AK, Neubeck B (1998) Virtual Reality as a Support System for Psychodynamic Treatment. CyberPsychology & Behaviour. 1(4): 341-345.

[60] Meyerbröker K, Emmelkamp, PMG (2008) Therapeutic Processes in Virtual Reality Exposure Therapy: The Role of Cognitions and the Therapeutic Alliance. Journal of CyberTherapy & Rehabilitation. 1(3): 247-257.

Chapter 5

Virtual Reality and Body Dissatisfaction Across the Eating Disorder's Spectrum

Annie Aimé, Karine Cotton, Tanya Guitard and Stéphane Bouchard

Additional information is available at the end of the chapter

http://dx.doi.org/10.5772/46414

1. Introduction

Body image dissatisfaction is very prevalent in women of all ages [1, 2]. Very early in life, girls are incited to pay attention to their hair, skin, or weight and to what they wear [3]. They learn that they can get praised for their body and appearance and that they have to watch their weight and shape in order not to become overweight [3, 4]. Girls and women are continuously exposed to displays of appearance and can put a significant amount of money, time, and effort in trying to achieve inaccessible beauty standards [5]. In some women, internalization of beauty norms can be so pervasive that they base their success and self-worth almost exclusively on their weight and shape.

Given the prevalence of body dissatisfactions and their consequences on the psychological functioning of women, assessment and treatment considerations must be pursued. Virtual reality represents an interesting way of enhancing motivation to work on body image. Over the years, a growing number of studies have highlighted the effectiveness of virtual reality for reducing body image disturbances in samples of women with eating disorders.

2. The concept of Body Image

Body Image (BI) is a multidimensional concept which includes cognitive, affective, behavioral, and perceptive components [6, 7]. It can also be conceived as a mental picture of the body [8]. The concept of BI implies two basic elements: investment and evaluation. While investment focuses on how important one's appearance can be, evaluation involves appearance appraisals [9, 10]. Evaluation gives rise to satisfaction or dissatisfaction with the body. Subjective satisfaction-dissatisfaction refers to a person's appreciation of his or her body (weight, shape, specific body sites). Body dissatisfaction is more frequently studied than body satisfaction. To be considered dissatisfied with one's body, an individual has to be seeking different body characteristics than its own, and must report negative thoughts and affects related to a discrepancy between his idealized and actual body [7, 11].

BI is subjective and can change over time, in response to new information or to situational and emotional factors [11, 12]. It develops through cultural socialization, interpersonal experiences, personality factors, physical characteristics, as well as physical changes [10]. Cultures convey information and standards regarding appearance [10]. Such standards are disseminated through media and advertisement. Cultural messages promote stereotypes and inaccessible beauty norms [13]. They dictate what physical characteristics are valued or not valued by people [10]. They also put a strong emphasis on means to achieve a body that better conforms to the beauty standards in vogue, such as dieting, exercising, or using beauty products. In addition to media exposure, family members, peers, and even strangers can exert a pressure to be thin and to meet social standards. While interacting with others, one is confronted to their opinions, expectations, comments, and potentially to their criticisms [10]. Weight loss is particularly encouraged and thinness is reinforced through compliments, approval manifestations, or one's own feelings of self-control or will power [14].

When they comment on their child's body or weight, and when they impose particular food rules in their household, parents are influencing their child's BI in a direct way. They can also have an indirect influence, especially through their own weight and shape preoccupations and through problematic eating behaviors aimed at loosing or controlling their weight [5]. Siblings can represent a source of social comparison and can also contribute to a child's negative self-appraisal [10]. Peer influence is particularly relevant as children move towards adolescence. Adolescent girls tend to report similar levels of body satisfaction/dissatisfaction and dietary restraint to those of their group of friends [5]. Peer influence includes weight and shape comments, appearance teasing, modeling of weight preoccupations and weight control behaviors, establishment of appearance and weight norms, and conversations about weight and shape [5]. In childhood and adolescence, BI evaluation can be closely linked to peer acceptance with regards to one's body and appearance [5, 10].

Physical and personality characteristics are important in BI development [10]. Women and girls who are overweight are more likely to be dissatisfied with their body and to try means of weight control [15, 16]. In fact, the more a woman's weight differs from what she believes to be the ideal weight, the more she is at risk of reporting elevated levels of body dissatisfaction [17]. As stated by Cash [10], how closely one's appearance conforms to cultural standards is essential in self-evaluation. Moreover, personality traits like perfectionism, public self-consciousness, and conformism can predispose an individual to put a strong emphasis on appearance, to internalize beauty standards, and to actively try to attain these standards [10, 18].

3. Body Image in women

Body shape and weight dissatisfactions are so widespread in the general population that they are collectively termed "normative discontent" [5, 19]. Physical appearance plays a critical role in women's gender identity. Women are significantly more vulnerable to social influences and they report higher body dissatisfaction than their male counterparts [20, 21]. Body image disturbances are so frequent in women that up to 80% of college women report some degree of body dissatisfaction [22]. According to the sociocultural perspective, in women, media exposure leads to internalization and social comparison with regards to appearance, which in turn increases body dissatisfaction and the likelihood of adopting problematic eating behaviors [5]. Body dissatisfaction and problematic eating are related through two possible pathways [23, 24]. The first possible pathway is one through which a negative evaluation of one's weight or body brings about dieting behaviors aimed at weight control or weight reduction. The second pathway pertains to depressive mood. It postulates that negative self-appraisal provokes depressive symptoms and intense feelings, which are regulated through overeating.

Once in place, problematic eating behaviors can lead to the development of an Eating Disorder (ED). Therefore, not only do body and weight dissatisfactions predict problematic eating, but they are also strong predictors of ED onset [25, 26]. In ED, the two basic elements of BI are present: first, the body (weight and shape) is intensively, almost exclusively invested, and secondly, over-evaluation is constantly observed [27]. In fact, in ED patients, body dissatisfaction transforms into over-evaluation of shape and weight. These patients' self-worth relies almost entirely on their shape and weight and even on their ability to control them [19]. Over-evaluation is believed to be the core psychopathology of ED. It affects socialization of ED patients and is manifested in many ways such as checking, scrutinizing, and behaviors of avoidance and comparison. It has a major impact on eating habits, with dietary restraint most likely playing a prominent role [19].

4. Virtual environment for assessing and treating Body Image in ED women

To this day, experimentation with Virtual Reality (VR) has been mainly done with samples of women with ED. Researchers started to use VR in the late 1990s with ED women in order to assess and treat their body dissatisfaction [12]. Riva's team in Italy and Perpiña's team in Spain both developed software packages aimed at assessing body image disturbances [28-30]. Both research teams used their software in treatment contexts, in order to measure changes in body dissatisfaction once treatment was completed.

Many types of virtual environments exist for evaluating and treating BI in ED. All of them share the basic goal of confronting ED women to their fears about weight and shape and to their dysfunctional mental representations of themselves [30]. VR appears particularly effective in treating misperceptions about one's body [31]. Through VR, it is possible to integrate new information about one's body and to acquire ways to differentiate the cognitive and affective perception from the real body. To achieve such distinction, a clinician can, for example, ask a woman who considers herself too big or too fat to use her virtual body to pass through a narrow door or a virtual space that looks too small to her. When she realizes she can pass through the narrow door, it creates an important doubt in her mind about the realism of her perception [32]. As a consequence, she becomes more likely to admit that her mental representation of her body is distorted, and to develop a new, more realistic and less distressing view of her body. VR allows for a complete BI experience, covering cognitive, behavioral, emotional, and perceptive components of BI [30].

Riva and his team were the first researchers to incorporate VR in the treatment of body image disturbances [33]. Their work was first applied to nonclinical samples, and since their results proved to be encouraging, they conducted the same treatment on samples of patients with ED. Their treatment program is based on what they refer to as Experiential Cognitive Treatment (ECT) [31, 34-36], which aims at challenging and changing ED women's perceptions toward their bodies. More precisely, the goal of the ECT program is to integrate VR as a way of challenging the idea that what looks like a perception (i.e. overweight shape) may in fact just be an assumption. To achieve this goal, they use the Virtual Environment for Body Image Modification 2 (VEBIM 2), which is an enhanced version of the original VEBIM, previously used with nonclinical samples. The VEBIM 2 is composed of five different zones that each correspond to a treatment component. In the first session (zone 1), participants explore the environment and learn how to use VR materials. They are then asked to weigh themselves on a virtual scale in order to focus their attention on the scale and elicit the importance of the weight dimension in the rest of the experimentation. The next zone, used in session 2, is composed of a kitchen, a closet, and a bedroom that each contain different food and drink. Subjects are encouraged to virtually consume what they would normally eat and drink. They do so by selecting the desired virtual stimuli. Once the food is "eaten", the matching calorie intake is computed in order to calculate the total calories ingested. This was created as a way to record the eating habits of the participants and to assess their reactions at the end of the session when confronted with their new weight on the virtual scale. In session 3, participants are immersed in a virtual environment in which they must choose two out of seven figures representing different body types, ranging from underweight to overweight. The two chosen figures represent their perception of their current body size and their ideal body. The discrepancy between the two figures is used as an indication of their level of body dissatisfaction. Session 4 represents a workplace environment in which participants must select food and drink as in session 2. This allows for the establishment of differences in eating behaviors while in the workplace rather than at home. In the fifth and final setting (zone 5), subjects are first immersed in an environment where pictures of models are painted across the room. This is done to elicit emotions that are then analysed by the clinician. Afterwards, the subjects are placed in front of a large mirror and then guided in a room with five doors of different dimensions. To exit this room, they must choose the door corresponding exactly to their width and height [37].

Perpiñá and his team used their own VR software exclusively with ED patients [38]. The VR part of their treatment program consisted of six different settings, including the training room, the kitchen, the poster room, the two-mirrored room, and the six-mirrored [12, 30]. They added 3D human images that can be altered to their software. This enables the participants to model their subjective and desired body, as well as the body shape they envisioned a significant person would have of them.

5. Efficacy of Virtual Reality focusing on Body Image in ED women

In their literature review, Ferrer-Garcia and Guitierrez-Maldonado [12] provide detailed explanations about the suitability of VR in the assessment and treatment of BI. They also provide an extensive list of the studies that have focused on the effectiveness of VR as an addition to the traditional treatment of body image disturbances, both in non-clinical and ED samples. They conclude that VR improves body image disturbances in both types of samples and that this improvement is maintained, even accentuated, at 6-month follow-up [12].

Research to date is mostly limited to the works of Riva's and Perpiñá's teams. Riva and his colleagues [31, 39] found that, in ED patients, a treatement including VR and ECT reduces binge eating, anxiety, and preoccupations having to do with being judged by others. It also improves body satisfaction, self-acceptance, self-esteem, self-efficacy, social functioning, as well as motivation to change. Perpiñá's team [30] also found that the addition of a VR component to a Cognitive-Behavioral Therapy increases its efficacy. ED patients who received the combined treatment condition showed better adherence to treatment, enhanced body satisfaction, decreased negative thoughts and attitudes towards their body, less intensive fears of weight gain following a meal, and a positive change in their beliefs concerning healthy weight.

Past research also shows that virtual stimuli can generate strong emotional responses such as anxiety or depressive symptoms in ED patients. For example, Gorini, Griez, Petrova and Riva [40] found that, for ED patients, virtual food was as anxiety provoking as real food and more so than pictures of food. Others have demonstrated that virtual environments involving high-caloric food elicit strong emotional responses in ED patients [41], reflecting a fear of gaining weight and an over-evaluation of weight and shape. ED patients also report higher anxiety and depressive symptoms in virtual environments where they virtually eat high-energy food [42]. Another significant environment for ED patients is the swimming pool environment. [41, 42]. According to Thompson and Chad [43], the high emotional reaction observed in the pool environment also reflects the over-evaluation of weight and shape in ED women. Additionally, it could be related to the tendency of ED patients to avoid situations in which they have to expose their bodies to other's evaluation, or in which they might compare themselves negatively to others.

6. Virtual Reality in women without ED

Few studies have used VR in women without ED (see Ferrer-Garcia & Guitierrez-Maldonado, 2011 for a review) [12]. Riva's research team showed that women without ED who were exposed to their VEBIM reported a significant decrease in body dissatisfaction following a VR session of 8 to 10 minutes [28, 29, 34, 35]. Such results indicate that VR represents an easy to use and time efficient prevention tool.

Scarce past research has also shown that body dissatisfaction might not be manipulated or treated exactly the same way across the eating disorder spectrum. The type of virtual environment could differ according to the intensity of weight and shape concerns. For example, Ferrer-Garcia and colleagues [42] found that, in participants without ED, a swimming pool environment is more anxiety provoking than any other virtual environment they used (high and low-calorie kitchen, high and low-calorie restaurant). In fact, the anxiety level of women without ED in the other four virtual environments was equivalent to that of the neutral environment.

Since the eating disorder's spectrum cannot be strictly confined to women with or without ED, our research team decided to evaluate the emotional reactivity of women preoccupied with their weight and shape who did not encounter the diagnosis criteria of ED [44]. In our experiment, we compared the emotional responsiveness of women with subclinical eating, weight, and shape concerns to that of a control group of women whose concerns were within the norms for women without ED [45]. Three virtual environments were used: an office, a swimming pool, and a buffet-style restaurant containing high and low-calorie food. Our results suggest that VR has the potential to generate substantial emotional responses across the eating disorder spectrum and that it could be useful as a prevention tool with individuals at risk of developing ED, or as a way to improve acceptance of one's own body in anyone who reports some body dissatisfaction.

In the subclinical group of our experiment [44], the anxiety level was significantly higher than in the control group during and after exposure to the challenging environments (swimming pool and restaurant), suggesting that women with eating, weight, and shape preoccupations share a tendency with ED women to be more strongly emotionally and cognitively activated by exposure to food (restaurant) and to other women (swimming pool), than women with normative weight and shape concerns. Therefore, as a woman's weight and shape concerns progress through the eating disorder spectrum, she becomes more likely to react to high-calorie food and social comparison.

Moreover, we found that for both groups of women, the swimming pool environment is significantly more effective in eliciting anxiety than the restaurant environment. This latest result is particularly interesting in light of the "normative" dissatisfaction so widely spread in our Western Society. It shows that, no matter the intensity of the eating, weight, and shape preoccupations, and no matter the positioning on the eating disorder spectrum, women are at risk of feeling anxious when they are in an environment where they must expose their body to others' evaluation, and where they have an opportunity to compare themselves to other women. The swimming pool environment has an interpersonal connotation other environments such as the restaurant do not have. It combines two kinds of social influences; the mass media pressure, in which women are exposed to unwanted slim beauty images, and the peer pressure, in which they have to tolerate the evaluation of average women. Since the buffet immersion was not as anxiety provoking as the swimming pool immersion in women with subclinical preoccupations with weight and shape, we believe social comparison might be more relevant than food exposure in women without ED.

7. Social comparison in VR

Social comparison appears to be a promising avenue to explore in VR, since every human being is using social comparison in order to assess their value and standing in life [45]. It has been recognized as an important component of ED treatment and is considered to be a form of shape checking that maintains the over-evaluation of weight and shape in ED patients [19]. ED patients tend to compare themselves to people of the same gender and age (or younger) who are thin and good-looking, and they fail to notice people who are not as good-looking or thin [19].

The Social Comparison Theory [45] offers an interesting framework to understand the increased emotional reactivity and potential body dissatisfaction following VR. This theory postulates that both intentional and unintentional comparisons with others are a common and basic phenomenon in humans [22]. Depending on its direction (upward or downward), social comparison has different affective consequences. When a woman compares herself to someone who appears to better meet beauty standards (upward comparison), she is more likely to experience deleterious effects from her comparison, such as feelings of discontent and dissatisfaction. However, positive emotional effects may occur when a woman compares herself to someone whose weight and appearance appear worse than her own weight and shape (downward comparison). VR can provide, in a controlled setting, standards against which women with different BI could compare themselves to other women or even to beauty icons. In VR, the characteristics of the comparison target can be manipulated in such ways to facilitate either upward or downward comparisons and to evaluate their affective consequences.

In order to test the social comparison theory, our research team [46] recently created a bar environment in which social comparison with a waitress was induced (Figure 1). Two different types of stimuli were used: an overweight (Figure 2) and a thin virtual waitress. Two very different body types were chosen as virtual humans as a way of not only inducing social comparison, but also of assessing the type of comparison prompted, whether it be upward or downward. Additionally, the bar environment included two virtual males sitting at a table, commenting on the waitress' appearance. This component was added as a way of further guiding the participants' attention towards the waitress and therefore facilitating the comparison process.

Before the immersions, participants were told to enter the bar with the intention of waiting for a friend that was supposed to meet them there. They were asked to pay attention to the waitress since she would inform them of the arrival of their friend. Once in the environment, participants were asked to take a few minutes to walk around the bar and then wait for their friend at the counter near the waitress. Once in position, the animation was activated and the male customers proceeded to comment on the waitress, their comments being adapted to each waitress. For instance, the thin waitress received more positive comments on her appearance, whereas the overweight waitress' appearance was commented on in a more negative light.

Body dissatisfied women were reporting even less satisfied feelings with their general appearance after being exposed to both types of virtual humans. This was particularly true for those who were first exposed to the overweight and then to the thin waitress. Therefore, exposure to a thin virtual human following an exposure to an overweight virtual waitress was associated with less satisfaction with one's appearance. These results partly support Festinger's theory of social comparison [45] and show that the order of exposure to virtual humans can modulate women's reaction. They suggest that a greater gap in the comparison is created when body dissatisfied women are first exposed to an overweight virtual human and then to a thin one. Social comparison could be exacerbated as participants went from an upward comparison to a downward comparison, subsequently creating a stronger reaction.

Although results obtained in our preliminary study weren't conclusive on all measures, these early findings do in fact concur with the idea that social comparison may play an important role in body dissatisfaction. Furthermore, it suggests that social comparison may be transposed to the virtual world. Findings like these are quite encouraging, and suggest that VR is a tool that offers a great alternative to traditional treatment programs used with body dissatisfied women.

Figure 1. Virtual bar environment

Figure 2. Overweight waitress virtual human

8. Conclusion and future directions in VR and BI

RV can be used in order to assess and treat body dissatisfaction across the spectrum of ED. It compares to real life situations in terms of its capacity to induce anxiety feelings and is often preferred over *in vivo* exposure. RV permits social interaction as well as social comparison in women with different levels of weight and shape preoccupations. While using VR, it seems possible to challenge the "normative discontent" view and better identify which characteristics lead to increased body dissatisfaction in certain people and not in others. Considering that, in our experiments, women without weight and shape preoccupations did not react as strongly as body dissatisfied women when immersed in a social comparison context, it seems that not "all" women similarly suffer from upward comparison.

Up to this day, in women with or without ED, VR has exclusively been used in treatement as an adjunct to cognitive-behavioral therapy. It has proved to be an effective tool that allows for exposure, desensitisation as well as cognitive retraining [47, 48]. Since it improves BI and body satisfaction when used in combination with cognitive-behavioral therapy [30, 39], some researchers have argued that VR should not be limited to this therapeutic approach but could also be considered beneficial in other types of psychotherapy, such as the interpersonal therapy or the psychodynamic therapy [33, 47, 48]. Given that VR is an experiential form of imagery that can induce emotions similar to those present in real life settings [47], it seems especially appropriate to use it in psychodynamic therapy. In fact, RV can facilitate the evocation and expression of emotions that would not necessarily be activated in a regular therapy setting (for example in the therapist's office), which in turn, increases the degree of closeness between the client and the therapist [47] and contributes positively to their relationship, an important focus in psychodynamic therapy. Moreover, VR provides a safe environment for eliciting and working on emotions in such a way that a corrective emotional experience becommes possible. Virtual humans can also represent an interesting tool to give access to repressed material [33] and to facilitate transference as well as resolution of past conflicts in clients [48, 49].

When considering future research in VR and BI, some drawbacks of the existing studies must be considered. Firstly, as others have pointed out, few controlled studies assessing the effectiveness of VR in body image disturbances have been conducted [12]. This is particularly true for samples of non-ED participants. Secondly, the ability of different virtual environments to evoke emotional reactivity in different types of participants across the eating disorder spectrum has to be evaluated, with new environments being tested and social comparison being considered. Thirdly, most studies using VR, including our own, used small samples of participants consisting almost exclusively of women. Fourthly, when studying treatment, follow-up results must also be obtained.

At a clinical level, psychotherapists interested in using VR to treat body dissatisfactions in women have to be aware of some possible adverse effects of VR with such clientele. First of all, VR in BI and other disorders has been linked to some side effects that can be classified in three types of symptoms: ocular problems (ocular fatigue, eye-strain, and blurred vision), disorientation, and nausea [50]. Clinicians must also be conscious that simple exposure to images of beauty models can increase body dissatisfactions in women. Therefore, they have to make sure their women clients will have time to discuss this possible increase in body dissatisfactions with them and they must set aside time to re-evaluate and counter the appropriateness of their perceptions about their BI. Clinicians should also assess the general psychopathology of their clients before using RV with dissatisfied or ED women. This seems particularly important considering that, in some women, dissociative episodes could be experienced as a consequence of a strong emotional response that cannot be tolerated by the client. Although, in most clients, the evocation of emotions is desirable and can accentuate their motivation to change, for those having a psychological disorder which can give rise to dissociative or even decompensating episodes, VR should be dosed accordingly to the results of the evaluation and to the diagnosis of their client.

Author details

Annie Aimé*, Karine Cotton, Tanya Guitard and Stéphane Bouchard
Université du Québec en Outaouais, Department of Psychology and Psychoeducation, Canada

9. References

[1] Bedford J.L, Johnson, C.S (2006) Societal Influences on Body Image Dissatisfaction in Younger and Older Women. Journal of Women & Aging. 18: 41-55.

[2] Neighbors L.A, Sobal J (2007) Prevalence and Magnitude of Body Weight and Shape Dissatisfaction Among University Students. Eating Behaviors. 8: 429-439.

[3] McKinley N.M (2011) Feminist Perspectives on Body Image. In: Cash T, Smolak L, editors. Body Image: A Handbook of Science, Practice, and Prevention. New York: The Guilford Press. pp.48-55.

[4] Dohnt H, Tiggemann M (2005) The Contribution of Peer and Media Influences to the Development of Body Satisfaction and Self-Esteem in Young Girls: A Prospective Study. Developmental Psychology. 42: 929-936.

[5] Tiggemann M (2011) Sociocultural Perspectives on Human Appearance and Body Image. In: Cash T, Smolak L, editors. Body Image: A Handbook of Science, Practice, and Prevention. New York: The Guilford Press. pp.12-19.

[6] Farrell C, Shafran, R, Lee M (2006) Empirically Evaluated Treatments for Body Image Disturbance: A Review. European Eating Disorders Review. 14: 289-300.

[7] Wertheim E.H, Paxton S.J (2011) Body Image Development in Adolescent Girls. In: Cash T, Smolak L, editors. Body Image: A Handbook of Science, Practice, and Prevention. New York: The Guilford Press. pp.76-84.

[8] Schilder P (1950) *The image and appearance of the human body.* NY: International Universities Press.

[9] Cash T.F (2002) Cognitive-Behavioral Perspectives on Body Image. In: Cash T, Pruzinsky T, editors. Body Image : A Handbook of Theory, Research, and Clinical Practice. New-York, NY: The Guilford Press. pp. 38-46.

* Corresponding Author

[10] Cash T (2011) Cognitive-Behavioral Perspectives on Body Image. In: Cash T, Smolak L, editors. Body Image: A Handbook of Science, Practice, and Prevention. New York: The Guilford Press. pp.76-84.
[11] Grogan S (2008) Body Image: Understanding body dissatisfaction in men, women, and children (second edition). New York: Routledge.
[12] Ferrerer-Garcia M, Gutiérrez-Maldonado J (2012) The use of Virtual Reality in the Study, Assessment, and Treatment of Body Image in Eating Disorders and nonclinical samples: A Review of the literature. Body Image. 9: 1-11.
[13] Vinette S (2001) Image corporelle et minceur: À la poursuite d'un idéal élusif. Reflets : revue d'intervention sociale et communautaire. 7: 129-151.
[14] Leonard T, Foulon C, Guelfi D.J (2005) Eating Disorders in Adults. Psychiatry. 2: 96-127.
[15] Davison K.K, Markey C.N, Birth L.L (2003) A longitudinal Examination of Patterns in Girls' Weight Concerns and Body Dissatisfaction from ages 5 to 9 years. International Journal of Eating disorders. 33: 320-332.
[16] Rukavina T, Pokrajac-Bulian A (2006) Thin-Ideal Internalization, Body Dissatisfaction and Symptoms of Eating Disorders in Coatian Adolescent Girls. Eating Weight Disorders. 11: 31-37.
[17] Cuadrado C, Carbajal A, Moneiras O (2000) Body Perceptions and Slimming Attitudes reported by Spanish Adolescents. European Journal of Clinical Nutritional. 54: S65-S68.
[18] Vartania L, Hopkinson M.M (2010) Social Connecteness, Conformity, and Internalization of Societal Standards of Attractiveness. Body Image. 7: 86-89.
[19] Fairburn C.G (2008) Cognitive Behavior Therapy and Eating Disorders. New York: The Guilford Press. 324 p.
[20] Hoek H.W, Van Hoeken D (2003) Review of the Prevalence and Incidence of Eating Disorders. *International Journal of Eating Disorders. 34:* 383-396.
[21] Babio N, Arija V, Sancho C, Canals J (2008) Factors Associated with Body Dissatisfaction in Non-Clinical Adolescents at Risk of Eating Disorders. Journal of Public Health. 16: 107-115.
[22] Fitzsimmon-Craft E.E (2011) Social Psychological Theories of Disordered Eating in College Women: Review and Integration. Clinical Psychology Review. 31: 1224-1237.
[23] Striegel-Moore R.H, Franko D.L, Thompson D, Barton B, Screiber G.B, Daniels, S.R (2004) Changes in Weight and Body Image over time in Women with Eating Disorders. International Journal of Eating Disorders. 36: 315-317.
[24] Stice E, Shaw H.E (2002) Role of Body Dissatisfaction in the Onset and Maintenance of Eating Pathology: A Synthesis of Research Findings. Journal of Psychosomatic Research. 53: 985-993.
[25] Aimé A, Begin C (2007) Modèle Conceptuel du Développement et du Maintien des Troubles des Conduites Alimentaires. Revue Francophone de Clinique Comportementale et Cognitive.12: 1-13.
[26] Neumark-Sztainer D, Levine M.P, Paxton S.J, Smolak L, Piran N, Werteim E.H (2006) Prevention of Body Dissatisfaction and Disordered Eating: What Next? Eating Disorders: The Journal of Treatment & Prevention. 14: 265-285.

[27] Thompson J.K (1990) *Body Image Disturbance: Assessment and Treatment.* New York: Pergamon.
[28] Riva G, Melis L (1997) Virtual Reality for the Treatment of Body Image Disturbances. In: Riva G, editor. *Virtual Reality in Neuro-Psycho-Physiology: Cognitive, Clinical and Methodological Issues in Assessment and Rehabilitation.* Amsterdam: IOS Press. pp. 95-111.
[29] Riva G, Melis L, Bolzoni M (1997) Treating Body-image disturbances. *Communications of the ACM, 40(8),* 69-71.
[30] Perpiñá, C, Botella, C, & Baños, R. M. (2003). Virtual reality in eating disorders. *European Eating Disorders Review, 11(3),* 261-278.
[31] Riva G, Bacchetta M, Cesa G, Conti S, Molinari E (2003) Six-Month Follow-up of In-patient Experiential Cognitive Therapy for Binge Eating Disorders. *Cyberpsychology and Behavior. 6:* 251-258.
[32] Riva G (2003) Virtual Environments in Clinical Psychology. *Psychotherapy. 40:* 68-76.
[33] Glantz K, Rizzo A, Graap K (2003) Virtual Reality for Psychotherapy: Current Reality and Future Possibilities. *Psychotherapy. 40:* 55-67.
[34] Riva G, Bacchetta M, Baruffi M, Rinaldi S, Molinari E (1998a) Experiential Cognitive Therapy in Anorexia Nervosa. *Eating and Weight Disorders. 3:* 141-150.
[35] Riva G, Bacchetta M, Baruffi M, Rinaldi S, Molinari E (1998b) Experiential Cognitive Therapy: A VR based Approach for the Assessment and Treatment of Eating Disorders. In: Riva G, Wiederhold B, Molinari E, editors. *Virtual Environments in Clinical Psychology and Neuroscience: Methods and Techniques in Advanced Patient-Therapist Interaction.* Amsterdam: IOS Press. pp. 120-135.
[36] Riva G, Bacchetta M, Baruffi M, Rinaldi S, Vincelli F, Molinari E (2000) Virtual Reality-Based Experiential Cognitive Treatment of Obesity and Binge-Eating Disorders. *Clinical Psychology and Psychotherapy.* 7: 209-219.
[37] Riva G (1998) Virtual Reality vs. Virtual Body: The use of Virtual Environments in the Treatment of Body Experience Disturbances. *Cyber Psychology & Behavior. 1:* 129-137.
[38] Perpiñá C, Botella C, Baños R, Marco J.H, Alcañiz M, Quero S (1999) Body Image and Virtual Reality in Eating Disorders: Exposure by Virtual Reality is more Effective than the Classical Body Image Treatment? *Cyberpsychology and Behavior. 2:* 149-159.
[39] Riva G, Bacchetta M, Baruffi M, Molinari E (2002) Virtual-Reality-Based Multidimensional Therapy for the Treatment of Body Image Disturbances in Binge Eating Disorders: A Preliminary Controlled Study. *IEEE Transactions on Information Technology in Biomedicine. 6:* 224-234.
[40] Gorini A, Griez E, Petrova A, Riva G (2010) Assessment of the Emotional Responses produced by Exposure to Real Food, Virtual Food and Photographs of Food in Patients affected by Eating Disorders. Annals of General Psychiatry. 9: 10p.
[41] Gutierrez-Maldonado J, Ferrer-Garcia M, Caqueo-Urizar A, Letosa-Porta A (2006) Assessment of Emotional Reactivity produced by Exposure to Virtual Environments in Patients with Eating Disorders. *Cyberpsychology and Behavior. 9:* 507-513.
[42] Ferrer-Garcia M, Gutiérrez-Maldonado J, Caqueo-Urizar A, Moreno E (2009) The Validity of Virtual Environments for Eliciting Emotional Responses in Patients With Eating Disorders and in Controls. Behavior Modification. 33: 830-854.

[43] Thompson A.M, Chad K.E (2002) The Relationship of Social Physique Anxiety to Risk for Developing an Eating Disorder in Young Females. *Journal of Adolescent Health. 31:* 183-189.

[44] Aimé A, Cotton K, Bouchard S (2009) Reactivity to Virtual Reality Immersions in a Subclinical Sample of Women Concerned with their Weight and Shape. Journal of CyberTherapy & Rehabilitation. 2: 111-120.

[45] Festinger L (1954) A Theory of Social Comparison Processes. SAGE Social Science Collections. 117-140.

[46] Guitard T, Aimé A, Bouchard S, Loranger C, Cotton K (2011) Body Dissatisfaction: Eliciting Emotions by Social Comparison in a Virtual Bar. Communication at the Cybertherapy Conference.

[47] Riva G (2005) Virtual Reality in Psychotherapy: Review. CyberPsychology & Behavior. 3: 220-230.

[48] Wiederhold B.K, Gavshon L, Wiederhold M.D (2010) A Psychodynamic View of Virtual Reality Exposure Therapy. Journal of CyberTherapy & Rehabilitation. 3: 395-403.

[49] Kahan M (2000) Integration of Psychodynamic and Cognitive-Behaviorla Therapy in a Virtual Environment. CyberPsychology & Behavior. 2: 179-183.

[50] Kennedy R.S, Stanney K.M (1996) Postural Instability induced by Virtual Reality Exposure: Development of a Certification Protocol. International Journal of Human Computer Interaction. 8: 25-47.

Chapter 6

A Discussion of the Use of Virtual Reality in Dementia

Linda Garcia, Adi Kartolo and Eric Méthot-Curtis

Additional information is available at the end of the chapter

http://dx.doi.org/10.5772/46412

1. Introduction

Dementia is a multifaceted disorder that impairs cognitive functions such as memory, language, and executive functions necessary to plan, organize, and prioritize tasks required for goal-directed behaviours. Ninety percent of persons with dementia present with dementia of the Alzheimer's type, vascular dementia, diffuse Lewy body dementia, or frontotemporal dementia [1]. All of these conditions affect the neurological functioning of the brain with different pathologies resulting in different clinical presentations [1]. In most, if not all cases, individuals with dementia experience difficulties interacting with their physical and social environments. The current chapter supports the use of virtual reality to explore the nature of these interactions.

2. Dementia and its impact on well-being

The most common cause of dementia is Alzheimer's disease, which leads to a physiological impairment in the functioning of neurons in the cerebral cortex [2], primarily in the hippocampus. The first stage of this type of dementia is marked by the individuals' inability to retain new information reliably and their difficulty in using cueing to enhance the retention of information. Typically, in the second stage, communication, behavioural and personality changes become more pronounced affecting daily life. In the last stage of Alzheimer's disease, most affected individuals lose their functional abilities to perform goal-oriented tasks which may include eating, grooming, and other self-care activities [3].

Vascular dementia is usually caused by a series of small strokes, resulting in an inadequate blood supply to the brain governing memory and thinking [2]. In addition to memory problems, individuals with vascular dementia may experience early gait disturbance, frequent falls, personality changes, depression, and issues related to emotional control [1].

Both diffuse Lewy body dementia and frontotemporal dementia present similar signs and symptoms manifested in both Alzheimer's disease and vascular dementia. The areas of the brain that are most affected in diffuse Lewy body dementia are very similar to those affected by Alzheimer's disease [4] but in frontotemporal dementia, the neuronal degeneration affects primarily the frontal and temporal lobes [5]. Diffuse Lewy body dementia has several hallmark features not common with Alzheimer's disease such as visual and/or auditory hallucinations, delusions, and fluctuation of cognition impairments, whereas frontotemporal dementia is marked by early behavioural changes such as aggressiveness [1] and more language-based deficits such as primary progressive aphasia [5].

Regardless of the form of dementia, many persons with dementia spend less time than their age-matched peers engaging in daily social activities. They are more likely to experience confusion and difficulties recognizing familiar environments or people at some point in the progression of the disease [2]. Even in the early stages, a simple fear of forgetting someone's name or getting lost in a new environment may impede persons with dementia from participating in social activities, as does their difficulty in communicating [1, 6]. Communication is critical to daily-life functioning as it provides a means to express one's needs or wishes [7]. Coupled with impaired memory, communication problems lead persons with dementia to have difficulties staying on topic [6, 8]. Conversational partners thus abandon their efforts for communication out of frustration, further isolating the persons with dementia.

In addition to memory and communication problems, persons with dementia have difficulty maintaining attention. When attending to a task, they may struggle to deactivate irrelevant stimuli in order to successfully perform cognitive tasks [9]. The result is that various important daily activities which require multi-tasking or undivided attention become challenging or impossible [9]. For instance, dementia might make it hard to socialize with several people at any given time [6].

All of these symptoms predispose individuals with dementia to social isolation and a progressive loss of autonomy. As such, they need to rely on individuals in their immediate environment to make sense of the world. The heavy reliance on others can ultimately be detrimental to their sense of self and well-being as they may be no longer able to exercise their own life choices.

3. The importance of environment

According to the Disability Creation Process model of functioning, a problem in life participation can be created when functional limitations (e.g. communication problems) interact with environmental factors (e.g. negative attitudes) that are non-facilitating [10]. Three interacting domains are used to describe the lived experience of individuals with functional limitations in creating successful engagement in life activities. The first domain, personal factors, includes organic systems and capabilities, which, together, entail all components of the human body that interrelate with the individuals' potential to perform mental or physical tasks; it also includes other variables such as age, sex and socio-cultural identity [10]. In the case of dementia, this domain would include the pathophysiological changes to the brain as well as their consequences on processes such as perception, memory and language. The second domain, environmental factors, encompasses the resources – which could be either facilitators or obstacles – within the individuals' living dimensions [10]. In dementia, for instance, this might include the physical design of long term care facilities, the attitude of the general public, or the access to health coverage. These factors influence the lived experiences as they can make life easier or harder as one transitions through the journey of dementia. Finally, the last domain, social (life) participation, refers to the result of the interaction of the first two domains as they impact the lived experiences valued by the individuals and their respective socio-cultural context [10]. This domain encompasses important life areas such as work, interpersonal relationships, community involvement and so on. Overall, this model of functioning gives a framework from which to identify the role of environmental factors in creating positive or negative lived experiences. For example, a simple goal-oriented task such as get-a-book-sit-and-read may be challenging for persons with dementia due to impaired executive functions. That is, they may simply not think of initiating the task of reading the book or may have difficulty planning the steps for executing the activity. An environmental cue such as a chair located near a bookcase might help persons with dementia improve their planning, organizing, and prioritization (i.e. executive functions) – to get a book, sit, and read. The impairment (i.e. dementia) in this example does not change, but the environment facilitates the realization of the activity. Fundamentally, the Disability Creation Process model offers a framework from which to understand the relationships amongst all the domains and help identify areas which might improve the lived experience through the modification of either or all of the domains.

If we are to improve the lived experiences of persons with dementia, attention must be allocated to altering the environment while other investigations continue to strive for a cure. Unfortunately, current pharmacotherapies for dementia act on the symptoms or succeed in slowing its progression but fall short of addressing the underlying pathology of dementia [11-13]. While environmental interventions have been the key to improved quality of life in dementia until a cure is found, it is important to not create overly facilitating environments. Environments that under-challenge persons with dementia may further deteriorate persons with dementia's quality of life by under-using their existing abilities. Lawton and Nehemow [14] describe this as 'environmental press'. In order to ensure an ideal zone of maximum comfort, the demands of the environment (weak to strong) should be balanced with the individual's level of competence (low to high). Situations where there is a strong environmental press on an individual with lower levels of competence will result in negative affect and maladaptive behaviour. Likewise, environments that are weak and demand little of the individual who has higher levels of competence will also result in negative and maladaptive behaviours. In the case of the get-a-book-sit-and-read task, placing a chair right beside the bookcase in addition to hiring an attendant who will get the book may prevent persons with dementia from using any residual planning processes, thereby fostering a sense of helplessness and loss of personhood.

The balance between the demands of the environment and the capabilities of the individual is difficult to maintain and study in a systematic fashion. One of the biggest limitations is the trade-off between experimental control and realism [15]. In other words, the more experimental control researchers have over an experience, the less ecologically valid it will be. The use of immersive virtual environment technology or virtual reality may be the answer to this shortcoming.

Immersive virtual environments allow researchers to create very realistic environments while maintaining a high level of experimental control [16]. Immersive virtual environments technology has been used in therapy for phobias [17, 18], stress [19-21], anxiety [22, 23], exercising [24], and memory problems [25, 26], yet it has hardly been used in the area of dementia. Several reasons might explain the dearth of publications on this topic. On the one hand, scientists might assume that persons with dementia cannot use the necessary paraphernalia to engage with virtual environments. On the other hand, there may be a concern that the cognitive difficulties associated with dementia prevent participants from attaining the level of presence necessary to engage with these environments.

4. Other media used in the area of dementia

As the numbers of individuals with dementia rise, there is an ever increasing interest in the use of technology to improve safety and quality of life. Smart Home Technology and environmental control systems are prime examples of emerging new technologies that have shown some success in facilitating the environment for individuals with disabilities, as well as improving their safety. Smart Home Technology is found in homes that are programmed to facilitate independent living. These homes include, for example, lights that turn on when the inhabitant gets out of bed and systems that remind indivudals with cognitive problems to turn off the stove. These are only a few examples of the large amount of technologies integrated within these homes to assist with everyday living [27]. While these types of technologies are seen as promising in assessing function, Brandt and colleagues, using a systematic review of the literature, found no solid evidence of their use in the area of dementia [27]. Only one group of authors in their selected studies included participants with cognitive difficulties (i.e. participants with traumatic brain injury) while most examined the performance of these technologies with participants with primarily physical limitations. Even in these populations with physical limitations, the evidence was inconclusive.

Bharucha and colleagues [28] searched the literature for strong evidence supporting the use of environmental assistive technologies for individuals with dementia. These systems use a large range of different sensors and artificial intelligence to assist the person with dementia as well as help ensure their safety. CareWatch consists of door opening and bed occupancy sensors that will alert the caregiver if the person with dementia exits the home at night. COACH is a system that can be used to guide people with dementia in hand washing. It provides verbal prompts and monitors to remind the person to wash their hands and assist them through the process [28]. These technologies allow for greater independence, security and overall quality of life for both the person with dementia and the caregiver, but strong evidence is still lacking. Bharucha and colleagues [28] found a dearth of studies focussed on dementia as a target population. When it came to individuals with cognitive problems, the search yielded studies whose participants were primarily young with nondegenerative conditions such as traumatic brain injuries. Again, the results were inconclusive as to the impact of these technologies. The authors call for more high quality investigations on the use of these technologies with dementia and for a need for more cost effectiveness data on these systems. In light of the cost of some of these systems, perhaps it would be worthwhile investing in experimental trial protocols that use technologies such as virtual reality. Rather than build these environments, virtual reality might help test their applicability in a virtual world prior to going to market.

On a more therapeutic note, technologies such as robot therapy and 'video respite' have shown some preliminary promise. One such example is the use of a seal robot named Paro which was designed for use with the elderly [29]. As an accompaniment and social stimulator to individuals with dementia living in long-term care facilities, its integrated sensors allow it to interact in various ways when it is touched. While the authors claim its widespread use, along with other robotic animals, the evidence for reducing levels of depression and improving quality of life in people with dementia needs more support. While the participants seemed to interact well with the 'virtual' animal, the evidence for diminishing stress and improving interactions is scarce for the population of individuals with dementia.

Similarly, 'Video Respite' has shown only preliminary success in regards to people with dementia [30]. Video respite is a videotape application that simulates the visit between an actor in the video and the person with dementia. For instance, people with dementia are seated in front of a television screen while an actor 'reminisces' with them about life in an earlier time. It is used mainly to provide respite for the caregiver as the person with dementia is occupied by the video. People with dementia watching the videos were subjectively found to be calmer and their disruptive behaviours were reduced.

The introduction of various types of technological media for use with people with dementia appears to be increasing. However, the evidence to support their use in helping people with dementia and their caregivers is scarce. Several interesting hypotheses can be generated however from these data. First of all, could media such as virtual reality be helpful in testing the preliminary effectiveness of these technologies prior to their development? With the advent of augmented reality, a person with dementia could potentially be immersed in a virtual apartment with simulated technological 'aids' to functioning. The impact of each environmental aid could be evaluated individually as well as in combination with other aids. Second, the preliminary evidence with robot therapy suggests that individuals with dementia may be able to obtain some level of presence with a 'non-real' animal. Is this indicative that they are able to obtain presence in other forms of media such as immersive virtual reality?

5. Immersive virtual environment technology and individuals with dementia

There is still insufficient evidence for concluding that immersive virtual environment technology can or cannot be used with persons with dementia. Flynn et al. [31] incorporated a user-centred-approach and found that it is feasible to immerse persons with dementia in a virtual environment via immersive virtual environment technology. The authors reported that persons with dementia felt a sense of control and enjoyed the interaction in virtual environments, as well as demonstrated little difficulty in manoeuvring a joystick [31]. There was also no significant deterioration in terms of simulator sickness and well-being [31].

Cushman, Stein, and Duffy (32) examined the feasibility of using immersive virtual environment technology to detect navigational deficits in people with Alzheimer's disease. Users had to navigate their ways through both a real hospital lobby and a virtual hospital lobby. By using eight navigation subtests, the authors found that the use of virtual environments was an effective method to assess navigational skills, and that quantifying a virtual-world navigational performance is easier and less time-consuming than quantifying a real-world navigational performance. In another study [33], Shaick, Martyr, Blackman, and Robbinson used virtual reality to observe how a physical outdoor environment could affect way finding. Persons with dementia were immersed within an outdoor virtual environment and were asked to go for a walk. Many facilitators and barriers which might affect orientation were incorporated such as larger street signs, street signs with pictures of uncommon objects, street signs with names only or unattractive street layouts. The authors found that if the street felt unappealing for the persons with dementia, it became a barrier to the overall success of the walk. In 2009, the same environments were used to examine spatial navigation in healthy older adults as opposed to older adults with dementia [34]. The users were asked to navigate a virtual neighbourhood and subsequently asked to recognize certain city buildings and objects. Young adults were quicker and more accurate than both older adults and individuals with Alzheimer's disease, while individuals with Alzheimer's disease had more difficulty with the recognition task than normally-ageing participants. In summary, these studies provide some evidence that, at least in the earlier stages, persons with dementia can use the paraphernalia and be successfully immersed in virtual environments. While these studies have shown some success with the use of virtual reality for assessing navigation, the question remains as to whether persons with dementia can obtain presence for all types of functional problems.

6. Presence and dementia

Early descriptions of presence define it as believing oneself to be in an environment while being physically located in another [35]. While presence is a necessary precursor for successful use of virtual environments, Riva, Waterworth, Waterworth and Mantovani [36] remind us that obtaining presence is not a media-dependent phenomenon. That is, the ability and belief that one can act upon one's environment and at the same time be influenced by this environment exists regardless of the medium of transmission [36]. In addition, attaining presence requires constant processing between human sensory and cognitive skills [37]; hence the concern that it might not be feasible with persons with dementia. Since there is very little literature pertaining to the use of virtual reality with persons with dementia, a better understanding of the phenomenon of presence will help with our understanding of how people with dementia might interact with their environments.

Riva et al. [36] suggest that presence is what links our volitional motivation to act upon our environment and our cognitive adjustment to changes in our environments. These interactions however can be explained as they evolve over several layers [38]. According to these authors, the first layer, proto presence, involves an unconscious sense of the self in space which is constructed using proprioceptive information. At this level the individual obtains a sense of how his body moves in the environment, and the challenge in virtual reality is to create technology that is sensitive to this proprioceptive feedback. The small literature reported above on the assessment of navigation suggests that individuals with early dementia are able to attain this layer of presence. Until the technology is able to capture propioceptive information more systematically, we may not be in a position to explore this level of presence in greater detail in the virtual environment [38]. Nevertheless, preliminary studies are promising that persons with dementia are indeed capable of obtaining proto presence.

In the second layer, core presence, the individual is able to determine what is real and what is not real based on the processing of perceptual information [38]. The perceptual information thus captured forms an integrated percept that is continually adjusted in real time according to current changes in the environment. The emphasis is on the here and now and the individual who obtains presence is influenced by these changes. The vast literature on environmental modifications to alter the expression of disruptive behaviours in dementia [35, 36] suggests that the actions of people with dementia can somewhat be mediated by a systematic control of the environment. For instance, persons with advanced dementia have been shown to improve real life navigation by the addition of visual environmental cues [41]. According to Riva et al. [36] the saliency of the stimuli is very important at this level since it is the perceptual information which creates this layer of presence. Once again the studies on navigation would suggest that persons with dementia are sensitive to the saliency of the stimuli as Schaik et al. [33] were able to create obstacles and facilitators by modifying the road signs. Persons with dementia were influenced by the visual nature of the information they received in the immersive virtual environment.

In the third layer of presence, extended presence, Riva et al. [38] argue that the individual uses cognitive processes to understand the meaningfulness of the environment. These processes require the individual to process the information received through levels one and two and compare it to information in memory to attach meaning and relevance. Based on what we know about the cardinal signs of dementia, persons with dementia should have considerable difficulties with this level of presence. According to Riva et al. [36], the individual in an immersive virtual environment will make hypotheses about what will happen in the environment as he makes sense of what he sees. It is plausible to assume that individuals with dementia may not be able to make these hypotheses or may make erroneous ones. If this is true, presence could only be achieved as high as the core level where information is processed online in the here and now.

Bouchard et al. [42] observed that activation of the parahippocampal cortex is correlated with a greater sense of extended presence [38]. This area of the brain is important in giving contextual meaning to scenes and locations [42]. According to these authors, the lateral surface of the occipital lobe, fusiform gyrus, and adjacent insula cortex are involved in providing virtual reality users with awareness in detecting the discrepancy between the virtual environment and the physical world; one might refer to this as core presence [38]. Since Alzheimer's disease can impair the functioning of either or all the brain regions stated above [43-46], one might argue that persons with dementia might not have the cognitive ability to analyze information within context and understand the meaningfulness of the environment. Hence, in dementia one might hypothesize that the processed information about the environment does not generate the same intention as was meant. It might be possible that individuals with dementia may find meaning in what they see but may be attaching the perceived information to wrong meanings in memory because they cannot access their memories as efficiently and effectively as individuals without dementia. It seems that the virtual environment might be ideal to test out this theory.

Current judgement theory indicates that our decision-making is governed by a two-system view in which the intuitive mode is monitored and corrected, if necessary, by the controlled mode [47]. The operation for the intuitive mode is automatic, and it is responsible for generating impressions [47]. In contrast, the operation for the controlled mode is more effortful as it is involved in deliberate reasoning [43]. In the event whereby the controlled mode is malfunctioning, as is often the case with persons with dementia, it is likely the intuitive mode would dominate. Persons with dementia might continue to believe in the impression they generated, regardless of whether it is right or wrong, given that the intuitive mode is not overridden by the controlled mode. In addition, the lack of a controlled mode reduces the ability for persons with dementia to experience doubt, a cognitive ability that allows the subject to deal with incompatible thoughts on the same subject [47]. This suggests that persons with dementia may be able to imagine, interact, and develop an emotional link towards the subjects within the virtual environment without a dilemma arising around whether it is real or not. Overall, this reasoning might suggest that persons with dementia may experience a greater sense of Riva et al.'s [38] core presence but a lesser extent – or erroneous – sense of Riva et al.'s [38] extended presence.

7. Potential of virtual reality as an assessment tool in dementia

Immersive virtual environment technology gives researchers the opportunity to study how persons with dementia interact with their environment and which elements are most facilitating or creating the least distress while capitalizing on residual cognitive resources. Virtual reality technologies offer the potential to capture the impact of physical environments, as well as social interactions by deconstructing these environments. Further research into this area will help develop a better understanding of how persons with dementia can act to influence their environments and how changes in the environment can affect them. The objective may not be to learn a new skill but rather to understand how to better design physical spaces or modify social environments for better quality of life.

One of the advantages of virtual reality is the potential to control and modify these environments when it cannot be easily, or completely, changed outside of the virtual world. For example, human interaction and social presence would be particularly useful to examine in persons with dementia using virtual reality. More detailed knowledge about how others influence interactions with persons with dementia would help offer advice for caregivers on how to interact with this population. For instance, one might manipulate the verbal and non-verbal aspects of communication of virtual humans as they interact with the user [16, 48]. As Garcia et al. [48] pointed out, if we want to study the impact of tone of voice on persons with dementia during social interaction, we could create experimental conditions where a virtual human modifies tone of voice while maintaining neutral facial expressions. This type of experimentation is impossible outside of the virtual environment since human conversational partners cannot voluntarily isolate these parameters. Knowing more about the impact of these changes in the environment may lead to remarkable changes in clinical approaches and improve quality of life. Immersive virtual environment technologies offers a promising avenue for testing these hypotheses as well as helping us to better understand the processes involved in presence.

8. Potential of virtual reality as a training tool in dementia

It is unlikely for persons with dementia to acquire new skills as they tend to experience rapid deteriorations in their brain mappings throughout the journey of dementia. Such progression, however, might be slowed down by medications and brain-stimulating activities based on the theory of neuroplasticity – the ability of brains and nervous systems to reconstruct new cellular synapses as a result of the interaction with enriched environments [49, 50]. Nonetheless, the perception of such environments may be different for individuals, and it would be financially impossible to construct a customized enriched environment for every individual. Virtual Reality has the potential to fulfil this requirement at a lower cost.

In addition, virtual reality would allow researchers to systematically study the various desired brain-stimulating activities as a function of the rate of neuroplasticity experienced in persons with different types of dementia. It is known that dementia of different types may experience deterioration in different regions of the brains; thus, it is very important to explore the possibility of conducting such 'targeted' brain-stimulating activities in light of deteriorating regions of the damaged brain.

9. Conclusion

It is surprising to see how little immersive virtual environment technology has been used with individuals with dementia, one of the largest growing clinical populations. The Disability Creation Process model of functioning suggests that each individual may experience the impact of diseases differently. It highlights the role of environmental factors in creating situations where social participation may be limited. Immersive virtual environment technology may prove to be useful in testing the role played by physical and social environments by immersing persons with dementia in virtual environments and observing the results following the process of virtual-environment-remodelling [33]. Immersive virtual environment technology can help us understand more about how persons with dementia interact with their environments – whether physically with the objects or socially with other people in their surroundings. Should the remodelling improve the lives of persons with dementia, such modification may be implemented outside the world of virtual environments.

Until a cure is found, our most promising interventions will rely on environmental modifications. Should we confirm that persons with dementia are unable to obtain extended presence, as argued in this paper, but can function at the level of core presence, this may prove to be a satisfactory level for improving quality of life. Persons with dementia typically function in the here and now and immersive virtual environment technologies might prove to be useful in creating pleasant virtual worlds where they may find it easier to interact while maximizing residual cognitive processes. These worlds may offer much needed respite to both formal and informal caregivers while allowing pleasant, meaningful activities in the here and now for persons with dementia. The literature is clear that success in lived experiences is dependent on both the personal and environmental factors. Immersive virtual environment technology can shed an important light on one of these domains.

Author details

Linda Garcia
Faculty of Health Sciences, Interdisciplinary School of Health Sciences,
University of Ottawa, EntourAGE Lab, Ottawa, Canada
Bruyère Research Institute, Ottawa, Canada

Adi Kartolo and Eric Méthot-Curtis
Faculty of Health Sciences, Interdisciplinary School of Health Sciences,
University of Ottawa, EntourAGE Lab, Ottawa, Canada

10. References

[1] Sounder E, Chastain JR, & Williams RD (2002) Dementia in the New Millennium. MEDSURG Nursing. 11(2): 61-70.

[2] Torpy JM, Lynm C, & Glass RM (2010) Dementia. Journal of the American Medical Association. 304(17): 1972.

[3] Clark C (2000) Clinical manifestations and diagnostic evaluation of patients with Alzheimer's disease. In C. Clark and J.Q. Trojanowski (Eds.). Neurodegenerative dementias (pp. 95-111). New York: McGraw-Hill.

[4] Brand M & Markowitsch HJ (2008) Brain structures involved in dementia. In GabrielaStoppe on behalf of the European Dementia Consensus Network (Eds).Competence Assesment in Dementia, Part 2, 25-34, DOI: 10.1007é978-3-211-72369-2_3

[5] Kirshner HS (2010) Frontotemporal Dementia and Primary Progressive Aphasia: An Update.Current Neurology and Neuroscience Reports. 10(6): 504-511.

[6] Svanström R & Dahlberg K (2004) Living with Dementia Yields a Heteronomous and Lost Existence. Western Journal of Nursing Research. 26(6): 671-687.

[7] Wright KB, Sparks L, & O'Hair HD (2008) Health communication in the 21st Century. USA: Blackwell Publishining Ltd

[8] Garcia LJ & Joanette Y (1997) Analysis of Conversational Topic Shifts: A MultipleCase Study. Brain and Language. 58(1): 92-114.

[9] Drzerga A, Grimmer T, Peller M, Wermke M, Siebner H, Rauschecker JP, Schwaiger M, & Kurz A (2005) Impaired Cross-Modal Inhibition in Alzheimer Disease. PLos Medicine. 2(10): 986-995.

[10] Fougeyrollas P, Cloutier R, Bergeron H, Coté J, Coté M, & Michel GS (1997) Revision of the Quebec Classification: Handicap Creation Process. National Library of Canada: International Network on the Handicap Creation Process.

[11] Carter MD, Simms GA, & Weaver DF (2010) The development of new therapeutics for Alzheimer's disease. Clinical Pharmacology Therapy. 88(4): 475-486.

[12] Kovács T (2009) Therapy of Alzheimer disease. Neuropsychopharmocol Hung. 11(1): 27-33

[13] Van Marum VJ (2009) Current and future therapy in Alzheimer's disease. Fundamental & Clinical Pharmacology. 22(3): 265-274.

[14] Lawton MP & Nahemow L (1973) Ecology and the aging process. In C. Eisdorfer and M.P. Lawton (Eds.). The Psychology of adult development and Aging, pp. 619-674. Washington, D.C.: American Psychological Association.

[15] Blascovich J, Loomis J, Beall AC, Swinth KR, Hoyt CL, & Bailenson JN (2002) Immersive Virtual Environment Technology as a Methodological Tool for Social Psychology. Psychological Inquiry. 13(2): 103-124.

[16] Persky S & McBride CM (2009) Immersive Virtual Environment Technology: A Promising Tool for Future Social and Behavioral Genomics Research and Practice. Health Communication. 24(8): 677-682.

[17] Côté S & Bouchard S (2008) Virtual Reality Exposure for Phobias: A Critical Review. Journal of CyberTherapy and Rehabilitation. 1(1): 75-91.

[18] Wiederhold BK, Jang DP, Gevirtz RG, Kim SI, Kim IY, & Wiederhold MD (2002) The Treatment of fear of flying: a controlled study of imaginal and virtual reality graded exposure Therapy. IEEE transactions on information technology in biomedicine. 6(3): 218-223.

[19] Bouchard S, Baus O, Bernier F, & McCreary R (2010) Selection of Key Stressors to Develop Virtual Environments for Practicing Stress Management Skills with Military *Personnel Prior to Deplyoment. Cyberpsychology and Behavior. 13(1): 83-94.*

[20] Riva G, Raspelli S, Algeri D, Pallavicini F, Gorini A, Wiederhold BK, & Gaggioli A (2006) Interreality in Practice: Bridging Virtual and Real Worlds in the treatment of Posttraumatic Stress Disorders. Cyberpsychology, Behavior & Social Networking. 13(1): 55-65.

[21] Villani D, Preziosa A, & Riva G (2006) Coping with stress using Virtual Reality: a new perspective. Annual Review of Cybertherapy and telemedicine. 4: 25-32.

[22] Harris SR, Kemmerling RL, & North MM (2002) Brief Virtual Reality Therapy for public speaking anxiety. CyberPsychology & Behavior. 5(6): 534-550.

[23] Repetto C & Riva G (2011) From virtual reality to interreality in the treatment of anxiety disorders. Neuropsychiatry. 1(1): 31-43.

[24] Bryanton C, Bossé J, Brien M, Mclean J, McCormick A, & Sveistrup H (2006) Feasibility, Motivation, and Selective Motor Control: Virtual Reality Compared to Conventional Home Exercise in Children with Cerebral Palsy. CyberPsychology & Behavior. 9(2): 123-128.

[25] Brooks BM & Rose FD (2003) The use of virtual reality in memory rehabilitation: Current findings and future directions. NeuroRehabilitation. 18(2): 147-157.

[26] Klinger E, Chemin I, Lebreton S, & Marié RM (2006) Virtual Action Planning in Parkinson's Disease: A Control Study. Cyberpsychology & Behavior. 9(3): 342-347.

[27] Brandt A, Samuelsson K , Töytäri O , & Salminen AL (2011) Activity and participation quality of life and user satisfaction outcomes of environmental control systems and smart home technology: a systematic review. Disability Rehability Assistive Technology. 6(3): 189-206.

[28] Bharucha AJ, Anand V, Forlizzi J, Dew MA, Reynolds CF, Stevens S, & Wactlar H (2009) Intelligent assistive technology applications to dementia care: current capabilities, limitations, and future challenges. The American Association for Geriatric Psychiatry. 17(2): 88-104.

[29] Shibata T & Wada K (2011) Robot Therapy: A new approach for mental healthcare of the elderly – a mini-review. Gerontology. 57: 378-386.

[30] Caserta MS & Lund DA (2003) Video Respite in an Alzheimer's Care Center. Activities, Adaptation and Aging. (27)1: 13-28.

[31] Flynn D, Schaik PV, Blackman T, FemCott C, Hobbs B, & Calderon C (2003. Developing a Virtual Reality-Based Methodology for People with Dementia: A Feasibility Study. Cyberpsychology & Behavior. 6(6): 591-611.

[32] Cushman LA, Stein K, & Duffy CJ (2008) Detecting navigational deficits in cognitive aging and Alzheimer disease using virtual reality. Neurology. 71(12): 888-895.

[33] Schaik PV, Martyr A, Blackman T, & Robbinson J (2008) Involving Persons with Dementia in the Evaluation of Outdoor Environments. Cyberpsychology & Behavior. 11(4): 415-424.

[34] Zakzanis KK, Quintin G, Graham SJ, & Mraz R (2009) Age and dementia related differences in spatial navigation within an immersive virtual environment. Medical Science Monitor. 15(4): 140-150.

[35] Witmer BG & Singer MJ (1998) Measuring Presence in Virtual Environments: A Presence Questionnaire. Presence. 7(3): 225–240.

[36] Riva G, Waterworth JA, Waterworth EL, & Mantovani F (2011) From Intention to Action: the role of Presence. New Ideas in Psychology. 29(1): 24-37.

[37] Lombard M, Reich RD, Grabe ME, Bracken C, & Ditton TB (2000) Presence and television: the role of screen size. Human Communication Research. 26(1): 75-98.

[38] Riva G, Waterworth JA, & Waterworth EL (2004) The Layers of Presence: A Bio cultural Approach to Understanding Presence in Natural and Mediated Environments. *Cyberpsychology & Behavior.* 7(4): 402-416.

[39] Calkins MP (2009) Evidence-based long term care design. NeuroRehabilitation. 25: 145-154.

[40] Fleming R & Purandare N (2010) Long-term care for people with dementia: environmental design guidelines. International Psychogeriatrics. 22(7): 1084-1096.

[41] Marquardt G (2011) Wayfinding for People with DementiaL A Review of the Role of Architectural Design. Health Environments Research & Design Journal. 4(2): 75-90.

[42] Bouchard S, Talbot J, Ledoux AA, Phillips J, Cantamesse M, & Robillard G (2009) The meaning of being there is related to a specific activation in the brain located in the parahypocampus. Proceedings at the 12th Annual International Workshop on Presence , Los Angeles (CA), November 11-13.

[43] Du AT, Schuff N, Kramer JH, Rosen HJ, Gorno-Tempini ML, Rankin K, Miller BL, & Weiner MW (2007) Different regional patterns of cortical thinning in Alzheimers disease and frontotemporal dementia. Brain. 130: 1159-1166.

[44] Bonthius DJ, Solodkin A, & Van Hoesen G (2005) Pathology of the Insular Cortex in Alzheimer Disease Depends on Cortical Architecture. Journal of Neuropathology & Experimental Neurology. 64(10): 910-922.

[45] Norfray JF & Provenzale JM (2004) Alzheimer's Disease: Neuropathologic Findings and Recent Advances in Imaging. American Journal of Roentgenology. 182: 3-13.

[46] Rombouts, SARB, Barkhof F, Veltman DJ, Machielsen WCM, Witter MP, Bierlaagh MA, Lazeron RHC, Valk J & Scheltens P (2000) Functional MR Imaging in Alzheimerès Disease during Memory Encoding. American Journal of Neuroradiology. 21: 1869-1875.

[47] Kahneman D (2003) Maps of bounded rationality: A perspective on intuitive judgment and choice. In T. Frangsmyr [Nobel Foundation], (Ed.), Les Prix Nobel: The Nobel Prizes2002 (pp.449-489). Stockholm, SE: The Nobel Foundation.

[48] Garcia LJ, Rebolledo-Leduc M, Metthé L, & Lefebvre R (2007) The Potential of Virtual Reality to Assess Functional Communication in Aphasia. Topics in Language Disorders. 27(30): 272-288.

[49] Kempermann G, Gast D, & Gage FH (2002) Neuroplasticity in old age: Sustained fivefold induction of hippocampal neurogenesis by long-term environmental enrichment. Annals of Neurology. 52(2): 135-143

[50] Clare L (2012) Cognitive rehabilitation and people with dementia. In: JH Stone, M Blouin, editors. International Encyclopedia of Rehabilitation. Available online: http://cirrie.buffalo.edu/encyclopedia/en/article/129/

Section 2

VR in Medical Applications

Chapter 7

Virtual Reality – A New Era in Surgical Training

C.E. Buckley, E. Nugent, D. Ryan and P.C. Neary

Additional information is available at the end of the chapter

http://dx.doi.org/10.5772/46415

1. Introduction

1.1. Current challenges in surgical training

The introduction of virtual reality and simulation into the world of surgery came about for a number of reasons. The most paramount reason being the adoption of laparoscopy which was first introduced in the 1990s. The advent of laparoscopy and minimally invasive procedures has created a necessity for more novel techniques to learn surgical skills. The skill set required in laparoscopy is very different compared with open conventional surgery(1). This is due to lack of tactile feedback, precise hand eye coordination, and a change from 3D to 2D visualisation as well as adaption to the fulcrum effect(2, 3). This skill set cannot be taught easily in the real life environment under the supervision of a senior surgeon. With the traditional open approach the supervising surgeon can directly guide the hands of the trainee and immediately intervene if a problem or difficulty arises. In laparoscopy however the expert surgeon has less control over what the trainee is doing. If a complication were to arise during the course of the surgery it would be more difficult for the expert surgeon to intervene and rectify the situation. The same can be said for endovascular procedures and endoscopy. The learning curve is also steeper in minimally invasive procedures than for open surgery as trainees have to learn not only new technology and overcome obstacles like the fulcrum effect but they also have to have a good fundamental ability for these procedures as well.

The early part of the learning curve is associated with a higher complication rate. Therefore it is intuitive that familiarity with surgical procedures should be taught outside the surgical environment in order to improve patient's safety. Laparoscopic cholecystectomy was the index procedure for laparoscopy. Although it was embraced with vigor it was also the procedure where problems and concerns with the minimally invasive approach were first highlighted. A higher than acceptable rate of bile duct injury in laparoscopic cholecystectomy when compared to open cholecystectomy became an important issue in the 1990s.

The Southern Surgeons Club study is an oft referenced paper (4). They found that 90% of common bile duct injuries occurred within the first 30 operations performed by the trainee surgeon. They also predicted that the surgeon had a 1.7% probability of causing a bile duct injury in their first operation, which reduced to 0.17% by the 50th case. The probability of injury was found to have dropped to a significantly safe level by the 10th case. This was one of the first articles to underline the significance of the learning curve in minimally invasive surgery.

As the complexity of the procedure increases so too does the learning curve. This has been demonstrated for laparoscopic fundoplication, where a significant reduction in complications has been reported to reduce only after the 50th case with the highest complication rate found within the first 20 cases (5). The learning curve for laparoscopic colectomy has been estimated to be even higher, with the highest rate of complications occurring during the first forty procedures(6). The initial learning curve has been shown to be associated with the period of greatest risk to the patient.

For these reasons the surgical community looked to virtual reality as a way of bridging this skill gap and providing a method of safely introducing new techniques into surgical practice.

However, these are not the only challenges surgical training has been faced with in recent times. Economic factors also affect training structures. The length of elective waiting lists and time pressures in the operating theatre play an important role in the amount of operative experience a trainee now receives. Coupled with the increasing cost of new operative technologies and instrumentation and the global economic recession the financial restraints on supervising surgeons is greater than before.

There has also been a change in the expectations of patients and the population as a whole. This has resulted from the publication of high profile medico-legal cases such as the 'Bristol Case' in the British Isles and the 'To Err is Human' report in the US. These cases have brought medical errors and the quality of surgical training to the forefront.

The last decade has seen considerable changes in the structure of healthcare delivery. There has been a steady shift towards a consultant-based service with reduced service activity by trainees. This shift is set to continue into the future. One factor propelling a consultant-led service is the fact that the complexity of surgery is increasing. This is due to advancements in the minimally invasive approach but it is also due to improvements in critical care services allowing increasing numbers of elderly and sicker patients to be operated on.

Fatigue due to an excessive workload and hours worked first became recognised as a significant problem following the landmark Libby Zion case (7). It has been demonstrated that fatigue is directly linked to medical error and a reduction in clinical performance. Fatigue has also been shown to effect mood and psychomotor performance.

The European Working Time Directive (EWTD) was a piece of legislation that demanded that doctor's work less hours per week (www.doh.ie). The aim of reducing hours worked was to ensure that patients receive high quality and safe care. The EWTD should have been fully implemented in the Irish healthcare system from 2004. It has been rolled out to a certain extent but as of 2012 has not reached the ascribed targets.

However with the introduction of the Calman reforms in the UK in 1993 and the implementation of the European Working Time Directive the surgical community has been forced to debate how best to train junior surgeons in a shortened period of time. Although the proposed reforms have received a cautious welcome from the medical community, there are significant worries about the impact of shortening the training time on trainees' experience.

All of these challenges pose a problem for trainees as they are no longer getting one on one teaching nor are they getting enough operative exposure, therefore the traditional Halstedian method no longer applies. However, most teaching in Ireland continues along this traditional apprenticeship route, where trainees are exposed to surgical procedures with the guidance of an experienced teacher. As a result teaching is quite unstructured and is very much dictated by the location of the hospital, the caseload, case variation and the enthusiasm of the supervising surgeon for teaching. Furthermore, the current training paradigm lacks objective feedback on trainee performance. The current structure is unlikely to change and therefore the approach to training must.

Training needs to be done in a more efficient manner to optimize the learning experience and surgical exposure of the trainee as well as combating the challenges we face in adapting the new skill set required with minimally invasive procedures. Virtual reality can offer the surgical community a solution.

1.2. How virtual reality offers a solution

In the past decade, various academic medical institutions have set up simulation laboratories. In the US, the American College of Surgeons has a specific training program for residents and also has introduced a process whereby they have accredited a number of institutions across the US and the UK. Ireland has followed suit by developing a surgical simulation laboratory in the Royal College of Surgeons, Dublin. A simulation laboratory is a space designated for trainees to practice various skills and procedures on a wide variety of available surgical simulators in a safe, controlled environment. Dedicated time is necessary in order to learn the required skills in a protected manner.

Surgical skill can be learnt very effectively on simulators. Simulation has much more to offer the trainee than the clinical environment alone as it allows for dedicated teaching which is focused and structured with specific learning goals. By mastering skills such as hand-eye coordination, counter intuitive fine movements and the ability to work with a 2D dimensional image in a 3D space on a simulator, the trainee surgeon can then focus on the critical steps of the operation when in the operating theatre. This is instead of trying to learn every aspect of a new skill set at once. One of the difficulties with acquiring these skills is due to the fulcrum effect of the body wall on instrumentation(2). This problem cannot be overcome with concentration; it requires practice until the process becomes automated(8).

Traditionally the skills required for minimally invasive surgery (MIS) have been attained in the operating theatre. It has been here that the steepest part of the learning curve has been battled out. It is obvious that this situation is not the ideal for either the patient or the trainee. However, the introduction of simulation to MIS has helped to address these issues. Simulation provides a safe environment for trainees to overcome the initial learning curve including the visual spatial, perceptual and psychomotor difficulties associated with minimally invasive techniques.

1.3. Background into simulation

Simulation has its roots in the commercial and military aviation industry. It was first considered in 1910 when student pilots trained in land-borne aircraft with reduced wingspans. The first rudimentary simulator was available in 1929 and was known as the Links Trainer (9). It consisted of a wooden fuselage mounted on an air bellows, which was able to represent the movements involved in flight. This allowed the pilot to train for hours. In 1934 the US purchased six Links simulators following a series of aviation accidents. At the time it was recognized that the current training programs were inadequate and simulation was a step towards improving the training system. World War II also had a dramatic impact on the uptake of simulation for training purposes. The war demanded that a greater number of pilots be trained and that skills such as the need to become proficient in instrument or blind flying were paramount. These factors led to simulator development and usage. Today, there are hugely sophisticated systems which replicate an aircraft environment precisely and can deal which a vast range of potential flight scenarios. Pilots must undergo ongoing annual training entitled "checking out" by the Federal Aviation Administration in order to ensure ongoing certification as well as additional training requirements if they wish to change to another aircraft. Astronauts are also required to follow similar procedures.

The first surgical simulator to use virtual reality technology was created at NASA by Rosen and Delp (10). It was an orthopaedic lower limb model that simulated tendon transfer. It was unique in that it allowed planning and therefore optimisation of operations. Virtual reality technology has evolved to the point today where actual patient data and radiological images can be inputted into the simulator allowing for a complete simulated run-through before operating on the patient; a process known as mission rehearsal.

The aviation industry paved the way for simulation, so we often look to their methods for guidance. However simulating the human body is a much more complex and unpredictable task and often we can only get close to elements of surgery rather than replicating it completely. If we take a look at the existing simulators today we notice they mainly involve the simulation of machinery (airplane, car, train, truck, bus, space vehicles). All these are perfect for reliable repetition of conditions and interface. The purpose is generally to allow the user to practice their skill in a controlled environment, with the additional benefit of having 'metrics' or computerised feedback on their performance.

There is an abundance of technology available today to help simulate these situations, so why are some areas better represented by simulation than others? As a general guide, if we made the object being simulated, then we can generally do a good job of simulating it. In flight simulation, airplanes are manmade so engineers understand every part of the airplane and its interaction with the physical environment. We can take the input from real (manmade) aeronautical instruments and interpret them perfectly. Flight simulators do not need to be any better than they are at present. They provide an excellent simulation of real flight. On the other hand, we have surgical simulation. We are trying to simulate interaction with the most complex system we know – the human body, coupled with trying to interpret the movement of laparoscopic instruments and the human hands. This is more challenging to the point that we may be generations away from being as satisfied with simulation as an optimal teaching tool when compared to aviation.

2. Virtual reality surgical trainers and types of assessment

2.1. VR Trainers

VR Trainers digitally recreate the procedures and environment of laparoscopy. The term "virtual reality" was coined by Jaron Lanier a philosopher and scientist in the 1980s. It is a phrase used to describe the concept of a virtual world which supports interaction instead of something that is passively visualised.

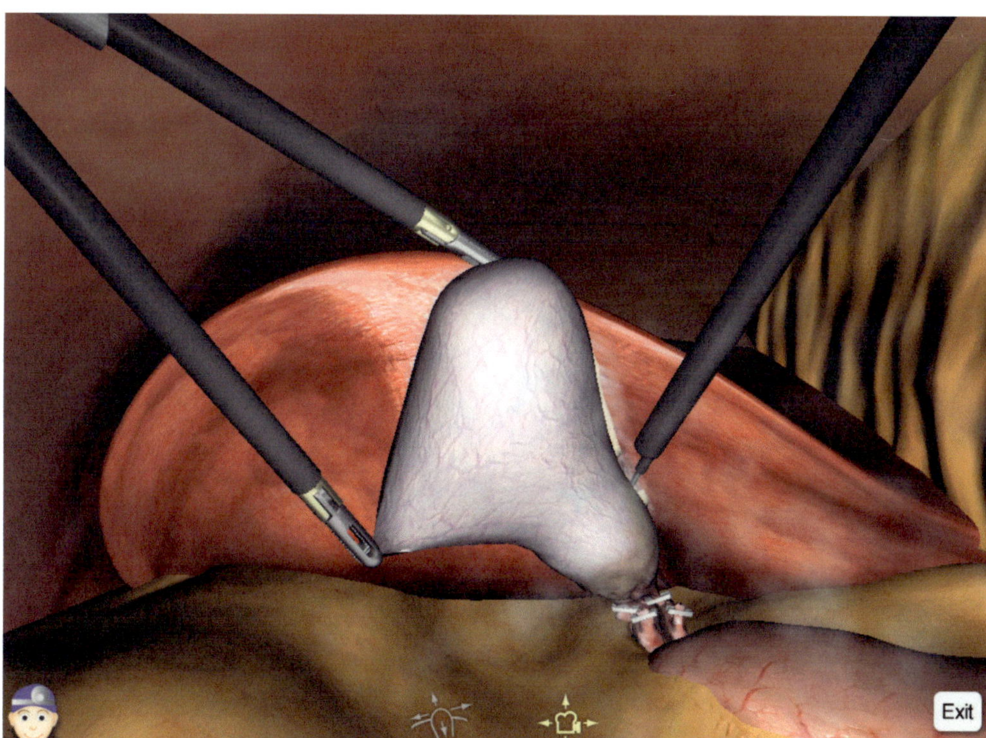

Figure 1. Screenshot of LapSim Laparoscopic Cholecystectomy

Here are some of the presently available commercial VR trainers:

LapSim (Surgical Science, Sweden): This system has practice sessions which can vary in complexity. Modules include basic laparoscopic skills, cholecystectomy, appendicectomy, suturing, anastomosis and laparoscopic gynaecological procedures. The metrics are specific to the task being performed. Time, instrument path length and procedure specific errors are measured. LapSim has been evaluated in many studies and construct validity has been established(11, 12).

LapMentor (Simbionix, USA): This has many modules including basic skills (camera navigation, clip applying, 2-handed maneuvers, hand-eye coordination drills, cutting, object translocation and suturing), laparoscopic cholcystectomy, laparoscopic ventral hernia repair, laparoscopic gastric bypass, laparoscopic nephrectomy, laparoscopic sigmoidectomy and a variety of laparoscopic gynaecological procedures. There are also several other platforms including URO mentor (urologic procedures), PERC mentor (percutaneous interventions), ANGIO mentor (catheter based interventions) and GI mentor (endoscopy). Metrics measured include time, economy of movement, safety and electrosurgical dissection, procedural errors, and procedure specific checklist items relating to knowledge of the procedure and handling of instruments. Studies have validated its validity (13, 14).

MIST-VR (Mentice, Sweden): The Minimally Invasive Surgical Trainer-Virtual Reality facilitates basic laparoscopy using two instrument handles, a computer, monitor and a foot pedal. Metrics measured include tool to tool contact, loss of tissue-tool contact, inappropriate "passing of the point" of the instrument through the tissue, inappropriate targert release, inappropriate cautery application and economy of movement. Mentice also make the Procedicus VIST which simulates catheter based interventions and the Procedicus COREP which simulates endovascular procedures. Several studies have demonstrated validity and transferability (15-17) of the MIST-VR and it is possibly the most established VR platform to date in terms of publications.

LapVR (Immersion Medical, USA): This offers simulation of basic skills (camera naviagtion, peg transfer, cutting and clip application), procedural skills (adhesiolysis and running of the bowel) and full laparoscopic cholecystectomy. Metrics measured include time and procedure specific errors. The LapVR system has only recently been validated (18).

SurgicalSim (METI, USA): This platform offers practice of core tasks (tissue manipulation, dissection, suturing and knot tying), transurethral resection of the prostate (TURP) and laparoscopic cholecystectomy. Metrics measured include time, instrument path length and procedure specific errors. The software enables the user to customize their own training programme which can be viewed by an administrator as well as practicing skills or procedures with virtual robotic arms and a 3D headset. No construct validity studies using the SurgicalSim were found to date.

Figure 2. LapVR, Used with permission by CAE Healthcare. All rights reserved

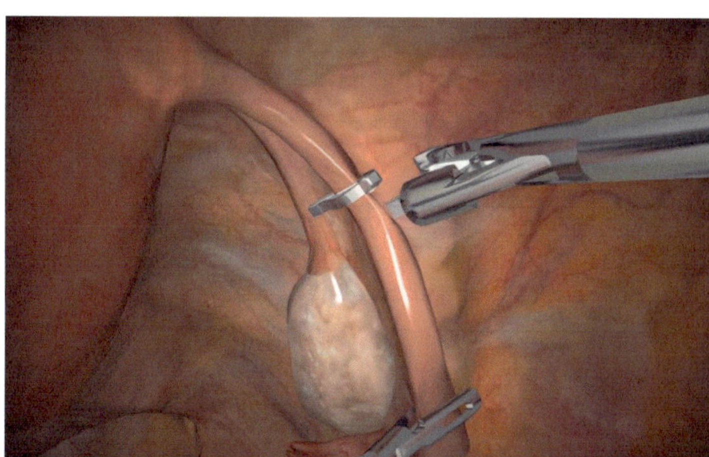

Figure 3. Screenshot of LapVR Laparoscopic Cholecystectomy, Used with permission by CAE Healthcare. All rights reserved

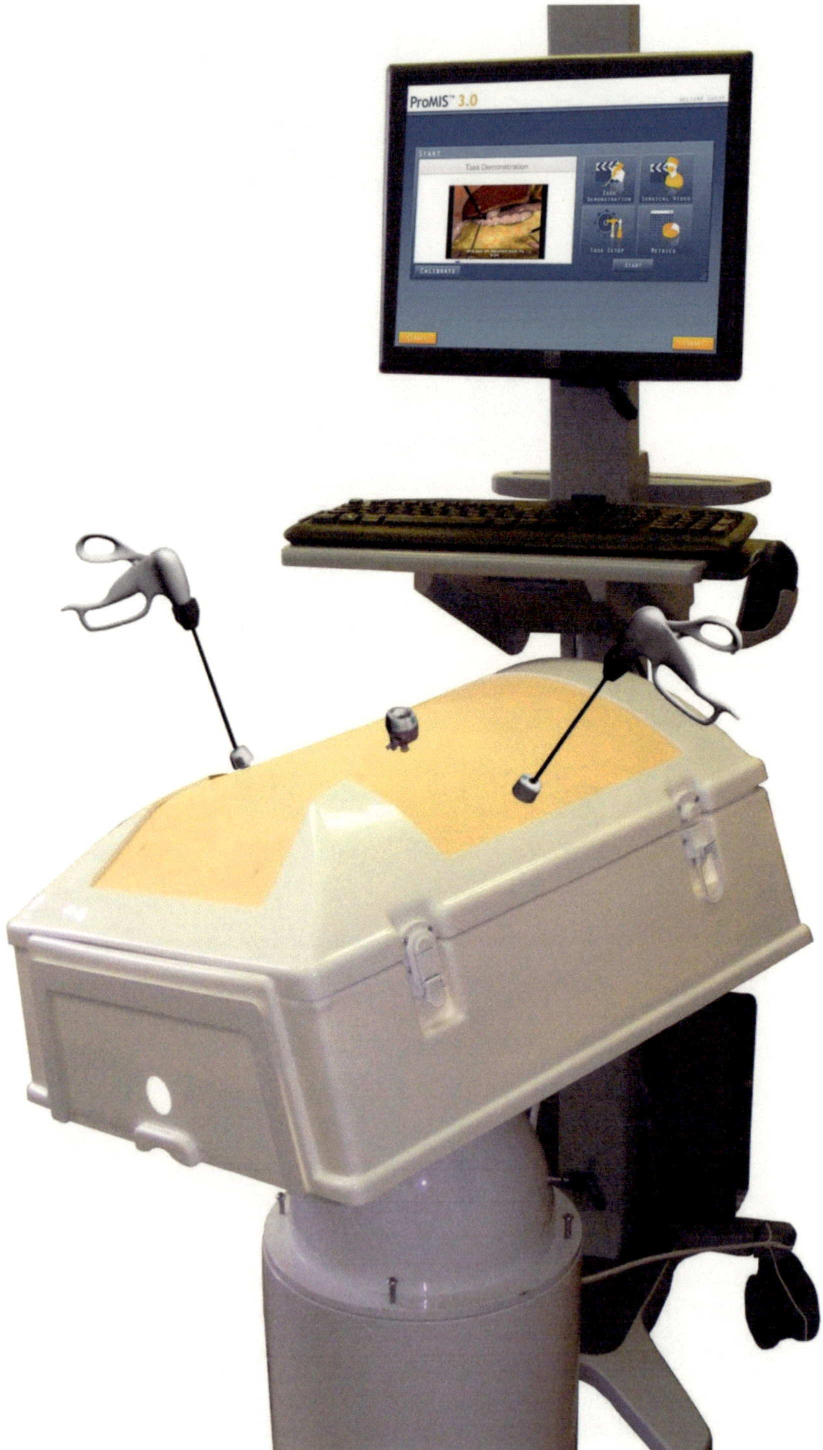

Figure 4. ProMIS, Used with permission by CAE Healthcare. All rights reserved

2.2. Hybrid trainers

VR Trainers have certain limitation due to lack of tactile feedback. In a response to these limitations, hybrid trainers were developed which combine computerized components with ex vivo synthetic parts to provide tactile feedback. Haptics experienced when using the simulator are real as you are interacting with real objects and instruments. The limitations of such physical models however include the increase in cost as the models can only be used once. Also complex human anatomy and physiology cannot be replicated precisely, for example bleeding vessels and leaking structures following trauma, and appropriate surrounding anatomy. Virtual reality surpasses physical models in this realm.

ProMIS Simulator (Haptica, Dublin) is a hybrid simulator which uses (a) 100% VR for certain tasks (b) Augmented reality that overlays graphics onto a task performed on a physical exercise. ProMIS supports both basic skills and a range of surgical procedures, including laparoscopic appendectomy and hand-assisted laparoscopic colectomy. ProMIS enables learners to practice on physical models to ensure appropriate tactile feedback, which is not easily replicated in VR simulators. The ProMIS open module is a perfect fit for surgical research and surgical procedure experimentation in that it allows you to insert any physical exercise into the simulator and by tracking the instruments, gives you full measurement and feedback on performance. Numerous studies have provided construct validity for this hybrid simulator(19-21).

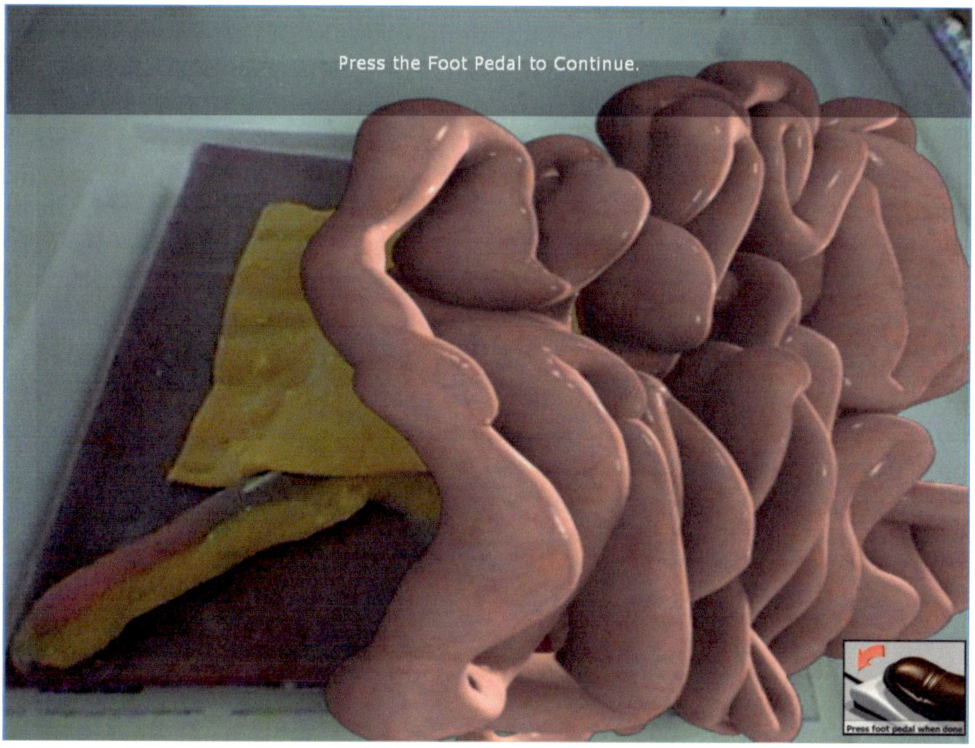

Figure 5. Screen shot of ProMIS with VR Bowel on an Appendix Model

2.3. Methods of assessment

Assessing improvement in surgical skill is essential to allow the development of surgical trainers, simulators and training programmes.

2.3.1. Metrics

Time is the most basic metric which may indicate progression in a task however it is not a real indicator of accurate performance. When time is combined with an error score (the amount of errors committed per task by the user), a trainee can also be assessed for accuracy. In order to use the metrics produced by a simulator as an assessment tool, they need to be validated. There are many different types of validity. Construct validity is the ability of a simulator to detect differences between groups with different levels of experience. Hence the simulator can measure what it claims to measure. Face validity is the extent to which simulation resembles the real task. Concurrent validity is defined as the concordance of a test to a known "gold standard".

Early box trainers lacked tracking systems which recorded errors and time; a simple stopwatch was used to access the speed at which a task was performed. With the advent of virtual reality simulators, we now have stand alone systems which can measure and record metrics. Simulators can generate a profile summary upon completion of a procedure or task which provides immediate feedback and an opportunity to see ones progress upon repeated practice. The easy to use nature of VR simulators along with practice sessions and step by step instructions provides the user with an opportunity for practice and attainment of proficiency.

Further to basic metrics (time, errors), more sophisticated markers of performance measurement have emerged over the years. An example of this is instrument path length which is the distance travelled by the instrument or the sum of deviations from a fixed point. When this is applied to laparoscopy, this suggests operative focus and greater overall performance and experience. A study by Smith et al used computer sensors on the tips of laparoscopic instruments to track motion paths. They found that speed did not equate to improved performance hence time can be a misleading if not used in conjunction with other metrics. Another metric used is economy of movement which is a score based on sudden changes in acceleration that works as an indication of smooth movement or instrument handling.

In order to use simple metrics to measure proficiency, appropriate scoring systems must be developed. The computer enhanced laparoscopic training system (CELTS) was developed by the Centre for the Integration of Medicine and Innovative Technology CIMIT and Harvard Medical School. They used a task trainer with a computer interface to form a task-independent scoring system against expert benchmark levels. Expert scores were calculated for suturing, peg transfer and knot tying using time, path length, smoothness, and depth perception as metrics. The user's score was then compared with an expert score which led to the development of a standardised scoring system. This scoring method provided a gold standard of comparing novices to experts. When ProMIS was later developed, it contains a similar system which can also compare the user's score in time, economy of movement and path length to expert proficiency scores. The scores need to be preset once they have been established for each module.

2.3.2. *Global rating scales*

Further to metrics, subjective rating of a surgical performance remains a very important tool. An approach to testing operative skills outside the operative setting led to the Objective Structured Assessment of Technical Skill (OSATS) which was introduced by Reznick et al in 1996 (22). This seven item table of technical performance on a fivepoint grading scale includes respect for tissue, time and motion, instrument handling, knowledge of instruments, flow of operation, knowledge of specific procedure and use of assistants. The OSATS tool has demonstrated high reliability and construct validity and is now used as a globally validated rating scale(23).

	1	2	3	4	5
Respect for Tissue	Frequently used unnecessary force on tissue or caused damage by inappropriate use of instruments		Careful handling of tissue but occasionally caused inadvertent damage		Consistently handled tissues appropriately with minimal damage
Time and Motion	Many unnecessary moves		Efficient time/motion but some unnecessary moves		Economy of movement and maximum efficiency
Instrument Handling	Repeatedly makes tentative or awkward moves with instruments		Competent use of instruments although occasionally appeared stiff or awkward		Fluid moves with instruments and no awkwardness
Knowledge of Instruments	Frequently used inappropriate instrument		Knew the name of most instruments and used the appropriate one for the task		Obviously familiar with the instruments required and their names
Use of Assistants	Consistently placed assistants poorly or failed to use assistants		Good use of assistants most of the time		Strategically used assistant to the best advantage at all times
Flow of Operation	Frequently stopped operating or needed to discuss next move		Demonstrated ability for forward planning with steady progression of operative procedure		Obviously planned course of operation with effortless flow from one move to the next
Knowledge of specific procedure	Deficient knowledge. Needed specific instruction at most operative steps		Knew all important aspects of the operation		Demonstrated familiarity with all aspects of the operation

Table 1. OSATS: Reznick et al., 1997

Global assessments are now widely used in the assessment of proficency during training and are used to study the effect that simulated surgical training has on operative skill. Studies by Scott et al, Hamilton et al, Traxer et al and Lucas et al demonstrating the transfer of skill from a simulated environment to the operating room have used a slightly modified version of OSATS with an included parameter of overall performance(24-27). In the study by Scott et al(26), the modified OSATS showed improvement in four of the eight parameters including the new parameter overall performance.

A study by Grantcharov(28) modified the scale so that a new parameter was created - economy of movement, which was a combination of time and motion (1= clear economy of movements and maximum efficacy; 5= many unnecessary moves) and instrument handling (1= fluent moves with instruments; 5= repeated tentative awkward or inappropriate moves). In Reznicks original scale, five was the best possible score and one was the worst. In this study, a parameter of error score was also created which is a combination of respect for tissue from Reznicks scale (1=consistently handled tissues appropriately with minimal damage; 5= frequently used unnecessary force on tissue or caused damage by inappropriate use of instruments) and precision of operative technique which is a new parameter (1= fluent, secure and correct technique in all stages of the operative procedure; 5= imprecise, wrong technique in approaching operative intentions)

Economy of movement					
	1	2	3	4	5
Unnecessary movements	Clear economy of movement and maximum efficiency		Some unnecessary moves		Many unnecessary moves
	1	2	3	4	5
Confidence of movements	Fluent moves with instruments and no awkwardness		Competent use of instruments but occasionally stiff or awkward		Repeated tentative awkward or inappropriate moves with instruments
Errors					
	1	2	3	4	5
Respect for tissue	Consistently handled tissue appropriately with minimal damage		Handled tissue carefully but occasionally caused inadvertent damage		Frequently used unnecessary force on tissue or caused damage by inappropriate use of instruments
	1	2	3	4	5
Precision of operative technique	Fluent, secure and correct technique in all stages of the operative procedure		Careful technique with occasional errors		Imprecise, wrong technique in approaching operative intentions

Table 2. Global Rating Scale: Grantcharov et al., 2002

The Global Assessment of Laparoscopic Skills (GOALS) tool was designed by Vassiliou et al(29) (based on Reznicks OSATS) for minimally invasive procedures. This five point scale assessed depth perception, bimanual dexterity, efficiency, tissue handling and autonomy. Results have shown that the tool is reliable and valid(30)

There is a trend towards using global rating tools in video analysis rather than direct observation in a live surgical setting due to time and cost resources. The advantage of simulation in this setting is the convenient storage of vast amounts of data. As there are so many available ways of rating surgical performance, the question of which is superior has been evaluated. A study by Aggarwal et al(31) assessed four different scales, OSATS, modified OSATS with four instead of seven parameters, a procedure-specific global rating scale and a procedure checklist using laparoscopic cholecystectomy. The generic global rating scales successfully distinguished between novices and experts unlike the procedure specific rating scale or checklist. An extensive systematic review was undertaken by van Hove and colleagues to examine the current evidence for objective assessment methods for technical surgical skills(32). It was concluded that OSATS is presently most accepted as the "gold standard" for objective skill assessment however it remains unknown whether OSATS can distinguish between different levels of performance. Furthermore cut off values have not been determined for OSATS. The same short comings apply to procedure specific checklists and currently there is only one checklist with a high level of evidence(33). The study also concluded that motion analysis devices can determine between operators with different levels of experience. An important point that was discusses in this study is that the value of a good assessment method can diminish when it is used in an appropriate setting.

3. The use of virtual reality in surgical training

3.1. The role of simulation in surgical training

The introduction and development of VR simulators has been one of the main innovations that have resulted in a change in training curricula in surgery. Satava was the first to recommend VR simulation as a complement to current training models(34).

The role of simulation in surgery is to provide our trainees with the opportunity to learn basic tasks in a safe and controlled environment. All movements the trainee makes can be recorded and therefore there is the facility for immediate and objective feedback. It is also possible to set a proficiency level on a simulator and therefore design a training program giving set goals that a trainee needs to accomplish before being allowed perform in the operating theatre. All of these factors contribute to skill learning, assessment, selection and credentialing. Simulators will also be invaluable in the teaching of the newer forms of surgery, single incision laparoscopy and natural orifice transluminal endoscopic surgery. The use of simulation should provide the setting in which challenges such as the use of new instruments and technology can be overcome. An example of this is in single incision laparoscopic surgery where it is difficult to have instruments working parallel to each other in a very narrow operative field.

Given that simulation is generally an education tool, there are two distinct parts to the delivery of a simulator. There is firstly the teaching aspect which is the way which we communicate or impart knowledge or information. Secondly there is the training aspect, which is the acquisition of psychomotor skill and cognitive skill(35). Furthermore, the learning of psychomotor and cognitive skill can become blurred. When a novice begins simulated training, they are naïve to both the fulcrum effect of laparoscopy and the steps of the surgical procedure, therefore it becomes unclear what rate each skill is learnt. The only way in which we can both teach and train in an effective way is through a carefully thought out, well-structured curriculum. Several studies(35-37) have proposed templates for this.

3.2. Mapping learning curves

A learning curve is a graphical representation of the changing rate of learning (figure 1.1). Typically the increase in retention of information is sharpest after the initial attempts. This increase gradually flattens out as less and less new information is retained after each repetition.

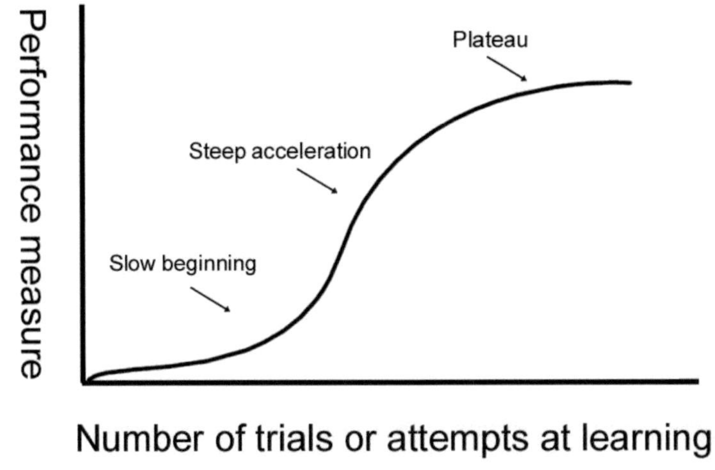

Figure 6. Graphical Representation of the Learning Curve

As mentioned earlier, simulation provides a protected environment for trainees to overcome the initial learning curve. This concept has been discussed and examined by researchers in several studies over the last ten years.

Gallagher and Satava carried out a study (8) which looked at using the MIST-VR trainer as a tool for assessing psychomotor performance. As an adjunct to this, they also looked at learning curves. Both senior (<50 laparoscopic operations) surgeons and junior surgeons (<10 laparoscopic operations) performed six tasks on the MIST-VR, by trial 10 there was a convergence of mean performance. This showed that juniors could potentially perform to the level of a senior surgeon with practice outside the operating theatre.

A study by Grantcharov (38) showed that different learning curves exist for surgeons with varying levels of laparoscopic experience. In this study, it was established that the MIST-VR was capable of differentiating between surgeons with different laparoscopic experience, which is, important for both construct validity and also for the potential development of internationally accepted norms of performance. If this was further developed then a trainee could use this as a reference point to establish where they currently are on the learning curve. Similar results were shown by Eversbusch (39). Three different learning curves were mapped for colonoscopy. The learning rate on the simulator was proportional to prior experience with endoscopy, which indicated that the simulator could assess parameters that are clinically relevant. Psychomotor training using the GI mentor compared with a control group who received no training demonstrated improved performance in the novice participants.

Aggarwal et al have produced several studies involving mapping learning curves. In a study in 2006(40), two different learning curves were mapped out using medical students who performed tasks of various complexity on the MIST-VR. All three parameters (time, economy of movement and error scores) plateaued at the second repetition for the twelve core skills and at the fifth repetition for the most complex two tasks. Another study in 2006(41) assessed the learning rate for dissection of Calot's triangle, a learning curve for novices was established as their performance plateaued at the fourth repetition. Learning curve data was established in a study in 2009 (36) to ensure that repetitive practice improved performance, as measured by the simulator. Moreover by applying a stepwise process to learning a laparoscopic cholecycstectomy, a whole procedure-training curriculum can develop. The learning curve for this procedure plateaued for all metrics between six and nine repetitions.

When laparoscopic suturing was examined in a study by Botden (42), the number of repetitions required to reach the top of the performance curve (defined as proficiency) was eight knots. Lin et al (43) evaluated the learning curve for laparoscopic appendicectomy and found that operative duration and complication rate decreased in proportion to the increasing experience of the resident.

Interestingly, Grantcharov's study in 2009 (44) assessed the learning curve patterns of acquisition of generic skills in laparoscopy. In this study it was hypothesized that the familiarization rate with laparoscopic technique is different depending on psychomotor ability. Four types of learning curves were identified, proficiency from the beginning (5.4%), ability to advanced with practice which was found to be between two and nine repetitions (70.3%), ability to improve but unable to reach proficiency (16.2%), and finally no tendency to improve and overall underperformance (8.1%). This data suggests a role for developing a proficiency-based curriculum based on innate psychomotor ability. Several studies have looked into aptitude tests, which may relate basic laparoscopic technical skill performance(45) Further to this research has attempted to ascertain the rate of skill acquisition in relation to innate ability (46).

3.3. VR-to-OR transfer

It is intuitive that training in a simulated surgical setting implies improved skill in a clinical environment; however this important concept requires definite clarification. There is little value to developing sophisticated training programmes in a simulated laboratory if laboratory training does not improve clinical performance. Transferability is often called VR-to-OR (a term coined by Professor Anthony Gallagher) and refers to the ability of simulation-based training to improve clinical performance. Transferability in clinical terms would imply predictive validity as discussed earlier.

Such trials are usually designed by using two groups who are randomised to either receive simulation based training or no training. Their performance is then compared in a specific laparoscopic procedure or task after simulation training or no training. The groups ideally have similar baseline psychomotor and visuospatial ability. Assessment in the operating room is performed by an examiner who is blinded to the status of the subject, using the methods described previously. Even with sound methodology human trials can have many logistical challenges therefore many investigators opt to conduct their trials using animal specimen's most commonly porcine models. Clinical transferability can be shown with animal models in suitable laboratorys as a bridge to the human setting. Transferability studies are essential in order to assess the ability of simulation based training to improve surgical performance in the operating room; they require approval from an intuitional review board.

The first study to demonstrate a transfer of simulator learned skills to the operating room was in Yale, 2001 (47).The control group had no simulation training and the trained group were taught to proficiency under supervision with emphasis on avoidance of errors. Candidates were assessed on dissection of the gallbladder from the liver edge both pre and post training or no training in the OR during human cholecystectomies. The scoring system used was a novel pre-defined eight error checklist; occurrence of these errors was recorded during each minute of the assessment. This was used instead of a global rating scale in an attempt to determine errors more accurately. A non-significant difference was detected in dissection time, with the trained group removing the gallbladder 29% faster than the non-trained group. In relation to error performance, the control group were five times more likely to burn the liver edge or injure the gallbladder and nine times more likely to fail to progress. Further evidence which supported this landmark research was in a study by Grantcharov et al(28) which assessed both a trained and a control group in the clipping and cutting of the cystic duct. Again both groups underwent pre and post testing in the OR during human cholecystectomies. Performance was measured using a modified OSATs scale by combining traditional parameters to create new parameters. It was found that the group who received simulated training on the MIST-VR performed faster, had greater economy of movement scores and lower error scores than the control group in the post-test assessment in the OR, hence the study demonstrated transferability.

Following on from this initial research, various other studies demonstrating transfer of skill have been published. Some of them have shown partial task transfer and some using whole laparoscopic procedures; the latter of which laparoscopic cholecystectomy form's the bulk. Three other studies (26, 48, 49) assessed the transfer of skill in laparoscopic cholecystectomy. Scott used OSATS and demonstrated a significant improvement in the trained versus control groups. McClusky and Ahlberg used total error scores; both studies showed that error scores were higher in the control groups.

Other studies have looked at the transfer of whole procedures. A study by Larsen et al(50) assessed the performance of an entire laparoscopic salpingectomy using an OSAT scoring system and found significant differences between trained and control groups. These same results have been shown with both laparoscopic hernia repairs(24) and laparoscopic nephrectomy(27). When laparoscopic appendicectomy was assessed on a porcine specimen, the results of this study did not show any difference between trained and control groups. In this study, training time was very short, with three hours training in total. Achieving proficiency in a shorter time frame may have been difficult and therefore could have affected the outcome of this study. The assessment method used was blinded rater analysis using a scale of bad, average and good, which had no previous validation in this setting.

One study(25) provided training for the novice group in laparoscopic cholecystectomy but assessed skill transfer in laparoscopic nephrectomy. The results showed that the group who received time based simulated laparoscopic cholecystectomy training outperformed the control group when a laparoscopic nephrectomy was performed in a porcine model. The students were assessed using OSATS. This shows not only the transfer of skill after simulated training but also that specific skills learnt for certain laparoscopic procedures are useful for other laparoscopic procedures.

Laparoscopic tasks as well as laparoscopic suturing have also been explored. Three studies(51-53) evaluated the transfer of laparoscopic suturing. Two of them (51, 52), used the same formula which was 600 – [(time + (10 x accuracy score) + (10x security error)]. This method awarded higher scores for the most accurate performance in the faster time. The purpose of this formula was to establish one value which if high implied a fast accurate performance and a good quality knot. By assigning one value to the user's performance as opposed to three, it gives results that are easy to compare and understand. Both studies showed significant improvements in the trained group compared with the control group. The third(53) study used an error scoring system which showed that the control group made more errors than the trained group and this study also performed blinded rater video analysis looking at economy of movement and error assessment (Table 2). Verdaadonk et al did not show any significant difference in the transfer of skill between the simulation-trained group and the control group.

4. Learning though virtual reality

Research has shown that training on simulators translates to the clinical environment but less is known about how best to integrate simulation into the surgical curriculum(37). In order to provide the ideal model for surgical training there are a number of factors to consider.

Firstly a structured curriculum needs to be developed(31). Wiggens and McTigue's backward design approach to curriculum development for technical skills is one approach that has been proposed for surgical simulation(54).

The second factor is that training should be carried out in a stepwise manner where the trainee begins on a simulator in the skills lab until predefined proficiency criteria are reached (55). An example of this is part-task training. Part-task training is a learning strategy whereby a complex task is deconstructed into smaller components for practice. Trainees gain proficiency in the individual components before progressing to the more complex task. It is thought that a higher level of skill can be attained if participants master individual components before integrating them into the whole task.

Thirdly, there needs to be clear criteria to determine the competence level of the trainee and skill mastery (56). The setting of training goals ensures that the trainee is required to reach a predefined standard and competence is not determined by time spent on the simulator or by performing a set number of repetitions. Standards should be benchmarked against both clinically established and simulator generated data. When this has been demonstrated and assessed in an objective manner then the trainee can progress to the real life operating room.

Training sessions should be spread out over a period of time in order to better augment and optimise learning. Previously it has been shown that one hour on a virtual reality simulator equates to two hours spent in the operating room. Other conditions to consider include the learning environment; for example is it in a quiet and relaxed setting or does it mirror the everyday stress of the operating theatre? Whether the trainee engages in purely self-directed learning or whether a mentor or trainer is present is also important.

Finally in order for any training programme to be effective the virtual reality simulator needs to demonstrate acceptability, validity, reliability and reproducibility in the real life operating environment.

4.1. Theories of learning

Historically observational learning has played a central role in surgical training, constituting the first step in the time-honored "see one, do one, teach one" model. However although there is a trend today away from this traditional approach to training, observation still has a role. No longer is observation limited to the operating room; many professional organizations now offer free Web-based videos of surgical procedures for training purposes. Such training videos are undoubtedly valuable resources. A study by Snyder and colleagues aimed to evaluate the use of video observation and to compare it to the real-life observation of procedures (57). They found that while instructional videos are useful they may however not be an adequate substitute for actual real-time observation in the minimally invasive surgery setting.

There are a number of theories of learning that have been discussed in the setting of surgical skills training. The learning model postulated by Fitts and Posner is one such theory. This theory has been discussed as being relevant in learning minimally invasive surgical skills

(58). Their theory states that there are three phases in the acquisition of skill; cognitive stage, associative stage and autonomous stage.

Cognitive stage: During this stage, learners need to know what the elements of the task are and what is expected in terms of performance. They will draw on their reasoning ability, past experience and instructions to use cognitive strategies that are subsequently modified as they gain experience with the task.

Associative stage: This stage involves working out how to optimize and integrate performance so as to greatly reduce major errors and make performance more efficient.

Autonomous stage: This stage refers to extremely advanced levels of performance where errors are greatly reduced and performance of the task seems to be almost automatic. At this stage, less attention is required to carry out the task and so can be allocated to other activities such as teaching, attending to anatomical anomalies, changes in instrument readouts and so on. Once a skill becomes automated, the learner has established a sequence of highly coordinated movements, which are integrated in time and are characterized by a rhythmic structure of their own.

The concept of deliberate practice, as proposed by Ericsson and Smith, has become the most popular learning theory of late(59). Their expert performance model comprises of three crucial stages which overall suggest that individual differences in performance can be explained by differences in deliberate practice. The first stage requires the identification of representative tasks of expert performance and their replication within a controlled laboratory setting. The second stage involves an empiric analysis to identify the mechanisms underlying experts' superior performance. The last stage examines the effect of specific practice activity to elucidate factors that might influence the acquisition of these expert performance mechanisms. Deliberate practise requires the individual to focus their training on defined tasks or drills. It involves repeated practice and immediate feedback delivered by experts. Because perceptual-motor tasks can be designed to capture the essence of specific surgical tasks, simulators lend themselves well to applying Ericsson's expert performance approach as they allow measurement and empiric analysis of representative tasks in a controlled setting and allow for repeated drills.

Crochet et al carried out a study to investigate deliberate practice in the simulated setting and compared it to the real-life clinical setting (60). They concluded that enhanced quality of surgical skills can be achieved with deliberate practice, both on simulated and realistic tissues.

4.2. Feedback

Feedback can be defined as the provision or return of performance-related information to the performer (61). Feedback that is delivered in a timely and regular manner has been recognised as an important part of the learning process in medical education. It can be intrinsic, where it is relayed directly by the sensory system of the trainee or extrinsic, where it is provided by an external source.

One of the dangers associated with the use of virtual reality simulation is a situation where the trainee is unaware of committing an error, and as a result persists in this error which in turn allows the simulator to reinforce undesirable behaviour. Therefore high-fidelity simulators run the risk of becoming ineffective as a training tool without feedback.

In spite of this virtual reality simulators confer a number of benefits with respect to feedback when compared to the real-life environment. Virtual reality simulation allows the delivery of immediate, objective and automated feedback. Tasks can be interrupted to highlight errors to the trainee and then repeated as required. Trainees can assess their own errors using the automated feedback provided or they can observe video playback of performance which is recorded by the simulator. Virtual reality simulation also allows for the delivery of feedback regardless of whether an expert is available. In fact it has been suggested that trainees may value virtual reality simulator feedback as being clearer and more objective than human expert feedback. Automated feedback has been demonstrated to have similar efficacy to live expert feedback(57)

Although the benefits of performance feedback are not debated, questions remain about the optimal way to provide this. Research is currently been conducted to analyse the optimal frequency and type of feedback. It has been shown that feedback delivered in a standardised and structured manner results in an improvement in simulator performance (61). It has also been found that providing feedback has resulted in a shortening of the minimally invasive surgery learning curve(55).

4.3. Limitations of virtual reality

Virtual reality is an acceptable way of simulating a surgical procedure however there are several challenges given the limitations of modern technology. Graphics can simulate anatomical structures visually however they are unable to model the physical properties of its real counterpart to an accurate enough degree therefore it cannot be manipulated in a realistic fashion. The biggest limitation however and future challenge of virtual reality simulation is haptic feedback. Currently none of the VR simulators are capable of providing any tactile feedback. There is ongoing research into this area however haptic technology is currently very basic with the phantom device being at this pinnacle of this technology. Technology could be ten or more years away from providing a solution to these challenges.

5. Technology's ability to deliver simulators

It is useful to compare the state of the flight simulator technology to the state of the surgical simulation technology. On one hand you can purchase a flight simulator today that will represent every single aspect of the flight experience right down to the chair you sit on. However, in no form does a similar setup exist for the surgeon. They cannot enter a virtual room and carry out whatever operation they like on a virtual patient and have everything behave exactly as it should. In order to understand why this is we need to break down the components of what makes a perfect simulator.

A perfect simulator will model a subset of the world (plane and terrain for the pilot, subset of human and procedure for the surgeon) and attempt to have you interact in your normal way with this simulated environment. We experience the world through our senses so to create the simulation we must be able to substitute for each of these experiences. For discussion sake we will concentrate on comparing a flight simulator with a cholecystectomy simulator.

1. Sight. For sight to be tricked, the objects we look at must look like their real counterpart – but must also behave like them.

Flight Simulation: The goal is to model and create a virtual terrain and have it behave correctly from a physics point of view: Because terrain, mountains, houses, in terms of a flight simulation need only be 'shells' and are generally far away in the distance (i.e.no physical interaction) then all we need to simulate is the effects of wind on an object (i.e. the airplane) that is a manmade object. Because it is manmade then we can easily understand and simulate all its aspects.

Cholecystectomy Simulation: Here the goal is to model the internals of a working body and have it behave correctly under physical interaction: Modeling a complex organic organ that is not fully understood is an incredibly difficult task. Unlike the 'shell' of the flight simulator, cutting into an organ must reveal a solid structure that behaves like a solid organic structure would – the weight of the separated tissue would help move it apart, blood would flow from blood supply and the combination of possible outcomes are infinite. With current technology we can do a reasonable job of modeling the outer shell, and have it behave in roughly an organic way, but any bleeding or cutting will be pre-scripted (fake). The model is just too complex.

2. Touch. For touch to work our sense must be fooled into thinking that we feel the same resistances as we would in the real situation.

Flight Simulation: Interaction is all through manmade interfaces (the cockpit is no more complex technology wise than any games console interface found in any home). They translate into increases/decreases in pressure and are very simple to interpret at a computer level.

Cholecystectomy Simulation: Interaction is physical and complex. Instruments must collide with each other, and their actions must cut/ burn/ grasp/ with organic material that will act realistically under pressure. If we press our virtual instrument against a liver it must stop in its tracks as a real liver would. We are now into the world of robotics. This technology is still in its infancy. For a surgeon to be able to pick instruments up and place them where they like in a simulated environment and have the instruments physically interact with every organ they come in contact with simply doesn't exist.

In summary the world of video games has given us vastly improved graphics (or visual representation of the real world). Their physics technologies (needed for accurate behavior) is excellent, but only in rigid solid objects. Everything else (organic soft body modeling/fluid dynamics) is too complicated. And touch (haptics), or recreating what our hand experiences, is just too complex. The human body is infinitely complex with so many interdependent systems that is too complex to replicate accurately – so we try to create pre-scripted experiences and in doing so we do our best with the current technical challenges.

5.1. The response from the simulation industry to these challenges

The industry's approach has been to try and combat the basic areas of surgery. The basic skills for example, can be practiced and measured in VR and Augmented Reality exercises with a high degree of accuracy and accountability. We are however, simulating manmade objects such as instruments, beads, suture needles and the consequences of the actions are easy to simulate. If a bead falls, it rolls somewhere. There is not a series of knock on effects; they are discrete pieces of simulation. After basic skills, the route has been to try to replicate the least complex operation, for example a laparoscopic appendicectomy or cholecystectomy. Several factors combine to create the level of complexity such as the number of organs involved, the number of structures involved and then the interaction between these. The rigidity of the organ being simulated varies; for example the small intestine would be more difficult to replicate than a gallbladder or even a liver. The level of interaction with the organ also varies; transecting the liver would require very complex physics as opposed to clipping the cystic duct during a laparoscopic cholecystectomy. Replicating the behavior of laparoscopic instruments interacting with tissue is infinitely less complex than replicating the experience of using your hands in open surgery when performing a procedure. We are essentially closer to the level of complexity of a flight simulator in that we are using a manmade interface to guide a rigid body through a known space.

5.2. Limitations of the simulation industry

The video games industry conquered their own limitations by delegation of complexity. The older model of a video game developer company involved one company creating every component to a video game. This is generally the case with surgical simulation companies. Simulator manufacturing may benefit greatly from having separate companies concentrate and perfect the various components needed for a more precise simulator, (e.g. interface, human body physics, anatomy rendering, fluid dynamics). These could all potentially run on a standardised platform designed by experts in the field internationally. By doing this, we may overcome the current restrictions that the simulation industry has regarding future developments.

5.3. The language barrier

A computer programming language is just a formal way of instructing a computer. If we leave what is possible out of the equation for a minute and assume we had limitless computer power, then to create a perfect simulator, we would need a surgeon to describe in perfect detail every aspect of a surgical procedure, describing at every moment how every cell reacted, every organ interacted, and the principles and physics behind every system such as blood flow. The computer programmers and artists would then have to understand every aspect of this with as much knowledge as the surgeon themselves. This doesn't happen so we end up with information being lost between the surgeon, the artist, the programmer and the limitations of how descriptive we can be in instructing a computer how to behave.

6. Summary

6.1. The future of VR in surgery

We still face a variety of challenges before we have a virtual patient that will behave in the exact same fashion as a real human. It is not only technical challenges, such as those concerning interface and complex system simulation, but the financial challenges such as hospital budgets and developer budgets. So what does the future hold for VR's role in surgical training? It would seem that excellent basic skills simulation and complete feedback is close. As we move towards procedure based simulation it may be a case of acknowledging the current technical limitations but expanding the material by improving and expanding the content using current platforms. Another step forward will be accessibility. The simulators need to become more integrated into surgical training programmes and should be onsite in teaching hospitals.

6.2. Conclusion

In an era of expanding minimally access technology, reduced working hours and increased awareness of patient safety; surgical simulation has helped to create a safe environment for surgeons to practice skills and procedures. The new ethos of proficiency based training programmes ensures that the surgical community can learn and perfect new skills. Further to this it has helped advance patient safety by battling out the steepest part of the learning curve.

Author details

C.E. Buckley, E. Nugent, D. Ryan and P.C. Neary
National Surgical Training Department, Royal College of Surgeons, Dublin, Ireland

7. References

[1] Figert PL, Park AE, Witzke DB, Schwartz RW. Transfer of training in acquiring laparoscopic skills. J Am Coll Surg. 2001;193(5):533-7. Epub 2001/11/16.
[2] Gallagher AG, McClure N, McGuigan J, Ritchie K, Sheehy NP. An ergonomic analysis of the fulcrum effect in the acquisition of endoscopic skills. Endoscopy. 1998;30(7):617-20. Epub 1998/11/24.
[3] Perkins N, Starkes JL, Lee TD, Hutchison C. Learning to use minimal access surgical instruments and 2-dimensional remote visual feedback: how difficult is the task for novices? Advances in health sciences education : theory and practice. 2002;7(2):117-31. Epub 2002/06/21.
[4] Moore MJ, Bennett CL. The learning curve for laparoscopic cholecystectomy. The Southern Surgeons Club. American journal of surgery. 1995;170(1):55-9. Epub 1995/07/01.

[5] Watson DI, Baigrie RJ, Jamieson GG. A learning curve for laparoscopic fundoplication. Definable, avoidable, or a waste of time? Ann Surg. 1996;224(2):198-203. Epub 1996/08/01.

[6] Bennett CL, Stryker SJ, Ferreira MR, Adams J, Beart RW, Jr. The learning curve for laparoscopic colorectal surgery. Preliminary results from a prospective analysis of 1194 laparoscopic-assisted colectomies. Archives of surgery (Chicago, Ill : 1960). 1997;132(1):41-4; discussion 5. Epub 1997/01/01.

[7] Asch DA, Parker RM. The Libby Zion case. One step forward or two steps backward? The New England journal of medicine. 1988;318(12):771-5. Epub 1988/03/24.

[8] Gallagher AG, Satava RM. Virtual reality as a metric for the assessment of laparoscopic psychomotor skills. Learning curves and reliability measures. Surg Endosc. 2002;16(12):1746-52. Epub 2002/07/26.

[9] LL K. As told to Robert B. Parke. The Pilot Maker. New York: Grosset and Dunlap; 1970.

[10] Delp SL, Loan JP, Hoy MG, Zajac FE, Topp EL, Rosen JM. An interactive graphics-based model of the lower extremity to study orthopaedic surgical procedures. IEEE transactions on bio-medical engineering. 1990;37(8):757-67. Epub 1990/08/01.

[11] Duffy AJ, Hogle NJ, McCarthy H, Lew JI, Egan A, Christos P, et al. Construct validity for the LAPSIM laparoscopic surgical simulator. Surg Endosc. 2005;19(3):401-5. Epub 2004/12/30.

[12] van Dongen KW, Tournoij E, van der Zee DC, Schijven MP, Broeders IA. Construct validity of the LapSim: can the LapSim virtual reality simulator distinguish between novices and experts? Surg Endosc. 2007;21(8):1413-7. Epub 2007/02/13.

[13] Andreatta PB, Woodrum DT, Birkmeyer JD, Yellamanchilli RK, Doherty GM, Gauger PG, et al. Laparoscopic skills are improved with LapMentor™ training: Results of a randomized, double-blinded study. Annals of surgery. 2006;243(6):854-60.

[14] Zhang A, Hunerbein M, Dai Y, Schlag PM, Beller S. Construct validity testing of a laparoscopic surgery simulator (Lap Mentor): evaluation of surgical skill with a virtual laparoscopic training simulator. Surg Endosc. 2008;22(6):1440-4. Epub 2007/11/01.

[15] Kothari SN, Kaplan BJ, DeMaria EJ, Broderick TJ, Merrell RC. Training in laparoscopic suturing skills using a new computer-based virtual reality simulator (MIST-VR) provides results comparable to those with an established pelvic trainer system. Journal of laparoendoscopic & advanced surgical techniques Part A. 2002;12(3):167-73. Epub 2002/08/20.

[16] Maithel S, Sierra R, Korndorffer J, Neumann P, Dawson S, Callery M, et al. Construct and face validity of MIST-VR, Endotower, and CELTS: are we ready for skills assessment using simulators? Surg Endosc. 2006;20(1):104-12. Epub 2005/12/08.

[17] Torkington J, Smith SG, Rees BI, Darzi A. Skill transfer from virtual reality to a real laparoscopic task. Surg Endosc. 2001;15(10):1076-9. Epub 2001/12/01.

[18] Iwata N, Fujiwara M, Kodera Y, Tanaka C, Ohashi N, Nakayama G, et al. Construct validity of the LapVR virtual-reality surgical simulator. Surg Endosc. 2011;25(2):423-8. Epub 2010/06/30.

[19] Broe D, Ridgway PF, Johnson S, Tierney S, Conlon KC. Construct validation of a novel hybrid surgical simulator. Surg Endosc. 2006;20(6):900-4. Epub 2006/06/02.

[20] Gilliam AD. Construct validity of the ProMIS laparoscopic simulator. Surg Endosc. 2009;23(5):1150. Epub 2009/03/06.

[21] Neary PC, Boyle E, Delaney CP, Senagore AJ, Keane FB, Gallagher AG. Construct validation of a novel hybrid virtual-reality simulator for training and assessing laparoscopic colectomy; results from the first course for experienced senior laparoscopic surgeons. Surg Endosc. 2008;22(10):2301-9. Epub 2008/06/17.

[22] Faulkner H, Regehr G, Martin J, Reznick R. Validation of an objective structured assessment of technical skill for surgical residents. Academic medicine : journal of the Association of American Medical Colleges. 1996;71(12):1363-5. Epub 1996/12/01.

[23] Martin JA, Regehr G, Reznick R, MacRae H, Murnaghan J, Hutchison C, et al. Objective structured assessment of technical skill (OSATS) for surgical residents. Br J Surg. 1997;84(2):273-8. Epub 1997/02/01.

[24] Hamilton EC, Scott DJ, Kapoor A, Nwariaku F, Bergen PC, Rege RV, et al. Improving operative performance using a laparoscopic hernia simulator. American journal of surgery. 2001;182(6):725-8.

[25] Lucas SM, Zeltser IS, Bensalah K, Tuncel A, Jenkins A, Pearle MS, et al. Training on a Virtual Reality Laparoscopic Simulator Improves Performance of an Unfamiliar Live Laparoscopic Procedure. Journal of Urology. 2008;180(6):2588-91.

[26] Scott DJ, Bergen PC, Rege RV, Laycock R, Tesfay ST, Valentine RJ, et al. Laparoscopic training on bench models: Better and more cost effective than operating room experience? Journal of the American College of Surgeons. 2000;191(3):272-83.

[27] Traxer O, Gettman MT, Napper CA, Scott DJ, Jones DB, Roehrborn CG, et al. The impact of intense laparoscopic skills training on the operative performance of urology residents. The Journal of urology. 2001;166(5):1658-61. Epub 2001/10/05.

[28] Grantcharov TP, Kristiansen VB, Bendix J, Bardram L, Rosenberg J, Funch-Jensen P. Randomized clinical trial of virtual reality simulation for laparoscopic skills training. The British journal of surgery [Internet]. 2004; (2):[146-50 pp.]. Available from: http://www.mrw.interscience.wiley.com/cochrane/clcentral/articles/215/CN-00460215/frame.html.

[29] Vassiliou MC, Feldman LS, Andrew CG, Bergman S, Leffondre K, Stanbridge D, et al. A global assessment tool for evaluation of intraoperative laparoscopic skills. American journal of surgery. 2005;190(1):107-13. Epub 2005/06/24.

[30] Vaillancourt M, Ghaderi I, Kaneva P, Vassiliou M, Kolozsvari N, George I, et al. GOALS-incisional hernia: a valid assessment of simulated laparoscopic incisional hernia repair. Surg Innov. 2011;18(1):48-54. Epub 2011/01/11.

[31] Aggarwal R, Grantcharov TP, Darzi A. Framework for systematic training and assessment of technical skills. J Am Coll Surg. 2007;204(4):697-705. Epub 2007/03/27.

[32] van Hove PD, Tuijthof GJ, Verdaasdonk EG, Stassen LP, Dankelman J. Objective assessment of technical surgical skills. Br J Surg. 2010;97(7):972-87. Epub 2010/07/16.

[33] Sarker SK, Chang A, Vincent C. Technical and technological skills assessment in laparoscopic surgery. JSLS : Journal of the Society of Laparoendoscopic Surgeons / Society of Laparoendoscopic Surgeons. 2006;10(3):284-92. Epub 2007/01/11.

[34] Satava RM. Virtual reality surgical simulator: the first steps. 1993. Clinical orthopaedics and related research. 2006;442:2-4. Epub 2006/01/06.

[35] Gallagher AG, Ritter EM, Champion H, Higgins G, Fried MP, Moses G, et al. Virtual reality simulation for the operating room: proficiency-based training as a paradigm shift in surgical skills training. Ann Surg. 2005;241(2):364-72. Epub 2005/01/15.

[36] Aggarwal R, Crochet P, Dias A, Misra A, Ziprin P, Darzi A. Development of a virtual reality training curriculum for laparoscopic cholecystectomy. Br J Surg. 2009;96(9):1086-93. Epub 2009/08/13.

[37] Kolozsvari NO, Feldman LS, Vassiliou MC, Demyttenaere S, Hoover ML. Sim one, do one, teach one: considerations in designing training curricula for surgical simulation. Journal of surgical education. 2011;68(5):421-7. Epub 2011/08/09.

[38] Grantcharov TP, Bardram L, Funch-Jensen P, Rosenberg J. Learning curves and impact of previous operative experience on performance on a virtual reality simulator to test laparoscopic surgical skills. American journal of surgery. 2003;185(2):146-9. Epub 2003/02/01.

[39] Eversbusch A, Grantcharov TP. Learning curves and impact of psychomotor training on performance in simulated colonoscopy: a randomized trial using a virtual reality endoscopy trainer. Surg Endosc. 2004;18(10):1514-8. Epub 2005/03/26.

[40] Aggarwal R, Grantcharov T, Moorthy K, Hance J, Darzi A. A competency-based virtual reality training curriculum for the acquisition of laparoscopic psychomotor skill. American journal of surgery. 2006;191(1):128-33. Epub 2006/01/10.

[41] Aggarwal R, Grantcharov TP, Eriksen JR, Blirup D, Kristiansen VB, Funch-Jensen P, et al. An evidence-based virtual reality training program for novice laparoscopic surgeons. Ann Surg. 2006;244(2):310-4. Epub 2006/07/22.

[42] Botden SM, de Hingh IH, Jakimowicz JJ. Suturing training in Augmented Reality: gaining proficiency in suturing skills faster. Surg Endosc. 2009;23(9):2131-7. Epub 2008/12/11.

[43] Lin YY, Shabbir A, So JB. Laparoscopic appendectomy by residents: evaluating outcomes and learning curve. Surg Endosc. 2010;24(1):125-30. Epub 2009/09/18.

[44] Grantcharov TP, Funch-Jensen P. Can everyone achieve proficiency with the laparoscopic technique? Learning curve patterns in technical skills acquisition. American journal of surgery. 2009;197(4):447-9. Epub 2009/02/17.

[45] Gallagher AG, Cowie R, Crothers I, Jordan-Black JA, Satava RM. PicSOr: an objective test of perceptual skill that predicts laparoscopic technical skill in three initial studies of laparoscopic performance. Surg Endosc. 2003;17(9):1468-71. Epub 2003/06/13.

[46] Stefanidis D, Korndorffer JR, Jr., Black FW, Dunne JB, Sierra R, Touchard CL, et al. Psychomotor testing predicts rate of skill acquisition for proficiency-based laparoscopic skills training. Surgery. 2006;140(2):252-62. Epub 2006/08/15.

[47] Seymour NE, Gallagher AG, Roman SA, O'Brien MK, Bansal VK, Andersen DK, et al. Virtual reality training improves operating room performance: results of a randomized, double-blinded study. Annals of surgery [Internet]. 2002; (4):[458-63; discussion 63-4 pp.]. Available from:

http://www.mrw.interscience.wiley.com/cochrane/clcentral/articles/307/CN-00398307/frame.html.

[48] Ahlberg G, Enochsson L, Gallagher AG, Hedman L, Hogman C, McClusky Iii DA, et al. Proficiency-based virtual reality training significantly reduces the error rate for residents during their first 10 laparoscopic cholecystectomies. American journal of surgery. 2007;193(6):797-804.

[49] McClusky DA, Gallagher AG, Ritter EM, Lederman AB, Van Sickle KR, Baghai M, et al. Virtual reality training improves junior residents' operating room performance: Results of a prospective, randomized, double-blinded study of the complete laparoscopic cholecystectomy. Journal of the American College of Surgeons. 2004;199(3):S73-S.

[50] Larsen CR, Soerensen JL, Grantcharov TP, Dalsgaard T, Schouenborg L, Ottosen C, et al. Effect of virtual reality training on laparoscopic surgery: randomised controlled trial. BMJ (Clinical research ed). 2009;338:b1802. Epub 2009/05/16.

[51] Korndorffer JR, Dunne JB, Sierra R, Stefanidis D, Touchard CL, Scott DJ. Simulator training for laparoscopic suturing using performance goals translates to the operating room. Journal of the American College of Surgeons. 2005;201(1):23-9.

[52] Stefanidis D, Acker C, Heniford BT. Proficiency-based laparoscopic simulator training leads to improved operating room skill that is resistant to decay. Surgical innovation [Internet]. 2008; (1):[69-73 pp.]. Available from: http://www.mrw.interscience.wiley.com/cochrane/clcentral/articles/702/CN-00638702/frame.html.

[53] Verdaasdonk EG, Dankelman J, Lange JF, Stassen LP. Transfer validity of laparoscopic knot-tying training on a VR simulator to a realistic environment: a randomized controlled trial. Surgical Endoscopy [Internet]. 2008; (7):[1636-42 pp.]. Available from: http://www.mrw.interscience.wiley.com/cochrane/clcentral/articles/210/CN-00637210/frame.html.

[54] Wiggins G MJ. Understanding by Design. Upper Saddle River, NJ: Prentice Hall, Inc; 2001.

[55] Grantcharov TP, Reznick RK. Teaching procedural skills. BMJ (Clinical research ed). 2008;336(7653):1129-31. Epub 2008/05/17.

[56] Sweet RM, Hananel D, Lawrenz F. A unified approach to validation, reliability, and education study design for surgical technical skills training. Archives of surgery (Chicago, Ill : 1960). 2010;145(2):197-201. Epub 2010/02/17.

[57] Snyder CW, Vandromme MJ, Tyra SL, Porterfield JR, Jr., Clements RH, Hawn MT. Effects of virtual reality simulator training method and observational learning on surgical performance. World journal of surgery. 2011;35(2):245-52. Epub 2010/11/19.

[58] Gallagher AG, Leonard G, Traynor OJ. Role and feasibility of psychomotor and dexterity testing in selection for surgical training. ANZ journal of surgery. 2009;79(3):108-13. Epub 2009/03/26.

[59] Ericsson KA. Deliberate practice and the acquisition and maintenance of expert performance in medicine and related domains. Academic medicine : journal of the Association of American Medical Colleges. 2004;79(10 Suppl):S70-81. Epub 2004/09/24.

[60] Crochet P, Aggarwal R, Dubb SS, Ziprin P, Rajaretnam N, Grantcharov T, et al. Deliberate practice on a virtual reality laparoscopic simulator enhances the quality of surgical technical skills. Ann Surg. 2011;253(6):1216-22. Epub 2011/04/26.
[61] Boyle E, Al-Akash M, Gallagher AG, Traynor O, Hill AD, Neary PC. Optimising surgical training: use of feedback to reduce errors during a simulated surgical procedure. Postgraduate medical journal. 2011;87(1030):524-8. Epub 2011/06/07.

Chapter 8

Virtual Reality Simulation: A Valuable Adjunct to Surgical Training

Bernadette McElhinney, Angela Beard, Krishnan Karthigasu and Roger Hart

Additional information is available at the end of the chapter

http://dx.doi.org/10.5772/50576

1. Introduction

Clinical experience has shown that surgeons need to perform a certain number of procedures to gain competency and continue performing a certain number of procedures to maintain these skills (1,2). More and more, it is becoming increasingly difficult to obtain an adequate amount of live operating, even for fully trained doctors. Reasons for this include reduced working hours, an increasingly consultant-led service, a better-educated patient body, with an increasing focus on their safety and rights. Previously junior doctors had ample opportunity to operate independently with indirect supervision from a more senior colleague; unfortunately this is becoming less common (3). Healthcare resources are becoming increasing scarce, which adversely affects the amount of theatre time that a trainee has access to (4). The European Working Time Directive led to a change in the working pattern of junior doctors in the UK with significant reduction in available hours and a greater proportion of their time spent in service provision. Furthermore, certain major operations are being replaced by less radical options such as a surgical/medical endometrial ablation replacing hysterectomy. Or alternatively, traditional surgery is being replaced by more sophisticated techniques, which experienced surgeons have to master prior to junior trainees having the opportunity to develop their skills. One example is robotic surgery and its incorporation into gynaecological minimal access surgery.

Surgical skills were traditionally acquired by practising on 'live' patients, but it is apparent that the operating room is not the ideal learning environment. Trainees are generally less time efficient than experienced surgeons with implications for theatre management and healthcare budget. The complication rate has been found to correlate with the experience of the surgeon (5-10) which is concerning with the ever-increasing emphasis on litigation. As surgeons become more experienced in laparoscopic surgery, the complication rate decreases and their ability to deal with complications in keeping with the minimal access approach increases (11). Even for experienced surgeons the learning curve for advanced laparoscopic procedures is fifty cases; total operative time for hysterectomies stabilised at approximately 95 minutes after fifty cases (12). The hypothesis that is being addressed is that training on a laparoscopic simulator shortens the learning curve, which has stimulated the development of simulation systems and their implementation into clinical practice.

2. Simulation in other professions

At present there is extensive knowledge about how to teach technical skills in professions where accurate and reliable performance is critical. High performance musicians and athletes on average invest 10 years of intense practice before they are considered experts (13). Surgeons, by comparison, are currently expected to 'perform' to a competent level without first practicing in a low-risk environment. Many doctors recall a familiar adage 'see one, do one, teach one'. Simulation-based training using flight simulators has been mandatory in the United States aviation industry since 1955 (14). All commercial and military pilots must train and be certified on a simulator before actual flight. Departments of Anesthesiology have applied principles similar to those used in pilot training with over 30 years of history in simulation-based training (15).

3. Physical simulation

Simulation can be described as an exercise that reproduces or emulates, under artificial conditions, components of surgical procedures that are likely to occur under normal circumstances (16). In the area of laparoscopic surgery, simulators fall into 2 broad categories: computer based simulators, in which the task is performed in a 'virtual' environment (17) and video-based simulators, in which the task is generally performed in a trainer box under videoscopic guidance. In the virtual reality simulator (VR), the student performs 'virtual' tasks in a computer-generated environment that allows sensory interaction. Unlike the box trainer, VR provides no tactile feedback (haptics). However, innate ability can be evaluated using computer-derived metrics; different aspects of performance can be analyzed at a later date (18). The main disadvantages of the VR simulator are the lack of portability, high start-up costs and ongoing maintenance. Physical simulators are widely available and include bench simulation, live animal model and human cadavers (19). The video or 'box' trainer (VT) is a basic training simulator in which users perform tasks with 'real' laparoscopic instruments under videoscopic guidance. Unlike the VR simulator, it is inexpensive, reproducible and provides 'haptic' feedback. Hybrid simulators combine both attributes of VT and VR simulators. Normally the hybrid simulation system incorporates a mannequin linked to a computer programme that provides visual images or feedback. This facilitates the creation of a realistic clinical environment where trainees can work as a team and respond to clinical situations.

4. Haptic technology

Haptic technology, or haptics, is a tactile feedback system that generates tactile sensations to the user. This mechanical stimulation can be used to assist in the creation of virtual objects in a computer simulation, to control such virtual objects, and to enhance the remote control of machines and devices (telerobotics). Haptics has been incorporated into VR simulators without compelling evidence that it adds benefit to training. This is significant because this technology costs a considerable amount of money in both the initial purchase of the equipment and the ongoing maintenance. Thompson et al (20) investigated the incorporation of haptics in virtual reality laparoscopic cholecystectomy training. Thirty-three laparoscopic novice students were placed into one of three groups: control, haptics-trained, or non-haptics trained group. The study found that haptics does not improve the efficiency or effectiveness of VR laparoscopic surgery training. They concluded that haptics should not be included routinely in surgery training. The strength of the study was weakened by the high attrition rate; more than 50% in the study groups but less than 10% in the control group. This was attributed to the time commitment involved and the technical difficulty encountered by the participants. Although the incorporation of haptic technology increases the financial burden, no clear benefit with respect to training has been demonstrated (21,22). A systemic review (23) reported that although the majority of results show a positive advantage from haptic technology in MIS, interarticle consensus is neither absolute nor firm. Furthermore, the general level of evidence was poor (level 3b). More objective study results based on valid end parameters need to be obtained to reliably report the value of haptic feedback.

5. Future developments

Telesurgery is a developing field for potential use in remote sites such as the battlefield and in space, once technology improves. Telesurgery will enable surgeons to operate on patients who are physically separated from them. Most of the research to date has been carried out on animals. A prototype telemanipulator has been used successfully to perform basic vascular and urological procedures in pigs (24). At present there is no role for the use of telesurgery in surgical practice due to the reliance on telephone line technology and telephone companies. Telesurgery will only become possible when surgery becomes digital with failsafe communications (25).

Another exciting area in laparoscopic surgery is the emergence of robotic assisted operations. The application of robotics provides surgeons with a remarkable three-dimensional image. The surgeon is able to sit comfortably and perform operations without the risk of soft tissue strain and fatigue, a common occurrence in laparoscopic surgeons who regularly partake in complex and prolonged operations. The consoles and instruments are very sensitive to movement and the awkward motions of minimally invasive procedures can be translated into natural hand motions from a surgical workstation. However these machines are extremely expensive to buy and require a large amount to space to operate because of the sheer size of the robotic arms. Entire operating theatres are taken up by the enormity of these robots. There are also some practical limitations, including not being able to change the position of your port sites once the operation has started.

These developments in surgical technology will influence and develop current practice. As these new technologies are validated there will be a new richness to surgery that will require even more surgical skills and training. The practice of surgery will not be replaced but will change and evolve. As a surgeon, the challenge is to be aware of the opportunities, rigorously evaluate the technologies and be willing to change if evidence-based outcomes demonstrate a clear benefit for the patient (26)

6. Simulation in surgery

Operative skill is a mixture of knowledge, clinical judgment and technical skill. As traditional surgical techniques are being replaced by minimal access approaches, surgeons in training need to adapt to this new technology. Minimal access surgery presents new challenges to the trainee surgeon such as operating in a 2-dimensional environment, reduced tactile feedback, new instrumentation and the 'fulcrum' effect (27). Fortunately minimal access surgery is amenable to simulator training and the benefits of simulator training are numerous. Laparoscopic simulators provide a safe, protected, unhurried environment where trainees can operate independently. Tasks can be presented consistently allowing the development of laparoscopic skills irrespective of prior surgical experience, sex or age (28). Laparoscopic skill can be measured on a simulator and performance improved with practice (most of this improvement was a result of speed rather than accuracy). The effect of repetition on performance overall and for each task individually was highly significant, confirming the simulator model as a valuable practice tool (29).

7. Acceptance of surgical simulation

Simulation-based training is becoming widely available to help trainees develop sound technical skills before they practice on real patients. Although it provides a nonthreatening, controlled environment, it is not being widely accepted into current clinical practice. An important issue is how to create optimal conditions for integration of simulators into the training curriculum. The willingness of twenty-one surgical residents to train on a voluntary basis was surveyed. Access was unrestricted for a period of 4 months, following which a competitive element was introduced. Free unlimited access to a VR simulator, without any form of obligation or assessment, did not motivate surgical residents to use the simulator; introducing a competitive element had only a marginal effect. The majority of residents (86%) stated that 'lack of time due to high working pressure' was the most important reason for not using the simulator. Therefore, the acquisition of expensive devices is probably only effective if it becomes a compulsory part of the training curriculum (30). Recent studies stated that trainees prefer video box trainers to virtual reality, citing better visualization and tactile feedback that made video box trainers more realistic; it should be the first choice if only one trainer was allowed (31,32).

8. The evidence for computer-based simulation

Virtual reality simulation allows trainees to interact efficiently with three-dimensional, deformable, computerized databases in real-time, using their natural senses and skills (33). Their application is more evident in laparoscopic as opposed to 'open' surgery. Surgical simulation provides the appropriate environment where very complex surgical procedures can be broken down into several simple tasks with the opportunity for mass and deliberate practice. Multiple repetitions of a skill, such as laparoscopic suturing, are needed to acquire the necessary hand eye coordination and muscle memory. There is evidence that computer-based surgical simulation leads to improved performance in complex laparoscopic tasks like suturing (34). This leads to decreased task completion time and increased accuracy. An important advantage of computer-based simulation is its ability to generate out-put data which reflects competence of the trainee and can be used for performance assessment.

A meta-analysis by Haque and Srinivasan (35) analysed 16 prospective and randomized studies for the effectiveness of VR simulation. The author's goal was to evaluate the effectiveness of surgical simulation and to assess the validity of current simulation. The authors found that surgical simulation was not superior to standard 'Heilsteidan' training methods. Their work suggested that training in VR simulators lessens the time taken for a given surgical task and clearly differentiates between experienced and naïve surgeons. However the authors sited several systematic problems as potential reasons for the failure of studies to show significant advantages of simulation technology including small sample sizes, low statistical power, lack of accepted validity measures, non blinded assessors and poor funding.

A recent Cochrane Review (36) of randomized, controlled trials investigated the effectiveness of simulation-based training interventions. The authors felt that until standards are adopted for establishing and reporting performance evidence from rigorous psychometric assessment instruments, the literature examining the efficacy of simulation-based surgical training will be limited. Although research of higher methodological quality is needed, the evidence would suggest that VR training improves standard surgical training with preliminary data supporting the concept that these skills translate into more effective operating room performance (17). A study by Larsen et al (37) showed that criterion based procedural training using a virtual reality simulator can help compensate for reduced working hours by bringing trainees to a higher level of performance more quickly.

9. Computer-based versus video box trainers

There is currently no universally accepted or recommended single model for laparoscopic simulation (38). Video box trainers seemed to be equally efficient as virtual reality simulators (39,40). In the systemic review by Sutherland et al (41), including 30 RCTs (760 participants), individuals trained in VR performed better than no training. The effect was less marked when compared with standard laparoscopic training; VR vs. VT no conclusive results. The Cochrane review by Gurusamy et al (42) that included 23 trials concluded that VR training can supplement standard laparoscopic training, and it is as least as effective as VT.

Youngblood et al (43) randomly assigned 46 surgically naïve medical students to three groups: tower training, VR (Lapsim) and the control group. The time and accuracy of three laparoscopic tasks in a living animal model were assessed; four experienced surgeons evaluated performance. Trained groups performed better compared to the control group but not for all outcomes measured. The authors reported that surgically naïve medical students (n=46) trained on a VR simulator performed better on three of seven outcome measures during live surgical tasks in a porcine model as compared with those trained with a box trainer (time, accuracy and global score; $p<0.05$).

Although training on both VR and VT effectively improves psychomotor skills, a trend towards greater improvement was found with the MIST VR that was transferable to the OR. Fifty surgical trainees were randomized to either a VT or VR trainer. The effect of task training was assessed via a pre- and post-test assessment on VT, VR and intraoperative assessment during laparoscopic cholecystectomy. Although both groups improved, operative performance improved only in the VR group ($p<0.05$). Furthermore, the VR group performed significantly better when tested on VT tasks suggesting that skills developed on one system appear to be transferable to the other modality (32).

The systemic review by Gurusamy et al (36) examined whether virtual reality training can supplement or replace conventional laparoscopic surgical training in trainees with little or no experience. Results were reported separately for trainees with no laparoscopic experience and for those with limited experience. The review included 23 randomised, controlled trials (612 participants). Four trials compared VR with VT, 12 trials compared VR with standard laparoscopic training (SLT), four trials compared VR, VT and no training and three trials compared different methods of VR training. Generation of allocation sequence, allocation concealment, blinding and follow-up were examined. Three trials that had adequate methodological quality in all four components were considered to have a low risk of bias. Five different parameters were examined in the VR vs. SLT group (limited laparoscopic experience) namely patient outcome, operating time, error score, composite score and economy of movements. Operating time was statistically significantly shorter in the VR group in two trials; five trials reported a statistically significantly lower error score in the VR group. Although there were methodological flaws with the majority of trials included, the author reported that virtual reality decreased time, decreased errors and increased accuracy compared with no training. The authors concluded that the advantages of VR over VT are not as evident as for VR over standard training. Virtual reality training should supplement standard laparoscopic surgical training. Common problems in studies to date include lack of universally agreed metrics, a variety of simulators, differing skill levels of participants, and small sample sizes. Despite this, most studies are in keeping with the positive impact of laparoscopic surgical simulation.

An important consideration in our era of financial restraints is the consideration of cost of surgical simulation. The hospital administration needs to be convinced that simulation will be cost effective before funding is made available. The main consideration is whether the low-tech, inexpensive video box trainer is as good as the considerably more expensive virtual reality trainer with the ability to provide haptic feedback for continual assessment? In a study performed at the University of Toronto (44), urology trainees were randomized to three types of training to extract a urethral stone. The first group received detailed instructions only, the second group was trained in a high fidelity virtual reality model and the third group was trained on a low-tech model using Styrofoam cups and straws placed in the anatomical orientation of the normal bladder. All participants were subsequently tested on the VR fidelity video endoscopic trainer. The two groups with hands-on teaching on either trainer did better than the group who received instructions only. Training in the low fidelity model conferred as much benefit as training on the high fidelity model. This evidence was backed up by a study performed by Goff et al (45) with respect to assessment of hysteroscopy skills; assessment in the low-tech trainer was actually better than assessment in the expensive virtual reality trainer.

10. Transfer of training

The main goal of any training method is the positive transfer of skills to the operating room. So does laparoscopic simulator training translate into improved operative performance? High-grade evidence on the effect of virtual reality simulator training on real operation performance was limited until now. The evidence that simulation training actually translates to improved surgical skills in the operating room is increasing and several studies now prove that laboratory based training improves surgical skills. Two studies compared simulator training and concurrent operating room performance in the porcine model (46,47). Grantcharov et al (46) assessed fourteen residents on an animal model with pre- and post-training on a VR model. The study demonstrated that in vitro scores for VR tasks are comparable to performance during operations on living animals. Although sample size was small and assessors were not 'blinded', the study suggests that the computer model shows promise as an aid to evaluate and assess trainee surgeons. Good correlation was found between performance in MIST-VR and cholecystectomy. A later study by Seymour et al (48) was one of the first studies to demonstrate a significant improvement in OR performance of residents. In a prospective, randomised, blinded study, sixteen surgical residents were randomized to VR training plus standard laparoscopic training (SLT) or control (SLT only). The training goal for the residents in the VR group was to perform as well as four experienced surgeons on the 'manipulate and diathermy' task on two consecutive trials. The assessors were blinded to training status. Gallbladder dissection was 29% faster in the trained group. The authors concluded that use of VR significantly improved OR performance of residents during laparoscopic cholecystectomy. The above evidence is also supported by Reznick et al (15) who showed that VR training significantly improves a resident's ability to perform a laparoscopic cholecystectomy with a reduced rate of errors, higher economy of movement scores and faster dissection than residents with no training.

There have also been studies in gynaecology training programs that show laboratory-based training improves technical skills in a clinical setting. A core curriculum of intensive video laparoscopic skills training improved not only technical but operative performance among residents A prospective randomised trial by Coleman & Muller (49) recruited obstetrics and gynaecology residents (skills cohort, 11; control cohort, 7) to laboratory based training for laparoscopic salpingectomy for treatment of ectopic pregnancy, compared to routine surgical training in residency. The aim of the study was to determine the effect and validity of an intensive laboratory-based laparoscopic skills training curriculum on operative proficiency. Study components included a baseline questionnaire, video skills testing, intraoperative skills assessment and resident skills perception. The residents that were assigned to a laboratory based skills curriculum had significantly higher ratings when performing a laparoscopic salpingectomy on patients. This study demonstrated that a short-term intensive laboratory-based video laparoscopic skills curriculum could translate into better individual operative proficiency. Banks et al (50) randomly assigned residents to a laboratory based surgical curriculum to teach laparoscopic tubal ligation versus routine surgical training. At baseline there were no differences in skills between the two groups. After completion of the curriculum, facility members blinded to the knowledge of which training the resident had received, assessed all residents in the operating room as to their ability to perform laparoscopic tubal ligation. Residents assigned to the simulation training obained higher scores compared with the control group.

There is also good evidence to support the positive transfer of surgical skills after training with VR simulation. A recent study by Larsen et al (37) proved that skills in laparoscopic surgery could be increased in clinically relevant manner using proficiency based virtual reality simulator training. These researchers performed a prospective, randomized, observer blinded, controlled trial. A group of junior gynaecology registrars were divided into a control group and an intervention group (trained to proficiency on a VR simulator). The intervention group was given seven hours of training outside the normal service setting and was found to perform their first laparoscopy on a patient up to intermediate level (20-30 cases). The control group performed at a novice level (0-5 cases) and took twice as long to complete the procedure. The results showed that the performance of novices was increased to the level of intermediate experienced laparoscopists and the operation time to complete the task was halved. They were able to show that VR training in laparoscopic salpingectomy, compared with standard clinical education, was associated with a clinically important improvement of operative skills during the actual procedure. The learning curve in the OT was also shorter. By using simulator training it might be possible to bypass the early learning curve, which is known to be associated with a higher number of complications. These results also show that criterion based procedural training on a VR simulator can help compensate for reduced working hours by bringing trainees to a higher level of performance before they start training. They concluded that simulator training should be considered before trainees carry out laparoscopic procedures. This is possibly the first well-designed trial to show benefit from simulation in surgical training, and therefore has huge implications for the future.

Recently there have been a few structured reviews published which appraise the current value of simulation, their incorporation into the surgical curriculum and aim to address the question regarding positive transfer of skills (19,35,41,42,49,51-53). A systemic review by Sturm et al (51) attempted to determine whether skills acquired by virtual-reality training are directly transferable to the operative setting. Eleven studies were included; ten RCTs and one non-randomised comparative study. In most cases, simulation-based training was in addition to normal training programs. In conclusion, there is an overall positive effect of simulation-based training on the actual OR performance, although for some parameters transference was not demonstrated. Other systemic reviews have shown that there is a positive transfer of skills from the 'simulated' to the 'actual' operating environment, but only for certain surgical procedures (cholecystectomy, colonoscopy and sigmoidoscopy). A recent systemic review concluded that VR training improves standard surgical training (36) with preliminary data supporting the concept that these skills translate into more effective operating room performance (17,50). The methodologies were flawed weakening the strength of the conclusion.

11. Practice distribution

Practice distribution refers to the schedule of practice that a trainee is given. 'Distributed practice' refers to a practice schedule in which periods of training are interspersed with rest periods; 'massed practice' refers to a continuous block of uninterrupted training (54,55). With regard to the effectiveness of laparoscopic simulator training, it is unclear whether it is preferable to undergo 'distributed' training or 'massed' training. Meta-analytic reviews indicated that distributed training resulted in better retention of motor skills than massed training, although this difference was dependent on the tasks trained (56). MacKay et al (55) examined the effect of practice distribution in the medical setting. Forty-one novice subjects were randomised into one of three groups to train on a VR simulator. Group A trained for 20 min continuously (n=14), group B trained for 20 min in 5 min blocks (n=14) and group C trained for 15 min in 5 min blocks (n=13). Post training, all groups had a rest followed by a retention test. The authors reported that distributed endoscopic training on MIST VR, with short breaks, was superior to continuous training within one single day (p=0.023), as determined by the retention test. A later study randomly assigned students with no endoscopic experience to distributed VR training on three consecutive days (n=10) or distributed training within 1 day (n=10). The training involved 12 repetitions of three different exercises in three differently distributed training schedules. All students performed a post-test on a VR simulator seven days after training; three technical parameters were measured. The group with training over several days performed faster (p=0.013), with the same number of errors and instrument path length used suggesting that 'rest' results in better consolidation of skills (54). It would appear that distributed training is more effective than massed training, and over several days rather than training on one day, potentially having implications for workshop based programmes.

12. Training methods

Harold et al (57) compared two methods of instruction in a randomised fashion for the teaching of laparoscopic intracorporeal knot tying. The intervention group in this study received instruction by lecture, video, and individual proctoring, which was compared with instruction by manual alone. The intervention group performed better than the control group in this study. Participants in the intervention group had the advantage of not only better understanding through the use of video, but also the advantage of practice and proctoring, which allowed their understanding to be translated into performance.

Other recent randomised trials further reinforce the point that conceptual understanding and technical performance are both important elements of laparoscopic proficiency. Stefanidis et al (39) reported that a combination of video tutorials and limited feedback were the most efficient way to reach proficiency in a laparoscopic suturing curriculum. Korndorffer et al (58) reported that participants who received video and practiced performed better than those who received instruction by video alone. Leung et al (59) tested the efficacy of video as an educational tool in laparoscopic training. This RCT compared text versus video alone for a laparoscopic procedure. The results showed that video is superior to text alone for achieving quicker and better understanding and greater competency at performing laparoscopic tasks.

Snyder et al (60) randomised 36 medical students into independent or proctored training groups (n=18); no significant differences in demographics. Simulator proficiency was reached after a median of eleven hours of training (range 6-21 hrs.). Trainees in the independent group achieved proficiency with significantly fewer hours of training (HR 2.62; 95% CI, 1.01-6.85; p=0.048). The authors concluded that for proficiency-based VR simulator training, an independent approach was just as effective and potentially less time consuming for trainees than a proctored approach.

13. Skill retention

It is widely accepted that laparoscopic skills improve after simulator training, however little is known regarding skill retention. Surgical competency depends on a combination of procedural knowledge and skill retention. A meta-analysis found that performance decay increased with longer retention intervals; a 92% skill loss at one year was documented (61). More recently, Maagaard et al (62) looked at two groups (novices and experts) who performed 10 sessions on the LapSim VR. Assessment of skill was based on time, economy of movement and error. The authors reported that, although novices showed retention of skills after 6 months, after 18 months, laparoscopic skills had returned to pre-training levels. Sinha et al (63) documented the retention of motor skills over time in 33 surgical residents who trained to established criteria (and passed an exam) on seven technical skills on a VR. Six months after training the residents underwent repeat testing. At retest, significantly more residents failed clip applying and cutting tasks ($p<0.05$). In failed tests, instrument and tissue handling skills deteriorated more than the speed with which a task was completed. Evidence of skill retention was present for some but not all tasks. Fine motor skills deteriorated more than skills needed for easier tasks. Residents were less likely to fail with increasing experience. Stefanidis et al (39) noted that there was a paucity of literature on skill retention and comparison of the durability of skill between VR and VT simulators. Fourteen surgical residents of varying levels were enrolled to train on VR and VT simulators until proficiency levels were achieved. VR scores were generated automatically and VT scores were based on completion time. Skill retention was evaluated by performing one task on both the VR (manipulate diathermy) and VT (bean drop) simulators. Skill acquisition was similar for both systems (Improvement:VR 59% vs. VT 56%). Despite an early performance decrement (VR 45% vs. VT 17%) the acquired skill persisted over a seven-month follow-up period. There was no correlation of skill loss with resident level, duration of training or any of the other parameters. The authors concluded that proficiency-based training on simulators results in durable skills, more so for VT than VR.

14. Assessment of Skills

The variable anatomy and different degrees of difficulty in live patients makes consistent assessment of technical skills in surgical trainees difficult. Traditionally trainee surgeons have been assessed by an 'expert' colleague, a process which is subjective and potentially prone to bias. A 'gold standard' for OR performance does not exist. Although improvement in surgical skill is usually reported, the extent of the improvement is hard to quantify. Therefore it is difficult to establish the effectiveness of simulation. More recently, studies have demonstrated the value of VR simulators for providing an objective assessment tool (64-67). Smith et al (68) developed a skills assessment device (SAD) incorporating VR and VT technology to quantify both speed and accuracy during laparoscopic skill performance. Untrained subjects performed ten repetitions of a standardised laparoscopic task. Task time improved dramatically during the first three repetitions and then stabilised. However, accuracy continued to improve. The authors concluded that although the time to perform a laparoscopic task improved more quickly than the accuracy of task completion, time alone is poor indicator of technical skill as it fails to account for the more protracted learning curve for accuracy. In their opinion, time was not a sufficient measure of proficiency.

Faster completion of a task does not presuppose accurate performance of the task (61,69). Both safety and accuracy need to be considered when assessing technical skills. Although a fast surgeon is not necessarily a safe surgeon, the idea that experience is related to greater efficiency of motion has face validity. Twenty-four subjects, with varying level of experience, were divided into three groups (naïve, junior, expert) depending on the volume of surgical experience. The results indicated that the 'time-error' scores are a valid measure of performance and improved significantly from baseline to final iteration in all groups. On all tasks, the 'expert' group performed significantly better than the naïve group (64). Furthermore, shorter times are indicative of familiarity and confidence with the instruments (28). Hyltander et al (70) found that students who performed tasks accurately also needed the least time. A technically skilled surgeon is one who executes a task quickly, is economical in movements as well as being precise (33). Studies have shown that performance, measured by either a subjective rating or time on a task, improve with practice (29,71,72).

15. Factors predicting performance

In laparoscopic surgery it is important to develop the ability to use both hands equally well. Many basic laparoscopic skills demand dexterity in both hands for successful completion of the task. Powers et al (73) assessed whether hand dominance had any effect on performance in a laparoscopic skills curriculum. Twenty-seven surgical residents underwent a four-week laparoscopic skills curriculum with pre- and post testing on six tasks during weeks one and four. During week two and week three, residents attended proctored practice sessions. The authors concluded that participation improved overall performance. The left-handed surgeons demonstrated better initial performance, but post-test comparison showed no difference.

Grantcharov et al (74) assessed impact of gender, hand dominance and computer games experience on psychomotor skills demonstrated with a VR simulator. Male surgeons were faster; no significant difference between genders in the number of errors and unnecessary movements was noted. Right-handedness was associated with fewer unnecessary movements. Computer game users made fewer errors than non-users. In a study by Derossis et al (29), forty-two surgeons viewed an introductory video, and then were tested performing seven laparoscopic tasks. Performance was measured using a scoring system rewarding precision and speed. Each candidate repeated all seven tasks and was rescored. Significant predictors of overall performance were level of training (p=0.002), repetition (p=0.0001) and interaction between level of training and practice (0.001). Construct validity was demonstrated by measuring significant improvement in performance with increasing residency training, and with practice.

As well as dealing with the stress of live operating, surgical trainees have to deal with many other stressors including unfavourable working patterns, sleep deprivation, large volumes of work and time pressures, concerns about patient outcomes, surgical emergencies or complications, team challenges, miscommunication and so forth. Andreatta et al (75) demonstrated that simulation provides an opportunity for trainees to manage stress in practice. They observed 27 medical students completing tasks using a laparoscopic simulator under two conditions; direct observation (stressor) and unobserved (no stressor). A simple stimulus of an evaluator observing the completion of a task incurred a stress reaction in terms of elevated heart rate and increasing performance errors. This has implications for training and assessment in the simulated context in that stressors imposed on the learner in a simulated environment may help support the acquisition of stress management skills that are necessary in the applied clinical setting. Exactly how these stressors influence surgical performance is not well understood, but simulation could be used to teach the trainee how to manage stress by developing coping mechanisms early in their training.

A primary aim for trainees is to practice skills in a safe and non-threatening environment. Evidence is accumulating which demonstrates a positive learning curve and improved surgical skills after training on surgical simulators. The availability of surgical simulators means that they can be incorporated into the surgical training curricula, and enable learning curves to be consigned to skills laboratories, away from live patients. The implementation of laparoscopy into residency training is difficult to achieve because of time and financial constraints. However, the benefits of simulator training seem to be greatest for the most inexperienced surgeons, in acclimatizing to the 2D environment, new instrumentation and the fulcrum effect (76). This would suggest that the ideal time to introduce the concept of surgical skills training to trainees is during their surgical attachments as a medical student. Bearing in mind that simulation is an adjunct to, not a replacement for, traditional methods of training. Supervision and feedback are essential (77).

16. In summary

Some studies support the role of VR in surgical skills training; others support VT and claim that a greater improvement in skills acquisition occurs. Other studies show no difference between the two methods. Overall it is unclear which method is superior. The bottom line is that research has demonstrated that practice in surgical simulators leads to improved performance. Furthermore, there is evidence to show that simulator training translates into improved operative performance, but in a limited number of procedures. Further good quality studies are needed to strengthen the evidence base in support of the various types of surgical simulation, and to establish to what extent simulation should be part of the surgical training program.

17. In conclusions

The increasing popularity of MAS makes it imperative that junior doctors have ample opportunity to master basic laparoscopic skills. At present, despite three decades of development, MAS training is still rather primitive. Worldwide, surgical simulators are playing an increasing role in the training of junior doctors. Evidence is increasing on the nature of the acquisition of surgical skill through the use of simulators rather than the traditional approach. The optimal timing and means of acquiring and retaining these skills to ensure optimal transfer of skill to the operating room is unknown.

Author details

Bernadette McElhinney
Department of Gynaecology, King Edward Memorial Hospital, Subiaco, Perth, Australia

Angela Beard
Department of Gynaecology, King Edward Memorial Hospital, Subiaco, Perth, Australia

Krishnan Karthigasu
Department of Gynaecology, King Edward Memorial Hospital, Subiaco, Perth, Australia

School of Women's and Infants' Health, University of Western Australia, King Edward Memorial Hospital, Subiaco, Perth, Australia

Roger Hart
School of Women's and Infants' Health, University of Western Australia, King Edward Memorial Hospital, Subiaco, Perth, Australia

18. References

[1] Davies RJ, Hamdorf JM. Surgical Skills Training and the Role of Skills Centers. British Journal of Urology. 2003;91:3-4

[2] Halm EA, Lee C, Chassin MR. Is volume related to outcome in healthcare? A systemic review and methodologic critique of the literature. Annals of Internal Medicine. 2002;137:511-520

[3] Rogers RM, Julian TM. Training the Gynecologic Surgeon. Obstetrics and Gynecology. 2005, 105; 197-200

[4] Reznick RK. Teaching and Testing Surgical Skills. The American Journal of Surgery. 1993, 165;358-361

[5] Shen C-C, Wu M-P, Kung F-T, Huang F-J, Hsieh C-H, Lan K-C, Huang E-Y, Hsu T-Y, Chang S-Y. Major complications associated with laparoscopic –assisted vaginal hysterectomy: ten-year experience. The Journal of the American Association of Gynecologic Laparoscopists. 2003;10:147-153

[6] Chapron C, Cravello L, Chopin N, Kreiker G, Blanc B, Duboisson JB. Complications during set-up procedures for laparoscopy in gynecology: open laparoscopy does not reduce the risk of major complications. Acta Obstetricia et Gynecologica Scandinavica. 2003:82;1125–1129

[7] Brosens I, Gordon A. Bowel Injuries during Gynaecological Laparoscopy: a multinational survey. Gynaecological Endoscopy, 2001;10:1441-1452

[8] Makinen J, Johansson J, Tomas C, Tomas E, Heinonen PK, Laatikainen T, Kauko M, Heikkinen A-M, Sjoberg J. Morbidity of 10110 hysterectomies by type of approach. Human Reproduction. 2001;16:1473-1478

[9] Jansen FW, Kapiteyn K, Trimbos-Kemper T, Hermans J, Trimbos JB. Complications of Laparoscopy: a prospective multicentre observational study. British Journal of Obstetrics and Gynaecology. 1997;104:595-600

[10] Chapron C, Querleu D, Mage G, MAdelenat P, Dubuisson JB, Audebert A, Erny R, Buhat MA. Complications of Gynaecologic Laparoscopy. Multicentre study of 7604 laparoscopies. Journal of Gynaecology, Obstetrics and Reproductive Biology. 1992;21:207-213

[11] Chapron C, Querleu D, Bruhat MA. Complications of gynaecological laparoscopy. A French collaborative study. Gynaecological Endoscopy 1998;13:867-72

[12] Lenihan JP, Kovanda C, Seshadri-Kreaden U. What is the Learning Curve for Robotic Assisted Gynecologic Surgery? The Journal of Minimally Invasive Gynecology. 2008;15:589-594

[13] Goff BA. Changing the Paradigm in Surgical education. Obstetrics and Gynaecology. 2008; 112 (1): 328-332.

[14] Kaufmann Cr. (2001), Computers in surgical education and the operating room. Annals Chirurgiao et Gynaecologiae, 2001; 90 (2): 141-146

[15] Resnick RK. Teaching and testing clinical skills. Am J Surg, 2003; 165: 358-61.

[16] Krummel TM. Surgical simulation and virtual reality: the coming revolution. Annals of Surgery. 1998;228:635-637

[17] Feldman LS, Sherman V, Fried GM. Using simulators to assess laparoscopic competence: Ready for widespread use? Surgery. 2004;135:28-42
[18] Oostema JA, Abdel MP, Gould JC. Time-efficient laparoscopic skills assessment using an augmented-reality simulator. Surgical Endoscopy. 2008;22:2621-2624
[19] Laguna MP, de Reijke TM, de la Rosette J. How Far Will Simulators be involved into Training? Endourology. 2009;10:97-105
[20] Thompson JR, Leonard AC, Doarn CR, Roesch MJ, Broderick TJ. Limited value of haptics in virtual reality laparoscopic cholecystectomy training. Surgical Endoscopy. 2011;25:1107-1114
[21] Salikini MW, Doarn CR, Kiehl N, Broderick TJ, Donovan JF, Gaitonde K (2010) The role of haptic feedback in laparoscopic training using the lapMentor 2. Journal of Endourology 24; 99-102
[22] Panait L, Akkary E, Bell RL, Roberts KE, Dudrick SJ, Duffy AJ (2009) The Role of haptic feedback in laparoscopic simulation training. Journal of Surgical Research 156;312-16
[23] Van der Meijden OAJ & Schijven MP. The value of haptic feedback in conventional and robotic-assissted minimal invasive surgery and virtual reality training: a current review. Surgical Endoscopy 2009;23:1180-1190
[24] Gorman PJ, Meier AH, Krummel TM. Arch Surg, 2009; 134: 1203-1208.
[25] Schlag PM, Moesta KT, Rakovsky S Graschew G. Telemedicine, The new must for Surgery, Arch Surg, 1999; 134: 1216-1221.
[26] Satava RM. Emerging Technologies for surgery in the 21st Century Arch Surgery, 1999; 134: 1197-1202.
[27] Gor M, McCloy R, Stone R, Smith A. Virtual Reality Laparoscopic Simulator for Assessment in Gynaecology. British Journal of Obstetrics and Gynaecology. 2003;110:181-187
[28] Rosser JC, Rosser LE, Savalgi RS. Objective evaluation of a laparoscopic surgical skill program residents and senior surgeons. Archives of Surgery. 1998;133:657-661
[29] Derossis, AM, Fried GM, Abrahamowicz M, Sigman HH, Barkun JS, Meakins JL. Development of a model for training and evaluation of laparoscopic skills. The American Journal of Surgery. 1998;175:482-487
[30] Van Dongen KW, van der Wal WA, Borel Rinkes IHM, Schijven MP, Broeders IAMJ. Virtual reality training for endoscopic surgery: voluntary or obligatory? Surgical Endoscopy 2008;22:664-667
[31] Madan AQ, Frantzides CT, Tebbit C, Quiros RM. Participants opinion on laparoscopic training devices after a basic laparoscopic training course. The American Journal of Surgery 2005;189:758-761
[32] Hamilton EC, Scott DJ, Fleming JB, Rege RV, Laycock RM, Bergen PC, Tesfay ST, Jones DB. Comparison of video trainer and virtual reality training systems on acquisition of laparoscopic skills. Surgical Endoscopy. 2002;16:406-411
[33] Moorthy K, Munz Y, Sarker SK, Darzi A. Objective Assessment of Technical Skills in Surgery. British Medical Journal. 2003;327:1032-1037
[34] Kothari SN, Kaplan BJ, DeMaria EJ, Broderick TJ, Merrell MD. (2002) Training in laparoscopic suturing skills using a new computer based virtual reality simulator

(MIST-VR) provides results comparable to those with an established pelvic trainer system. Journal of Laparoendoscopic & Advanced Surgical Techniques, 2002; 12: (3): 167-169.

[35] Haque S & Srinivasan S. A meta-analysis of the training effectiveness of virtual reality simulators. IEEE Transactions on Information Technology in Biomedicine. 2006;10:51-58

[36] Gurusamy KS, Aggarwal R, Palanivelu L, Davidson BR. Virtual Reality training for surgical trainees in laparoscopic surgery. Cochrane Database of Systemic Reviews. 2009, Issue 1. Art. No: CD006575

[37] Larsen CR, Soerensen JL, Grantcharov TP, Dalsgaard T, Schouenborg L, Ottosen C, Schroeder TV, Ottesen BS. Effect of virtual reality training on laparoscopic surgery: randomised controlled trial. BMJ. 2009;338:1-6

[38] Feldman LS, Sherman V, Fried GM. Using simulators to assess laparoscopic competence: Ready for widespread use? Surgery. 2003;135:28-72

[39] Stefanidis D, Korndoffer JR, Sierra R, Touchard C, Dunne B, Scott DJ. Skill retention following proficiency based laparoscopic simulator training. Surgery. 2005;165:165-170

[40] Munz Y, Kumar BD, Moorthy K, Bann, Darzi A. Laparoscopic Virtual Reality and box trainers: is one superior to the other? Surgical Endoscopy. 2004;18:485-494

[41] Sutherland LM, Middleton PF, Anthony A, Hamdorf J, Cregan P, Scott D, Maddern GJ. Surgical Simulation A systematic Review. Annals of Surgery. 2006;243:291-300

[42] Gurusamy KS, Aggarwal R, Palanivelu L, Davidson BR. Virtual Reality Training for Surgical Trainees in Laparoscopic Surgery. The Cochrane Database of Systemic Reviews. 2007, 3:CD006575

[43] Youngblood PL, Srivastava S, Curet M, Heinrichs WL, Dev P, Wren SM. Comparison of training on two laparoscopic simulators and assessment of skills transfer to surgical performance. Journal of the American College of Surgeons. 2005;200:546-551

[44] Matsumoto ED, Hamstra SJ, Radomski SB, Cusimano MD. The effect of bench model fidelity on endourological skills: a randomized controlled study,. Am J Surg, 2003; 185: 378-385.

[45] Goff BA, VanBlaricom A, Mandel L, Chinn M, Nielsen P. Comparison of objective structured assessment of technical skills with virtual reality hysteroscopy trainer and standard latex hysteroscopy model. J Reprod Med, 2007: 52: 407-12.

[46] Grantcharov TP, Rosenberg J, Pahle E, Funch-Jensen PM. Virtual reality computer simulation-an objective method for evaluation of laparoscopic surgical skills. Surgical Endoscopy. 2001;15:242-244

[47] Fried GM, Derossis AM, Bothwell J, Sigman HH. Comparison of laparoscopic performance in vivo with performance measured in a laparoscopic simulator. Surgical Endoscopy. 1999;13:1077-1081

[48] Seymour NE, Gallagher AG, Roman SA, O'Brien MK, Bansal VK, Andersen DK, Satava RM. Virtual Reality Training Improves Operating Room performance: Results of a Randomized, Double-Blinded Study. Annals of Surgery. 2002;236:458-464

[49] Coleman RL & Muller CY. Effects of laboratory-based skills curriculum on laparoscopic proficiency: A randomised trial. American Journal of Obstetrics and Gynaecology. 2002;186:836-842

[50] Banks EH, Chudnoff S, Karmin I, Wang C, Pardanani S. Does a surgical simulator improve resident operative performance of laparoscopic tubal ligation? Am J Obstst Gynecol , 2007; 197: 541, e1-5.
[51] Sturm LP, Windsor JA, Cosman PH, Cregan P, Hewett PJ, Maddern GJ. A Systematic Review of skills transfer after surgical simulation training. Annals of Surgery. 2008;248:166-178
[52] Gurusamy K, Aggarwal R, Palanivelu L, Davidson BR. Systemic review of randomised controlled trials on the effectiveness of virtual reality training for laparoscopic surgery. British Journal of Surgery. 2008, 95; 1088-1097
[53] Paisley AM, Baldwin PJ, Paterson-Brown S. Validity of surgical simulation for the assessment of operative skill. British Journal of Surgery. 2001;88:1525-1533
[54] Verdaasdonk EGG, Stassen LPS, van Wijk RPJ, Dankelman J. The Influence of different training schedules on the learning of psychomotor skills for endoscopic surgery. Surgical Endoscopy. 2007;21:214-219
[55] MacKay S, Morgan P, Datta V, Chang A, Darzi A. Practice distribution in procedural skills training: a randomised controlled trial. Surgical Endoscopy. 2002;16:957-961
[56] Lee TD & Genovese ED. Distribution of practice in motor skill acquisition: learning and performance effects reconsidered. Research Quarterly. 1988;59:277-287
[57] Harold KL, Matthews BD, Backus CL, Pratt BL, Heniford BT. Prospective randomised evaluation of surgical resident proficiency with laparoscopic suturing after course instruction. Surgical Endoscopy. 2002;16:1729-1731
[58] Korndorffer JR, Dunn JB, Sierra DR, Stefanidis D, Touchard C, Scott D. Simulator training for laparoscopic suturing using performance goals translates to the operating room. The American Journal of Surgery. 2005;201:23-29
[59] Yeung P, Justice T, Paya Pasic R. Comparison of text versus video for teaching laparoscopic knot tying in the novice surgeon: A randomised controlled trial. The Journal of Minimally Invasive Gynaecology. 2009;16:411-415
[60] Snyder CW, Vandromme MJ, Tyra SL, Hawn MT. Proficiency-based laparoscopic and endoscopic training with virtual reality simulators: a comparison of proctored and independent approaches. Journal of Surgical Education. 2009;66:201-207
[61] Arthur W, Bennett WJ, Stanush PL. Factors that influence skill decay and retention: a quantitative review and analysis. Human Performance. 1998;11:57-101
[62] Maagaard M, Sorensen JL, Oestergaard J, Dalsgaard T, Grantcharov TP, Ottesen BS, Larsen CR. Retention of laparoscopic procedural skills acquired on a virtual-reality surgical trainer. Surgical Endoscopy 2011;25:722-727
[63] Sinha P, Hogle NJ, Fowler DL. Do the laparoscopic skills of trainees deteriorate over time? Surgical Endoscopy. 2008;22:2018-2025
[64] Sherman V, Feldman LS, Stanbridge D, Kazmi R, Fried GM. Assessing the learning curve for the acquisition of laparoscopic skills on a virtual reality simulator. Surgical Endoscopy. 2005;19:678-682
[65] Grantcharov TP, Rosenberg J, Pahle E, Funch-Jensen PM. An objective method for the evaluation of laparoscopic surgical skills. Surgical Endoscopy. 2001;15:242-244

[66] Grantcharov TP, Bardram L, Funch-Jensen P, Rosenberg J. Learning curves and impact of previous operative experience on performance on a virtual reality simulator to test laparoscopic surgical skills. The American Journal of Surgery. 2003;185:146-149

[67] Jordan J, Gallagher, AG, McGuigan J, McGlade K, McClure N. A Comparison between Randomly Alternating Imaging, Normal Laparoscopic Imaging, and Virtual Reality Training in Laparoscopic Psychomotor Skill Acquisition. The American Journal of Surgery. 2000;180:208-211

[68] Smith CD, Farrell TM, McNatt SS, Metreveli RE. Assessing laparoscopic manipulative skills. The American Journal of Surgery. 2001;181:547-550

[69] Aggarwal R, Moorthy K, Darzi A. Laparoscopic Skills Training and Assessment. British Journal of Surgery. 2004;91:1549-1558

[70] Hyltander A, Liljegren E, Rhodin PH, Lonroth H. The transfer of basic skills learned in a laparoscopic simulator to the operating room. Surgical Endoscopy. 2002;16:1324-1328

[71] Chung JY & Sackier JM. A method of objectively evaluating improvements in laparoscopic skills. Surgical Endoscopy. 1998;12:1111-1116

[72] Rosser JC, Rosser LE, Savalgi RS. Skill Acquisition and Assessment for Laparoscopic surgery. Archives of Surgery. 1997;132:200-204

[73] Powers TW, Brentem DJ, Nagle AP, Toyama MT, Murphy SA, Murayama KM. Hand dominance and performance in a laparoscopic skills curriculum. Surgical Endoscopy. 2005;19:673-677

[74] Grantcharov TP, Bardram L, Funch-Jensen P, Rosenberg J. Impact of hand dominance, gender, and experience with computer games on performance in virtual reality laparoscopy. Surgical Endoscopy. 2003;17:1082-1085

[75] Andreatta PB, Woodrum DT, Birkmeyer JD, Yellamanchilli RK, Doherty GM, Gauger PG, Minter RM. Laparoscopic skills are improved with Lap Mentor training: results of a randomised, double-blinded study. Annals of Surgery. 2006;243:854-860

[76] Scott DJ, William NY, Tesfay ST, Frawley WH, Rege RV, Jones DB. Laparoscopic Skills Training. American Journal of Surgery. 2001;182:137-142

[77] Undre S & Darzi A. Laparoscopy Simulators. Journal of Endourology. 2007;21:274-279

Chapter 9

Virtual Rehabilitation and Training for Postural Balance and Neuromuscular Control

Kristiina M. Valter McConville

Additional information is available at the end of the chapter

http://dx.doi.org/10.5772/50133

1. Introduction

Virtual Reality (VR) has been used in an increasingly wide range of applications, and technological advances have made it accessible to the everyday consumer. Virtual reality games have become the norm in today's consumer marketplace. This makes it possible to consider such devices for more productive use of leisure time, such as in exercise and health maintenance. Gaming systems such as the *Nintendo® wii-fit™* have already introduced the concept of VR guided balance and fitness training.

Studies have shown that virtual reality has potential for assisting recovery from illnesses such as stroke and vestibular disorders and improvements in cerebral palsy. Custom designed VR systems in the laboratory have been successful with balance training [1]. Active video game systems, like the *Sony PlayStation® 2* with the *Sony EyeToy®*, a camera accessory, have also been used in studies of rehabilitation. Flynn et al. used this system in a case study of a stroke patient and documented the types of training for each game [2]. The *Nintendo® wii™* system was designed for in-home training as well as gaming. This system was tested on a 13-year-old patient with cerebral palsy who showed improvements in visual perceptual processing, postural control and functional mobility [3]. Therefore, these gaming systems show promise in rehabilitation, and have the added benefit of entertainment.

Other custom-designed games and virtual environments have also been developed to address specific disabilities and user requirements. Why do these systems work? In order to answer this question, the basics of human postural balance control have to be considered. The balance control system is a complex integration of multiple sensory inputs along with feedback from the motor component. The sensory inputs include visual, vestibular, auditory, tactile, and proprioceptive signals. The motor aspect is the "plant", driven by neuromuscular control. Figure 1 demonstrates the complex interaction between the various sensory systems in the central processing of incoming signals. Different inputs are assigned weights based on motor learning and experience, i.e. the reliability of the incoming cues.

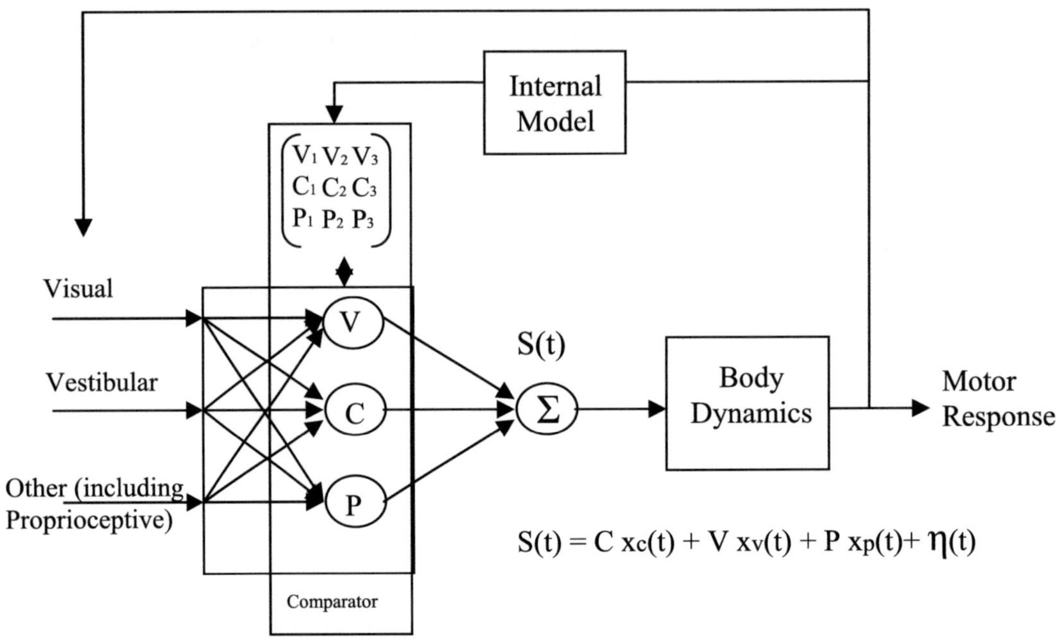

Figure 1. Proposed comparator model of postural control: Sensory inputs are compared to each other for consistency, and a matrix of gains [V,C,P] is assigned accordingly. The weighted inputs are combined as S(t) which is the total sensed self-orientation x as a function of time t, with noise $\eta(t)$ added. This results in a compensatory motor output through body dynamics. Feedback is provided through the senses, but also internally to update a memory or database of previously experienced combinations of sensory inputs. In addition to comparing the sensory inputs to each other, this database is also referenced.

Visual inputs are generally dominant among human sensory inputs for balance control as well as a general source of information, and virtual reality systems have evolved with an emphasis on visual displays. Balance improvement has been demonstrated through both static and dynamic visual stimuli in VR systems. [4] This results in a natural fit for visual-modality systems for both cognitive and neuromuscular control training along with entertainment value. However, other channels of information cannot be discounted, including the vestibular system, which is located in the inner ear and senses head acceleration.

The visual and vestibular systems are very closely linked, for example eye movements respond to vestibular inputs. Since the vestibular system senses head movements, it can be hypothesized that head movements together with visual stimuli should be used in training the balance control system. [5] Cohen et al. showed that purposeful head movement exercises improved balance in vestibular patients. [6] Therefore active games that track player head and body movements could demonstrate potential for balance improvement.

Other inputs include auditory and tactile modalities. Auditory signals provide orientation cues based on sound localization. Tactile cues provide pressure feedback and external inputs regarding contact with surfaces. Proprioceptive inputs are also important, providing us cues about our body position in space, through joint angle feedback signals. These signals are directly relevant to the control of ankles, knees, hips and upper body to maintain balance. Furthermore, the arms can be used for stabilization control and fall prevention.

In this chapter, we will investigate both commercial and custom-designed virtual reality interfaces for balance training and rehabilitation, considering the various inputs that are important for balance control. We can then begin to evaluate the important elements of virtual reality games, so that they can be developed more effectively for rehabilitation applications.

2. Arm movement importance and potential training system

Arm movements have been shown to be very important in balance maintenance and recovery [7]. Both unilateral and bilateral arm movements generate anticipatory and reactive responses in postural leg muscles [8] and hip and trunk muscles [9]. Furthermore, reaching and grasping actions are important in balance recovery and prevention of falls [7].

In a recent study we showed the importance of free arm movements in a series of classical balance and mobility tests in which three out of four tests showed significant improvement with the free use of arm movements [10]:

1. Maximal step length test (length and timing of a single maximally long step)
2. Step test (number of steps up and down onto an elevated surface in one minute)
3. Timed get-up-and-go test (time to get up from a chair, walk to a target, and return to sitting)
4. Walk along an elliptical line

Only test (4) did not show signficant improvement, possibly because there is less change of support motion of the body relative to the task of walking, compared to launching a single large or elevated stepping motion, and certainly compared to rising from a chair. These results show the importance of arm movements in controlling body balance.

Arm movements are affected by stroke and progressive disease, such as Parkinson's disease (PD). Given the importance of arm movements for balance, such an effect could result in decreased balancing ability or steadiness of stance in these patients. Arm movement training is thus an important part of rehabilitation. We have developed a virtual reality arm movement training system [11], in which different games can be programmed to suit user abilities and interests. This system uses the *Microsoft Kinect*™ interface, along with pressure sensors attached to the index finger and thumb as an object selection tool. This system shows the actual image of the user on the screen, and we have developed games, which appear overlaid on the user's image. This allows the user to reach out to the target letters and "select" them by making contact between the thumb and index finger. Initial testing on healthy subjects provided a baseline scores, so that we can now proceed to evaluate patients' performance.

3. Balance training: *Sony PlayStation®* 2 with the *Sony EyeToy®* and the Harmonix *AntiGrav™* game [12]

Based on the importance of head and arm movements along with visual inputs in balance training, the Harmonix *AntiGrav™* game was selected for evaluation. The purpose of this study was to to characterize balance-related motor learning effects through game performance and to evaluate the effectiveness and specificity of the video game for balance training in healthy subjects. [12]

3.1. Game description

Sony's EyeToy® used video capture technology, in which it imaged the user with a small video camera mounted above the display and used face-recognition and motion-tracking technologies (no body markers were required). [13, 14] The system used two-dimensional user motion in the frontal plane as inputs. Each user was individually calibrated before each session, to account for height differences and range of head motion. During this procedure, the subject's own image was shown on the screen overlaid with a calibration figure, instructions and controls. The subject had to move to a position such that his/her face was centered on the figure, and then was instructed to wave his/her hand over the "lock" control. In the subsequent screen, the subject was required to bend to the left and right such that the head was aligned with target positions. If this procedure was not properly executed, it became very apparent in the early stages of game-play that the game was not responsive (the avatar movements did not respond to user movements), and the calibration was repeated. A diagram of the system and screenshots is shown in Figure 2.

The user was portrayed in the game as an avatar and the user's image was not shown after the calibration procedure. The avatar raced on a hoverboard along roadways and rails, over jumps, and when airborne, through rings. The user, standing in front of the display, used head movements to guide the avatar, and was required to jump or duck to avoid obstacles. The user also had to reach with either arm or both to strike targets as the avatar passed them. The arm motions were directed diagonally up or down or to the side, requiring additional head/upper body motions to contact the targets. The game was fast-paced and challenging, and included sound effects and "life or death" consequences. It presented unpredictable situations requiring maintenance of balance and decision making at the same time. As the user gained experience in the game, anticipation of upcoming movements became easier. Upon achieving a pre-set score, the user was advanced to playing both difficulty Level 1 and Level 2 in each training session.

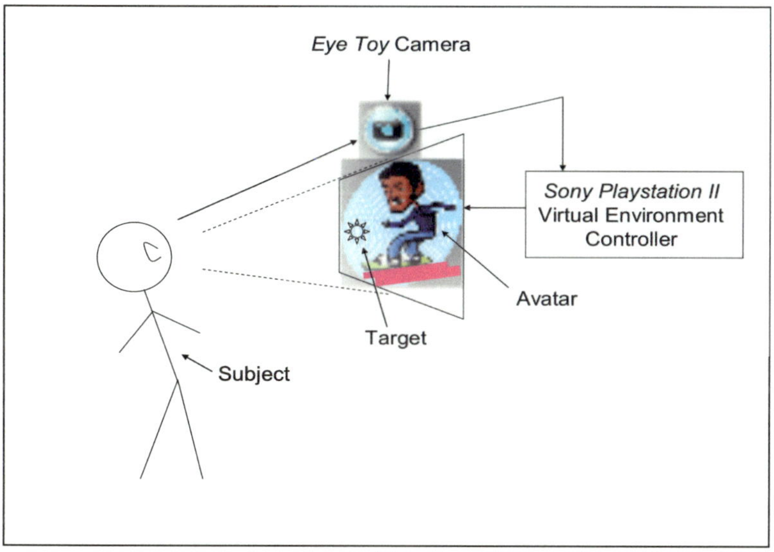
Eye Toy Camera
Sony Playstation II
Virtual Environment
Controller
Avatar
Target
Subject

(a)

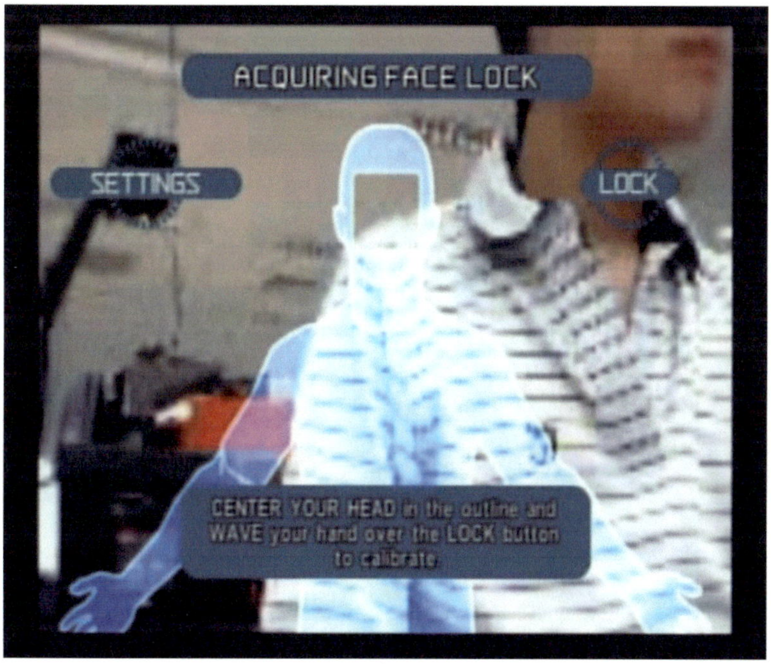
ACQUIRING FACE LOCK
SETTINGS
LOCK
CENTER YOUR HEAD in the outline and
WAVE your hand over the LOCK button
to calibrate

(b)

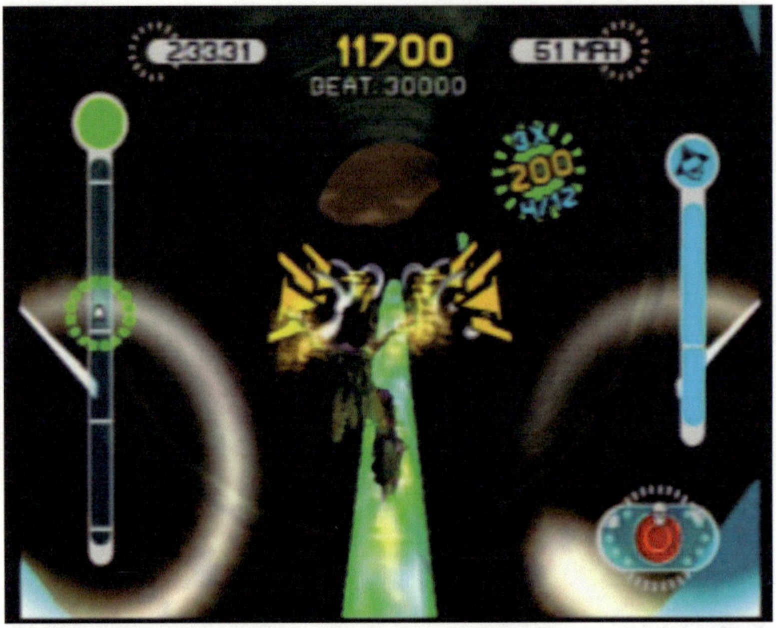

(c)

(d)

Figure 2. a) Conceptual schema of the *PlayStation®2*, *EyeToy®* and hoverboard game (not an actual screenshot) with the user.

b) Screenshot of the first step in calibration. The actual image of the user is shown superimposed (off to the side in this picture to highlight the two images) on an outline of a human frame with an opening at the face.

c) Screenshot of the game in progress. The avatar seen in the middle of the screen is riding on a "rail", which constrains the path while presenting targets for arm movements. While elsewhere, head movements steer the avatar, here head movements are required to lean toward targets (if presented on one side) to be struck with rapid arm movements and to duck or jump over obstacles. The lower right corner has an indicator of the user's head and arm positions in real time. These are also reflected in the avatar actions. In the center, the accumulated points are shown along with the target number of points to beat. The time remaining is shown in the top left corner. The bar on the left side indicates the current position along the racetrack.
d) Screenshot of the game in progress. Here the user has successfully struck a pair of targets with both hands, and the user has been given a credit of extra time to complete the race. [12]

3.2. Experimental protocol

The control (no training, n=7) and experimental (video game training, n=7) groups completed the balance tests (as described below) at the beginning and at the end of a three-week period. Both the control group and the experimental group were free to conduct their daily activities as normal, and thus, the control group controlled for the effect of the specific experimental video game training. The experimental group trained 9 times during the three-week period, and the control group received no training. For details, see [12].

Game performance was measured by the number of tokens or targets struck with arm movements and the speed and accuracy in navigating the avatar through the race course using head movements. Additional points could be achieved by doing tricks, executed by arm movements while airborne. The scores were provided after each game, and points were displayed when awarded, which reinforced the immediate feedback provided during game play.

Three balance tests (Balance Board (BB), Tandem Romberg (TR), and One-leg Standing (OL)) were selected to evaluate transfer of the video game training to the real world. These tests evaluated transfer of learning to the real world because they were conducted without the virtual environment and thus tested for changes in balance outside of the game. Many balance-related measures were found in the literature ranging from the Dynamic Gait Index to the Sensory Organization Test to various functional measures and clinical tests. Our tests were selected based on common usage and relevance to overall balance [15-17] rather than factors such as sensory contribution or functional outcomes such as locomotion.

The average scores from two trials for each test were used. To minimize a "ceiling effect" (which occurs if most subjects pass a balance test given for a specified time, e.g. 30 seconds), the time was recorded when balance was lost. This gave a finer measure to detect and quantify improvement.

3.3. Results

3.3.1. *Game scores*

As expected, the game scores in both levels of difficulty improved over the 9 training sessions. The performance in the game was based on the game scores in both levels of difficulty. All subjects except one achieved the target score for Level 1 in the first session, and were moved to training at both Levels 1 and 2 by the second session. Overall, there was a gradual increase in the total scores for both Levels over the period of the experiment.

3.3.2. Model of motor learning through game scores

The improvement in game scores was consistent with our common understanding of improved performance with practice and with studies of motor learning. [18] The game scores were averaged across subjects and fit to an exponential given in Equations 1 and 2 for Level 1 and Level 2 games respectively [19]:

$$y(n) = 57{,}500\text{-}41{,}000e^{-0.26n} \quad (1)$$

$$y(n) = 57{,}500\text{-}43{,}500e^{-0.087n} \quad (2)$$

where n is the session number and y is the curve fit value. All three variables were initially used to obtain curve fits. It was found that constraining the DC term to be the same for both curves did not affect the quality of the fit, and it represented the average performance plateau after a large number of sessions. These curves are shown together with the averaged data in Figure 3. The curve fits reflected the increase and plateau of motor learning as a function of the session number for the first level (Level 1), and suggested that the subjects had not yet mastered Level 2 because of the absence of the plateau.

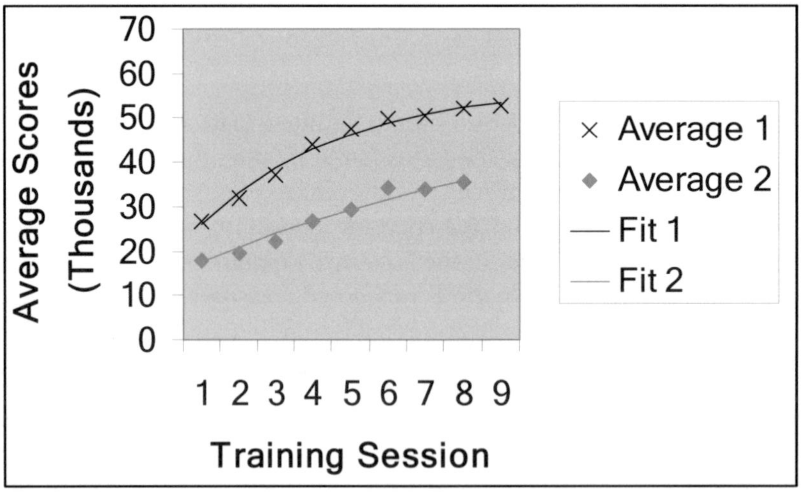

Figure 3. Averaged scores over all 7 training subjects for Level 1 and Level 2 training along with corresponding exponential curve fits (see text). [12]

3.3.3. Balance tests

Performance on balance tests improved over the training period for the training subjects as compared to controls. During pilot trials, it was found that some volunteers were balancing for longer than three minutes on the balance board and two minutes on the Tandem

Romberg tests. The BB and TR tests were discontinued after three and two minutes respectively because of the potential for muscle fatigue, which would interfere with the measure of balancing ability. [20] Of the 7 experimental subjects in each group (training and control), three control subjects reached the maximum time in the BB pre-test and one control subject in the TR pre-test.

Figure 4 shows the balancing times in seconds averaged for training and control subjects for each test. Trained subjects showed significant improvement in the BB test, moderate improvement in the TR test, and no significant difference in the OL test, while control subjects did not show significant difference on any of the tests.

3.3.4. *Simulator sickness*

Virtual Reality games have the potential for simulator sickness, which is related to motion sickness, with many of the same symptoms. Kennedy et al. have established a Simulator Sickness Questionnaire (SSQ) based on work with the US Navy, which is often used in evaluating simulator sickness. [21] It contains 16 items on which the subjects rated their responses (ranging on a 4 point scale; 0 (none), 1 (slight), 2 (moderate), 3 (severe)).

In this study, simulator sickness symptoms were recorded in every training session. Figure 5 shows the incidence of all symptoms across the 7 subjects and 9 training sessions. Most subjects who experienced symptoms reported slight sweating. No severe symptoms were present, and the symptoms subsided over the course of the training sessions. The presence of the symptoms suggested potential for vestibular rehabilitation, and their mildness demonstrated the safety of this system for the general public.

3.4. Discussion

The results showed that game scores improved over the sessions as expected. The average game scores were modeled, and showed an almost linearly increasing component followed by a plateau. Motor learning analysis showed that Level 1 was mastered by the participants while Level 2 was still showing learning. We developed a game difficulty parameter, which could be used in rehabilitation game design to provide appropriate increases in difficulty for continued learning and training.

Pre- and post-test results showed that subjects who played the game had an improvement in balance board and tandem Romberg tests over the 9 sessions. One-legged standing tests did not show improvement, which could be the result of training specificity, since the training was not performed on one leg.

The simulator sickness questionnaire showed the presence of several symptoms, especially in the earliest sessions. These were mild and decreased over the training period. The symptoms indicated that the game had a physiological effect on the participants, and the decrease in symptoms paralleled the motor learning.

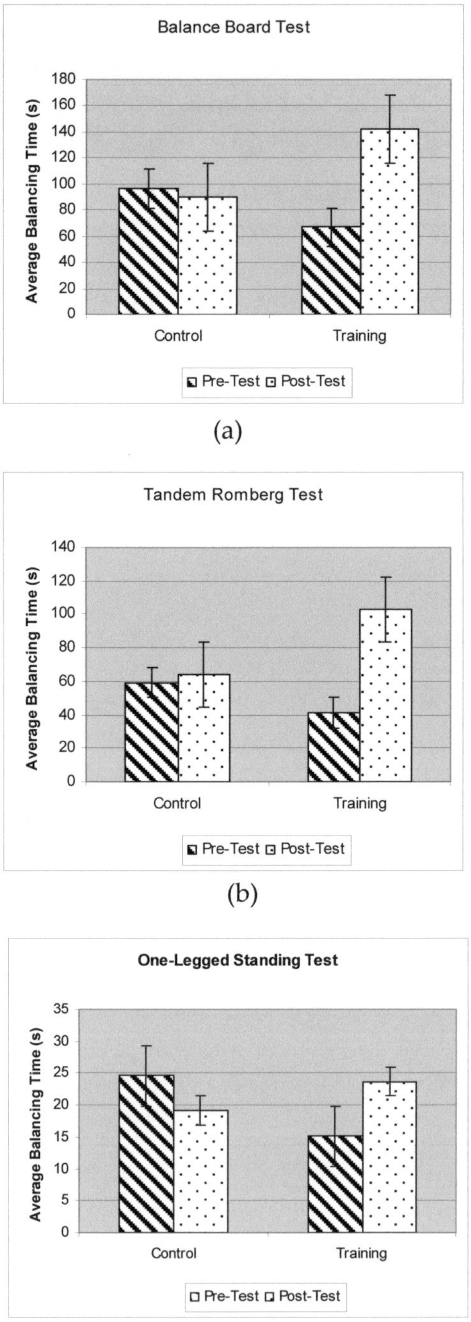

Figure 4. a) Time that the subjects were able to maintain balance on a Balance Board (BB) Pre- and Post-training. The performance for the training subjects (n=7), mean=67 and 142 s, is compared to the performance of the control subjects (n=7) who did not receive training during the 3-week period; mean=97 and 90 s. The trained subjects had a significant difference in balancing times ($p<0.005$).

b) Time that the subjects were able to maintain balance in the Tandem Romberg (TR) test Pre- and Post-training. The performance for the training subjects (n=7), mean=41 and 102 s, is compared to the performance of the control subjects (n=7) who did not receive training during the 3-week period; mean=59 and 64 s. The trained subjects had a significant difference in balancing times ($p<0.05$).
c) Time that the subjects were able to maintain balance in the One-Leg (OL) standing test Pre- and Post-training. The performance for the training subjects (n=7), mean=15 and 24 s, is compared to the performance of the control subjects (n=7) who did not receive training during the 3-week period; mean=25 and 19 s. The differences in balancing times were not significant for either group. [12]

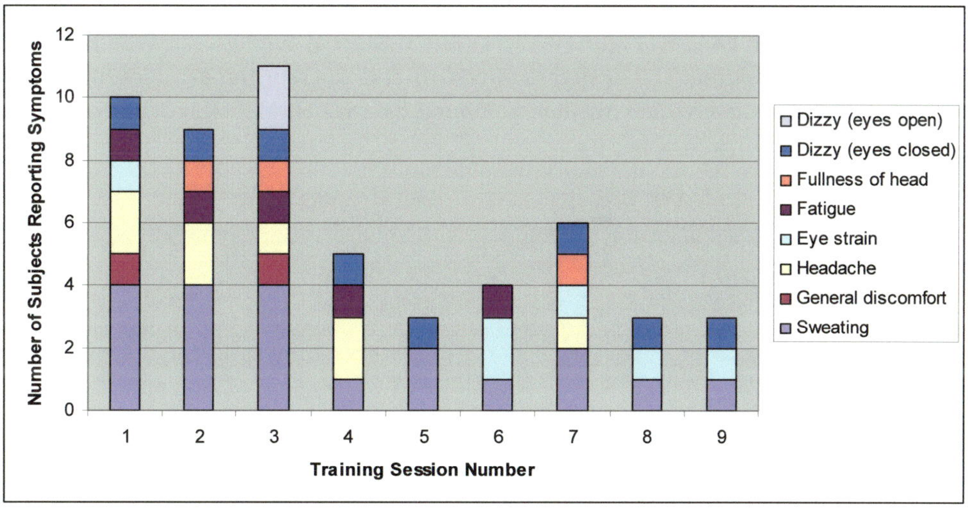

Figure 5. Simulator Sickness symptom occurrences over 7 subjects and 9 training sessions during a three week period. Some subjects had more than one symptom on a given training day, and some subjects reported no symptoms.

4. Auditory inputs

The auditory system also provides important cues to an individual's self-orientation in space due to sound-localization capability. This ability is based on the difference in arrival time of a sound to each ear. We have used this concept in a prototype balance training aid, in which both visual and auditory feedback were provided to four healthy subjects. The system was tested with the subjects standing on a balance board. The auditory feedback was provided through stereo headphones worn by the participant. Both the Inter-aural Intensity Difference and the Inter-aural Timing Difference were controlled based on accelerometer signals from the tilt of the balance board. The combined audio-visual feedback resulted in significant improvement in the subjects' balancing ability [22]. Future studies will test for the relative contribution of each of the two feedback modalities.

5. Tactile inputs

Tactile inputs can be utilized in VR systems with the increasing availability of haptic feedback. Multimodal feedback increases the sense of immersion in the virtual environment, and different sensory inputs can be harnessed for sensory substitution, additional information channels and most importantly, redundancy. Redundancy is a means of bringing important signals to the attention of the user, and alerting the user in emergencies. For this reason, we considered it important to include tactile feedback in the design of an alerting system to inform subjects about potential loss of balance.

We developed a Wireless Vibrotactile Rehabilitation System (WVRS), which provides vibration feedback during body tilt. [23] The system incorporates three vibrating disks on each arm, which are activated when an accelerometer, mounted on the mid-back detects body tilt within certain ranges. Figure 6 shows the design of the system, which includes a wireless link to a computer. The system can record and display the accelerometer signals. The three vibrating disks on each arm are activated as a function of the degree of body tilt and tilt direction.

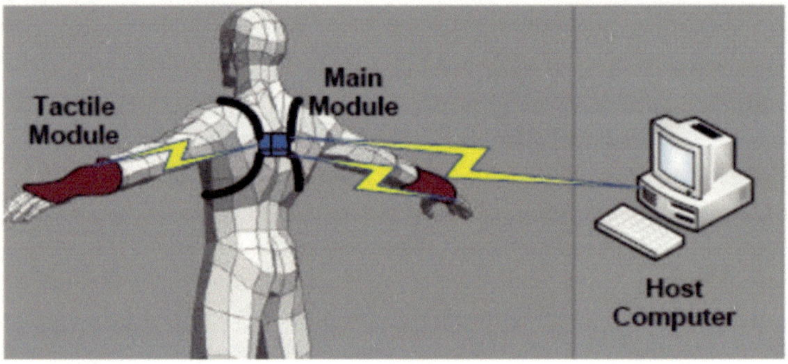

(a)

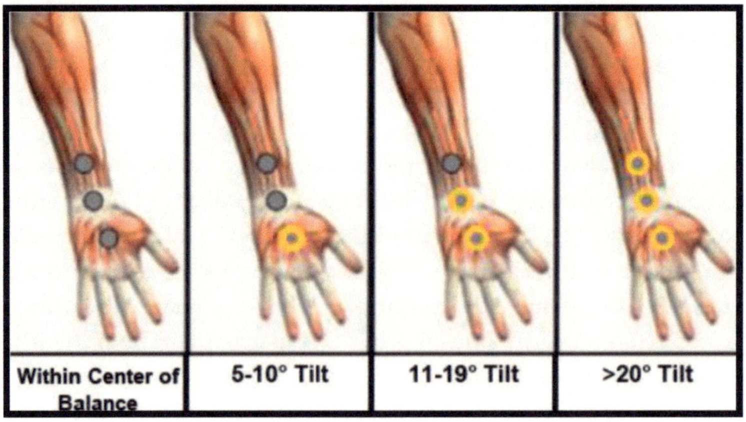

(b)

Figure 6. System design of a vibrating tactile balance feedback system. a) The main module includes a tri-axial accelerometer, Arduino microprocessor, and ZigBee wireless system, incorporating a Wireless Personal Area Network ID to distinguish the communication with the arm subsystems and the communication with the host computer. b) A system of three vibrating disks was provided on each arm, activated in each arm as shown, based on the amount of body tilt in each direction respectively. The vibration was applied at 200 Hz. [23]

The arms were selected for the vibration transducers to naturally produce a reaction in the appropriate arm, and by eliminating unnecessary processing steps, the arm movement could be rapid enough to help stabilize posture. An example result from one subject is shown in Figure 7.

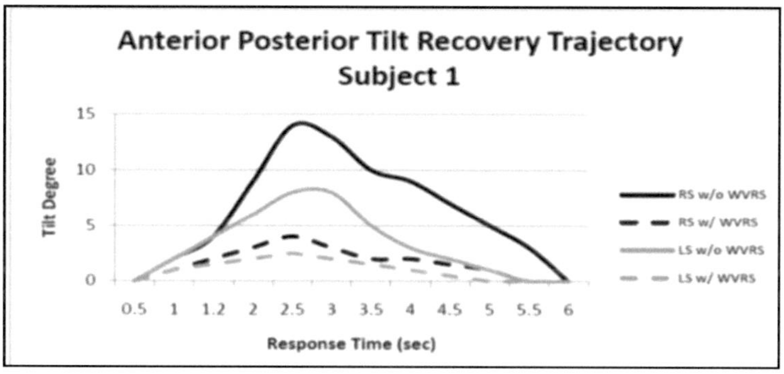

Figure 7. The subject was perturbed in the antero-posterior direction, and left and right body tilt were recorded with and without the use of the WVRS system. In both directions, the amount of body tilt was reduced. [23]

Such a system can be used as a VR rehabilitation tool, utilizing multimodal feedback with the visual display and providing progress tracking with the data recording capability. It is also anticipated that in the next design phase, the recording could also be housed in the main module, making the system portable. Such a wearable device could be worn as an everyday balance maintenance aid. This is an example of taking the virtual reality into the real world, or augmented reality.

6. Ankle training

In addition to head/body and arm movements, balance also depends on the quality of control in the joints of the lower limbs. In a balance training program, these joints can be involved in general training, or targetted specifically. When a joint injury occurs, for example in the knee or ankle, then a phased and progressive rehabilitation program is essential.

We have demonstrated ankle VR rehabilitation systems in which ankle movements in up to two degrees of freedom can be used to control a custom-designed VR game. The program was motivated by a sports injury, which can often take a long time to heal without properly designed rest and training. The games tracked the progress of the training and could be used as an inexpensive home-based training tool that motivates exercise adherence. The system had the following objectives [24]:

a. the training has a range of difficulty levels to suit different patient abilities, including isometric contraction when no joint rotation is desired
b. the system has the ability to constrain movement in both magnitude and direction so that specific muscle groups can be trained without the risk of re-injury, and
c. the game provides visual feedback through a game, which provides scores of the ankle control ability related to game performance.

These objectives have been satisfied with the incorporation of either a balance board simulating a 2 degree-of-freedom foot pedal interface [25], or a boot with a frame and locking pins that either prevent motion, or constrain motion to the plantarflexion/dorsiflexion plane [26]. Force sensors and accelerometers measured the muscle activity and ankle motion and provided inputs to the game. Visual interfaces were developed for calibration, as shown in Figure 8. Game scores were tracked and recorded across the training sessions, so that the patient and also the caregiver can monitor progress.

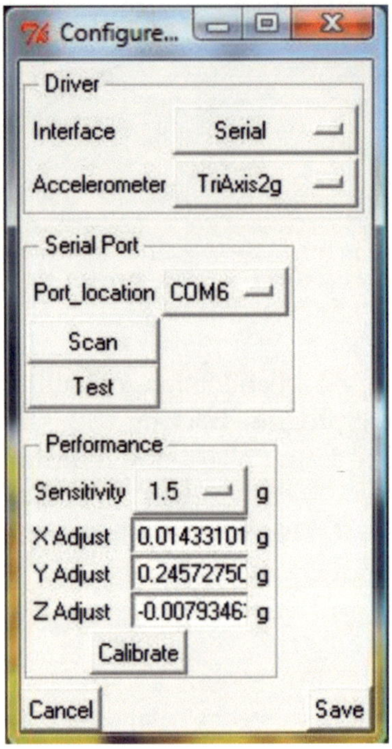

Figure 8. Windows interface for accelerometer calibration [25]

The important side-benefit of such a visual gaming interface is the motivation to improve scores, and the entertainment value of the game. Sayenko et al. have shown that a custom-designed VR game provides useful feedback and motivation for a spinal cord-injured patient to prevent muscle atrophy through functional electrical stimulation training. [27] The game encouraged adherence to the exercise program. It is important to conduct further studies on patients to develop optimal training protocols.

7. Occupational health and safety: Training for correct lifting posture

Very specific aspects of postural balance can also be isolated and trained through VR. In the field of occupational safety, VR can be a very important training tool and real-time aid in injury prevention. We have developed a lifting technique trainer, in which an avatar shows the correct motion on the screen, based on published positioning guidelines. [28] The trainee lifts a training object wearing a harness instrumented with accelerometers as shown in Figure 9.

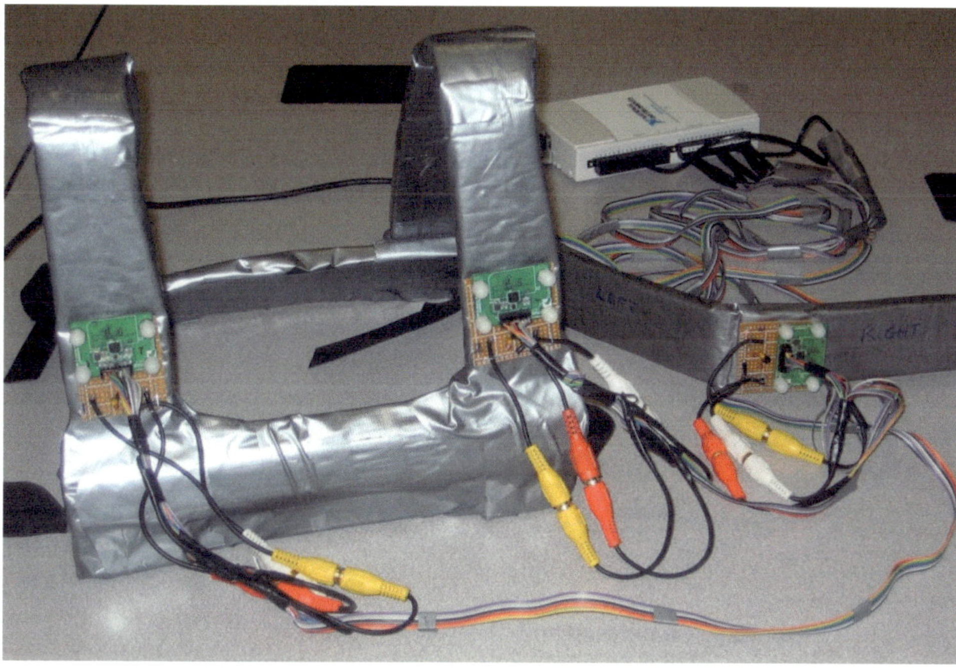

Figure 9. Instrumented harness providing input to the VR training system. Accelerometers attached to the shoulder harness are positioned on the trainee's shoulder blades, and the belt contains a third accelerometer, which is positioned on the lumbar region of the low back. [29]

The user can view the image of the avatar reflecting actual measured values from his/her upper body during the lifting motion. The discrepancy between the ideal and actual lifting motions is displayed, which guides the trainee to improve his/her lifting technique. Such a VR training aid provides instant visual feedback to allow the user to learn the body proprioceptive feedback signals associated with the correct postural positioning.

8. Conclusions

Different virtual reality training environments were evaluated or developed for improving postural balance and neuromuscular control. Balance training was shown to transfer to the real world from commercial video-game based training, and because balance is related to overall fitness, these training interfaces represent an important means of combining leisure and health maintenance. The results of such training could reduce the likelihood of falls and injury in everyday activities.

The benefits of VR training include the ability to isolate specific sensory inputs and motor outputs. Clearly, VR also provides immediate feedback to the user, which enhances learning. Further, VR provides an automated way to guide and track the progress of the training. Adaptive systems can sense the user's performance and adjust parameters to provide the appropriate level of difficulty in each training session. Finally, VR provides a relatively safe environment for training compared to many real-world environments.

Author details

Kristiina M. Valter McConville
Electrical and Computer Engineering, Ryerson University,
Institute of Biomaterials and Biomedical Engineering, University of Toronto, Canada
Toronto Rehabilitation Institute, Toronto, Canada

Acknowledgement

I would like to thank Dr. Ken Norwich, University of Toronto for his insight and wisdom and endless encouragement. I am also grateful to all of my students who have helped me immensely in this work, especially Matija Milosevic and Sumandeep Virk. I also thank the Natural Sciences and Engineering Research Council of Canada for funding the research.

9. References

[1] Keshner AE, Kenyon VR. Using immersive technology for postural research and rehabilitation. Asst Technol 2004;16(1) 27-35.

[2] Flynn S, Palma P, Bender A. Feasibility of using the Sony Playstation 2 gaming platform for an individual poststroke: a case report. J Neurol Phys Ther 2007;31 180-89.

[3] Deutsch JE, Borbely M, Filler J, Huhn K, Guarrera-Bowlby P. Use of a low-cost, commercially available gaming console (Wii) for rehabilitation of an adolescent with cerebral palsy. Phys Ther 2008;88(10) 1196-207.

[4] Suárez H, Suárez A, Lavinsky L. Postural adaptation in elderly patients with instability and risk of falling after balance training using a virtual-reality system. Int Tinnitus J 2006;12(1) 41-4.

[5] Foster CA. Vestibular Rehabilitation. Baillieres Clin Neurol 1994;3(3) 577-92.

[6] Cohen H, Kane-Wineland M, Miller LV, Hatfield CL. Occupation and visual/vestibular interaction in vestibular rehabilitation. Otolaryngol. - Head & Neck Surgery 1995;112(4) 526-32.

[7] Maki BE, McIlroy WE. Control of rapid limb movements for balance recovery: age-related changes and implications for fall prevention. Age Ageing 2006;35(Suppl 2) ii12-ii18.

[8] Mochizuki G, Ivanova TD, Garland SJ. Postural muscle activity during bilateral and unilateral arm movements at different speeds. Exp Brain Res 2004;155 352-61.

[9] Yamazaki Y, Suzuki M, Ohkuwa T, Itoh H. Maintenance of upright standing posture during trunk rotation elicited by rapid and asymmetrical movements of the arms. Brain Res Bull 2005;67 30-9.

[10] Milosevic M, McConville KMV, Masani M. Arm Movement Improves Performance in Clinical Balance and Mobility Tests. Gait Pos 2011;33(3) 507-9.

[11] Seevaratnam S, Attalla R, El-Falou A, Valter McConville KM. Arm Biofeedback System for Parkinson's Disease Patients. Inst Biomat Biomed Eng Sci Day. 7 May, 2012. University of Toronto; 2012. p12.

[12] McConville KMV, Virk S. Evaluation of an Electronic Video Game for Improvement of Balance. Virtual Reality 2012;In Press. DOI: 10.1007/s10055-012-0212-7.

[13] Weiss P, Rand D, Katz N, Kizony R. Video capture virtual reality as a flexible and effective rehabilitation tool. J NeuroEng Rehabil 2004;1-12.

[14] Wang S, Xiong X, Xu Y, Wang C, Zhang W, Dai X, Zhang D. Face-tracking as an augmented input in video games: enhancing presence, role-playing and control. SIGCHI Conference on Human Factors in Computing Systems. 22-27 April, 2006, Montréal, Québec, Canada. New York: ACM; 2006. p1097-1106.

[15] Quesada PM, Durham MP, Topp RV, Swank AM, Biton D. Quantitative assessment of balance performance on a passively unstable surface. Occupat Ergonom 2007;7(1) 3-10.

[16] Vereeck L, Wuyts F, Truijen S, Van de Heyning P. Clinical Assessment of Balance, Normative data and Gender and Age Effects. Int J Audiol 2008;47 67-75.

[17] Curb JD, Ceria-Ulep CD, Rodriguez BL, Grove J, Guralnik J, Willcox BJ, Donlon TA, Masaki KH, Chen R. Performance-based measures of physical function for high-function populations. Am Geriatr Soc 2006;54(5) 737-42.

[18] Newell KM, Liu Y-T, Mayer-Kress G. Time scales in motor learning and development. Psychol Rev 2001;108(1) 57-82.

[19] McConville KMV, Virk S. Motor learning in a virtual environment for vestibular rehabilitation. Proceedings of the 3rd Int IEEE EMBS Conference on Neural Engineering. 2-5 May, 2007, Kohala Coast, Hawaii. IEEE Press; 2007. p600-3.

[20] Gandevia SC. Spinal and Supraspinal Factors in Human Muscle Fatigue. Physiol Rev 2001;81(4) 1725-89.

[21] Kennedy RS, Lane NE, Berbaum KS, Lilienthal ML. Simulator sickness questionnaire: an enhanced method for quantifying simulator sickness. Int J Aviat Psychol 1993;3 203-20.

[22] Milosevic M, McConville KMV. Audio-visual biofeedback system for postural control. Int J Disabil Hum Dev 2011;10(4) 321-4.

[23] Tino A, Carvalho M, Preto NF, McConville KMV. Wireless Vibrotactile Feedback System for Postural Response Improvement. IEEE Eng Med Biol Soc Conf. 30 Aug – 3 Sept., 2011, Boston, MA. IEEE Press; 2011. p5203 – 6.

[24] Holme E, Magnuson SP, Beeher K, Bieler T, Aagaard P, Kjarr M. The effect of supervised rehabilitation on strength, postural sway, position sense and re-injury risk after acute ankle ligament sprain. Scand J Med Sci Sports 1999;9 104-9.

[25] Ching A, Guirguis R. Haptic Interface For Virtual Reality Training Game Using The Ankle. Thesis. Electrical and Computer Engineering Ryerson University Toronto; 2009.

[26] Battaglia B, Brukson A, Nassery M. Ankle Rehabilitation. Thesis. Electrical and Computer Engineering Ryerson University Toronto; 2012.

[27] Sayenko DG, Masani K, Milosevic M, Robinson MF, Vette AH, McConville KMV, Popovic MR. Video game-based functional electrical stimulation system for calf muscle training: A case study. Med Eng Phys 2011;33(2) 249-55.

[28] Genaidy, A. M., & Houshyar, A. Biomechanical Tolerance Limits For Manual Lifting Tasks: A Tool To Control Back Injuries. IEEE Eng Med Biol Soc Conf. 9-12 Nov., 1989, Seattle, WA. IEEE Press; 1989. 3:803-5.

[29] Feizizadeh S, Baretto E, Joackhim R. Virtual Reality Biofeedback Training Interface For Upper Body Movements In Lifting Technique. Thesis. Electrical and Computer Engineering Ryerson University Toronto; 2008.

Chapter 10

Applications of Virtual Reality Technology in Brain Imaging Studies

Ying-hui Chou, Carol P. Weingarten, David J. Madden,
Allen W. Song and Nan-kuei Chen

Additional information is available at the end of the chapter

http://dx.doi.org/10.5772/48445

1. Introduction

Virtual reality is an evolving technology that allows for the possibility of interactive environments with stereoscopic, three-dimensional (3D) visual displays, auditory input, haptic feedback, and immersive interaction from a first person perspective [1]. Thus virtual reality technology makes it possible to simulate an environment with better ecological validity and control than previously possible for brain research and clinical applications. Virtual reality is compatible with many brain imaging methods and this has allowed researchers to evaluate typical and atypical brain function when users are immersed in a virtual reality environment. Virtual reality is also being developed as a therapeutic tool for a wide range of clinical populations, and because the brain is a primary mediator of effects of virtual reality treatments brain imaging is an important method for assessing some types of treatment effects. In a very different type of therapeutic application, virtual reality is also being developed to augment clinical use of brain imaging results for presurgical planning. This chapter provides an overview of these kinds of studies that employ both virtual reality and brain imaging technologies. In the first part of the chapter we will describe brain correlates of a few examples of tasks that are very difficult, if not impossible, to employ inside a brain imaging scanner without virtual reality technology. These tasks are spatial navigation, car driving, and social interactions. In the second part of this chapter we will survey examples of the application of virtual reality and brain imaging methods to clinical populations. These applications are in early stages of development and will require further studies to assess the value of using a virtual reality versus conventional task and to provide specific evidence of a therapeutic benefit from a virtual reality treatment. Nonetheless these examples will indicate some innovative ways being explored to apply virtual reality and brain imaging in the clinical sphere. We will conclude with a prospective view on using real-time magnetic resonance imaging (MRI) technology combined with virtual reality environments for research or therapeutic applications.

Many imaging modalities have been employed in virtual reality studies reviewed in this chapter. These imaging modalities include structural MRI [2] and diffusion tensor imaging (DTI) [3] that are used to measure brain anatomical structure, and positron emission tomography (PET) [4], functional MRI (fMRI) [5], electroencephalography (EEG) [6], and magnetoencephalography (MEG) [7] that are used to assess brain functional activity.

2. Measuring brain function in virtual reality environments

Brain function can be measured using imaging methods in a straightforward fashion for many cognitive and sensorimotor performances, such as attention, memory, facial recognition, finger tapping, or language. However, during these measurements of brain function participants usually must be imaged individually and must remain immobile (especially the head). Therefore it has been very difficult to measure brain function associated with tasks that require participants to move through the environment or to have face-to-face social interactions. Virtual reality technology makes it possible to "walk" or "drive" or to perform social interactions in simulated environments during measurements of brain function. Here we will review selected topics in spatial navigation, car driving, and social cognition for which attempts have been made to relate brain function to these behaviors using virtual reality and brain imaging techniques in healthy adults. We will also describe some studies that probed if the brain responds differently to tasks with two-dimensional (2D) versus 3D stimuli or real objects, as steps towards a more complex understanding of the neural correlates of virtual reality tasks.

2.1. Spatial navigation

"Where am I?"

"Where are other places with respect to me?"

"How do I get to other places from here?"

These are the main questions associated with spatial navigation [8, p.305]. Spatial navigation is the process of determining and maintaining a course or trajectory to a goal location accurately and efficiently [9] and is a requirement of daily life. Behavioral studies of spatial navigation indicate two important aspects for navigation involving allocentric or egocentric representations [10, 11]. Allocentric representations (e.g., South or North) are linked to a reference frame based on the external environment and independent of one's current location in it, while egocentric representations (e.g., left or right) reference spatial locations in the external world with respect to individual body space. Different cognitive strategies have also been observed during spatial navigation [12]. One is called spatial strategy, which involves a more Euclidean representation of space allowing a target location to be reached in a direct path from any given location. The other is a non-spatial strategy, which is related to using environmental information, such as turning left or right at certain points, without knowing the relationships between the start and the target locations.

In addition, gender- and age-related differences in spatial navigation have been widely reported from behavioral studies [13, 14]. It has been suggested that the gender-related differences may result from disparate cognitive strategies [12, 15-17]. For example, in a self-report study [12], men are more likely to report using a spatial strategy whereas women are more likely to report using a non-spatial strategy. Men perform best when using instructions indicating the directions (e.g., North or West) and metric distances (e.g., 100m), whereas women performed best when using instructions indicating the salient landmarks (e.g., the purple doors) and egocentric (e.g., left or right) turn directions [17].

Age-related deficits in spatial navigation have also been studied extensively in animals and humans. For example, older animals show decreased performance in a water maze task [18-20]. In this task, rats are placed in a large circular pool of opaque water. Younger rats quickly learn to escape by finding and climbing onto a small platform hidden beneath the water surface if this platform remains in a fixed location over a series of trials. However, older rats take longer to find the hidden platform, travel a longer distance in locating the platform, and may require more trials before reaching a designated criterion performance. Moffat and Resnick [21] developed a virtual water maze task for human application and found that, compared to younger adults, older adults traversed a longer linear distance to locate a hidden platform. Older adults also showed evidence of impairment in cognitive mapping as revealed by their poorer map constructions of the previously visited virtual environment and their impaired ability to locate the platform on the experimenter-provided 2D maps of the virtual environment.

With the development of 3D virtual reality environments it has now become possible to probe human brain function involved in spatial navigation using brain imaging methods. Virtual environments, rendered in a first-person view, have been created for spatial navigation brain imaging research. Participants could navigate through the virtual environment with the use of keyboard, control pad, joystick, or mouse. The most frequently used virtual environment is the virtual maze, in which participants were instructed to learn and recall the topographical information to locate a target object or to find their way out of the maze [22-27]. Another frequently used virtual environment is a complex virtual town, in which participants were asked to freely navigate the environment first and later they were required to either head directly toward the goal location or follow a trail of arrows [28-31]. In addition, a water maze task used in animal studies to examine the brain correlates of spatial navigation [32] has also been employed in many virtual reality studies [33-35]. In this task participants are required to search for a hidden target in a large pool of water.

Based on one PET [29] and five fMRI [24-26, 31, 36] virtual reality studies, the hippocampus (especially in the right hemisphere) is one of the most consistently identified substrates associated with spatial navigation in these virtual reality studies. Involvement of the hippocampus has usually been implicated in allocentric representations of space that allow the computation of the direction from any start location to any goal location [29]. The parietal cortex is another brain region that has been frequently identified as being involved in spatial navigation [24, 26, 29, 31, 36] and has been implicated in providing complementary egocentric representations of locations [37]. Various brain correlates were attributed to different cognitive strategies during spatial navigation. The hippocampus was specifically more involved in spatial strategy, and the caudate nucleus more associated with a non-spatial strategy [23, 25].

In addition to the hippocampus many other brain regions, including parahippocampal gyrus, parietal cortex, caudate nucleus, frontal cortex, posterior cingulate cortex, cerebellum, putamen, thalamus, and retrosplenial cortex, have also been identified as possible elements of a navigation system from numerous PET [29, 38] and fMRI virtual reality findings [22, 24, 26, 31, 36]. For example, the parahippocampal gyrus has been implicated in allocentric representations with a different role than the hippocampus. When objects can be used as specific landmarks for navigation there is significant activity of the parahippocampal gyrus, but no activity of the hippocampus, suggesting a role for the parahippocampal gyrus in object-location associations [36, 38]. In addition to its role in non-spatial strategy, activation of the caudate nucleus is also associated with getting to the target locations quickly [29]. Activations in the frontal areas (especially the inferior frontal gyrus and frontal eye fields) may be involved in planning, decision making, and attention during spatial navigation [29].

Virtual reality studies using fMRI to explore the brain mechanisms underlying gender-related differences in cognitive strategies during spatial navigation have been reported [24, 27, 34]. In a study by Gron et al. [24] participants were scanned as they searched for the way out of a complex, 3D, virtual reality maze. Men were significantly faster than women at finding the way out of the maze. Women coped with the task by engaging a right parietal and a right prefrontal area, whereas men recruited the left hippocampal region. Two later fMRI studies matched spatial navigation performance between men and women and consistently found that women showed significantly increased activation of parahippocampus in comparison with men [27, 34]. In addition, in one of the studies [34] women also showed increased activation in the hippocampus and cingulate cortex while in the other study [27] men showed increased activation of posterior cingulate cortex and retrosplenial cortex relative to women. These findings suggest that even when men and woman are well-matched on navigation performance they appear to use different brain mechanisms to achieve the same behavioral end point, and the distinct functional correlates of spatial navigation in women versus men may be related to the differential use of cognitive strategies during spatial navigation.

Virtual reality studies using fMRI to study age-related differences in spatial navigation have found that, compared to younger adults, older adults showed reduced activation in the hippocampus, parahippocampal gyrus, and parietal cortex [26, 39, 40]. For both age groups, level of activation in hippocampus and parahippocampal gyrus was positively correlated with navigation accuracy [26]. In addition, age-related attenuation in functional activation in prefrontal and parahippocampal areas was accompanied by brain volume reduction [39]. Findings from these studies suggest that hippocampus, parahippocampal gyrus, parietal cortex, prefrontal cortex, and caudate nucleus play a critical role in age-related decline in spatial navigation.

Collectively, these findings are consistent with previous animal [32, 41] and human lesion [42, 43] studies on spatial navigation. When virtual reality technology was not incorporated into functional brain imaging paradigm, many brain imaging studies could only measure brain correlates of "mental navigation" by asking participants to imagine and make decisions about the routes and landmarks from their mental representations of the environment [44, 45]. Although the feasibility of virtual reality technology in understanding the neuronal underpinnings of spatial navigation in human beings has been demonstrated, future studies investigating similarity and disparity in brain correlates of spatial navigation between viewing virtual reality and conventional 2D visual display will be needed.

2.2. Car driving

Car driving is a complex behavior involving interrelated cognitive elements, which may include attention, perception, visuomotor integration, working memory, and decision-making. Simulated driving environments using virtual reality have been used to understand the cognitive elements involved in driving behavior and to study how driving performance is affected by factors such as alcohol intoxication [46] and distraction [47, 48]. Understanding the neuronal underpinnings of these cognitive elements and factors degrading driving performance is crucial for safe driving. The combination of virtual reality and brain imaging techniques is a useful approach for studying the brain during driving.

Virtual environments created for simulated driving are relatively homogeneous across studies. Participants are presented with an 'in-car' view of a road and a readout of speed. Most studies included both driving and observing conditions. During driving conditions, participants were instructed to drive within a predetermined speed range [49-57], to maintain a constant distance from a preceding car traveling at varying speeds [58], or to drive without deviating from the road [59]. Different input modalities, such as a brake pedal plus accelerator and a steering wheel, joystick or a controller with buttons, were used to brake, and to adjust speed and driving direction. During observing conditions, participants passively viewed a simulated driving scene. Some studies added traffic signs (e.g., speed limit, stop signs, and yield for pedestrians) and/or a stream of oncoming traffic and asked participants to abide by all conventional traffic rules [49, 53, 54, 56, 57].

Using these virtual environments, fMRI studies of healthy adults showed that a network of brain regions, including cerebellum and fronto-parietal areas, was more active during simulated driving than passive viewing conditions [51, 55-59]. The increased activity in these brain regions was common to all prepared movement executions (such as starting, stopping and turning), with the cerebellum more specifically linked to fine-control during movement execution and the fronto-parietal region related to visual attention [57, 58]. Involvement of other brain regions, including occipital cortex, basal ganglia, thalamus, and amygdala, has also been reported [51, 55, 56, 58, 59], although without clear consensus regarding what cognitive elements were associated with these other brain regions.

FMRI studies have also examined the alcohol dose effects on brain networks during simulated driving in healthy adults [52-54]. Participants received single-blind individualized doses of beverage alcohol designed to produce moderate, high or placebo (i.e., sober baseline) blood alcohol content. Compared to the placebo condition, many brain networks, including cerebellum and fronto-temporal-basal ganglia circuits, were significantly affected by the high levels of alcohol, which was associated with unstable motor vehicle steering, increased driving speed, and opposite white line crossings during simulated driving [52-54].

Distraction is another factor that impacts driving performance reported in behavioral studies [47, 48]. However, there is scant evidence showing how distraction during driving modulates brain function, mainly due to difficulties in applying ecologically valid driving hardware to the investigation. Recently, an fMRI study examined distraction effects on brain correlates of simulated driving in a virtual reality environment [56]. In this study, healthy participants were instructed to respond to audio tasks consisting of simple true or false questions, such as "a triangle has four sides", by pressing a left (true) or right (false) buttons on the steering wheel during simulated driving. Driving with this type of distraction, increased activations in auditory cortex, precuneus, frontal cortex, and cerebellum and slightly decreased activations in the visual cortex were observed suggesting that distraction might draw resources away from the primary task of driving. This compromise in resources also indicates the potential risk for danger when distracted. This finding may have implications for screening drivers with a history of head injuries or pre-existing cognitive impairment.

Together, these findings have demonstrated that virtual reality technology allows simulations of driving tasks that would be impractical, dangerous, unethical, or even impossible in real contexts and enables the measurement of brain function during simulated driving [60].

2.3. Social cognition

Social cognition is a key topic in brain imaging studies. Almost all brain imaging studies in social cognition have been performed on single subjects who are isolated from a real social presence or interaction relevant to the study. Thus studies of theory of mind or social emotions may employ stories or pictures of others, attachment studies may use photographs or texts about loved ones, etc. This isolation is due in part to limitations imposed by brain imaging technologies, such as MRI and PET scanners that do not allow for most types of direct social contact or interactions during imaging. Although an individual can engage alone in important types of social cognitive processing, such as social memories, imagination, prospection, etc., by its very nature social cognition depends on the dyad or group of individuals [61, 62]. Virtual reality is a means by which some degree of simulation of social interactions within the constraints of many brain imaging methods is possible. A few examples of how virtual reality has been employed in MRI studies of social cognition are described.

Eye gaze is an important social act. Virtual reality methods have been used in several innovative ways to study brain responses when an individual observes the gaze of another's eyes. Pelphrey et al. [63] used fMRI to assess brain activations while adult participants used virtual reality goggles to observe an animated man who shifted the gaze of his eyes either toward or away from the participant. Activations were observed in the superior temporal sulcus, known for its role in social cognition, and the fusiform gyrus. A similar study was then conducted on seven to ten year old children, which also showed the importance of the superior temporal sulcus and several other regions including the fusiform gyrus, middle temporal gyrus, and inferior parietal lobule [64]. Another study has examined brain activations when participants observed virtual characters for eye gaze and facial expressions and pointed to the importance of the medial prefrontal cortex [65]. All of these studies were based on passive observance of a virtual character. However, in naturalistic settings social acts such as eye gaze are socially responsive and interactive acts. Thus a more technically advanced methodology for interactive study of eye gaze during MR imaging has been developed [66]. The method employs an eyetracking system with real-time data transmission that can modulate a visual stimulus, such as the gaze of a virtual character, "allowing the participant to engage in 'on-line' interaction with this virtual other in real-time" [66, p.98]. Using this approach a virtual "joint attention task" was examined. Results included increased activations in the medial prefrontal cortex, posterior cingulate cortex, and anterior temporal poles. Note that some of these regions overlap the default mode network [66]. Regions of the default mode network typically decrease during cognitively demanding tasks [67]. However, regions in the medial prefrontal cortex can also be involved in outcome monitoring, understanding intentionality, and triadic relationships (joint attention toward a third object) [66, p.105].

Joint action is a key type of behavioral aspect of social interactions. Because joint actions are difficult to study during MRI studies, virtual actions have been employed to probe the neural correlates of joint action. In one approach, the participant performed a task of lifting a virtual bar while balancing a ball on the bar during MR imaging and did so either alone or with the assistance of another person (who was not being imaged) [68]. Joint action versus solo action showed increased activations in regions of the mirror neuron system of the inferior frontal gyrus and inferior parietal lobule.

Finally, there is an important emerging brain imaging technique of hyperscanning for which the use of virtual reality environments has also been proposed [69]. Hyperscanning employs simultaneous brain imaging of more than one participant interacting with a task [70]. Simultaneous scanning of two (or more) interacting participants allows for observation of the complex interplay of brain responses that emerge in real-time during an interpersonal interaction. Among these responses some of the most interesting new observations made possible by hyperscanning are inter-brain synchronization or coherence [71, 72]. For example, interbrain synchronization of EEG or near infrared spectroscopy (NIRS) brain signals has been observed when two participants imitated hand movements [72] or played a computer game together [71] respectively. FMRI hyperscanning has also been conducted [70]. In one study fMRI hyperscanning of two participants was conducted simultaneously while the participants worked together or alone to "drive" through a maze displayed on a video screen [73]. One result was that highest activation in the caudate, putamen, and orbitofrontal cortex – regions in the reward system – was observed when the two participants worked together to successfully complete the task, indicating the importance of cooperation in activating the reward system. Note that in all of the above hyperscanning studies, whether fMRI, EEG, or NIRS, the participants being simultaneously scanned did not watch the other participant but, instead, watched a video screen of a maze, televised hand movements, or computer game respectively. Thus virtual reality paradigms could be useful for hyperscanning. Preliminary reports of hyperscanning employing virtual reality include a study in which participants competed to increase regional brain activations [74].

2.4. Measuring brain function using 3D, 2D, or real stimuli

Many types of virtual reality tasks in studies cited in this chapter could have conventional counterparts. "Driving" could be performed using a virtual environment [49-59] or a nonvirtual environment [73]. Pain distractors could comprise virtual reality protocols [75-77] or 2D equivalents. Would the use of virtual reality 3D versus conventional 2D versions of task stimuli lead to observable differences in brain responses? Such questions will be important not only for understanding the neural substrates of virtual reality tasks but can also contribute to testing and understanding the efficacy of virtual reality versus conventional therapies. From another perspective it is also the case that, although 3D virtual reality appears more consistent with the 3D nature of reality than 2D representations, virtual reality is not identical to reality. Would differences be observed for brain responses to stimuli comprising 2D or 3D representations versus real objects, such as response to a 2D televised display of someone's hands moving [72] versus direct observation of the person's hands; or to 2D versus 3D renditions of televised displays of someone's hands?

At this time these types of questions are only beginning to be addressed. Virtual reality protocols are still relatively infrequent in brain imaging studies and our most productive brain imaging methods for neurocognitive studies, namely MRI and PET, have usually precluded the use of real objects, environments, and interpersonal interactions for comparisons. Studies that have been conducted so far, however, suggest that there can be significant differences in how the brain responds to 2D, 3D, and real stimuli. A recent fMRI study on the neural substrates of visual motion processing included observations of brain regions involved in 2D versus 3D flow stimuli [78]. The cingulate sulcus visual area, a region in the dorsal posterior cingulate cortex, showed stronger activation for 2D stimuli, while regions in the occipital and temporal lobes (V5/middle temporal and middle superior temporal areas) were more involved in 3D stimuli. There has also been an innovative fMRI study of brain responses to repeated observation of 2D pictures of objects versus the real objects [79]. Repeated observation of 2D pictures of objects showed well-known repetition effects in activations in two regions of the lateral occipital complex, a posterior dorsal portion and anterior ventral portion in the posterior fusiform sulcus. The repetition effect is an attenuation of signal referred to as "fMRI adaptation" or "repetition suppression" [79, p.1]. Repeated observation of the real objects, however, showed a surprisingly weak repetition effect. Overall, results from these two studies indicate that neural processing of 2D, 3D, and real stimuli may differ in important ways. Although this topic will require many further studies these results [78, 79] suggest that there can be significant differences from use of 2D and 3D representations and real objects as stimuli in brain imaging studies. Understanding these differences will contribute to an understanding of the neural correlates of virtual reality and can facilitate the development of more ecologically valid virtual simulations for research and therapeutic applications.

3. Applications to studies of clinical disorders

Virtual reality has been integrated into non-invasive brain imaging in hopes of improving some kinds of clinical assessments and treatments. Specifically, virtual reality tasks have been employed as cognitive paradigms in brain imaging assessments of several neurological and psychiatric disorders. This approach is being developed to understand pathophysiological mechanisms underlying the diseases or to evaluate treatment effects. Virtual reality treatments have also begun to be developed for some clinical populations and brain imaging is being explored to assess effects of treatment. Finally, use of virtual reality to augment imaging procedures for presurgical planning is also being developed. In this section we will describe some of these applications for different clinical populations. All of these applications are in early stages of development and will require further studies to determine if and how these applications of virtual reality add clinical benefits and to explicate the neural mechanisms of action. Nonetheless the applications are interesting examples of approaches being explored for assessment and treatment of clinical disorders.

3.1. Addiction

Successful recovery from additions is often hampered by craving or strong desire for the substance of addiction. Craving can be initiated or enhanced by cues of the substance and associated factors. One approach to treatment for nicotine addiction and cigarette smoking is Cue Exposure Treatment (CET), which attempts to extinguish cue-induced craving by use of repeated exposure to cues [80]. CET, however, has limits on the nature of environmental cues that can be simulated, the degree of sense of "immersion" and reality, and has also shown decreased efficacy over time. Thus a virtual reality variation of cue exposure treatment, the Virtual Environment- CET (VR-CET), has been developed as an alternative to CET [80, 81]. To explore VR-CET treatment effects fMRI was conducted on eight adolescents who were cigarette smokers before and after they received VR-CET [81]. (No comparison treatment was assessed.) The fMRI task was to observe photographs of smoking cues versus neutral images during the scan. Participants did not report significantly decreased craving for cigarettes after treatment with VR-CET. However, fMRI showed decreased activation in the inferior frontal and superior frontal gyri in response to smoking cues after treatment with VR-CET. In a previous study, the inferior frontal gyrus region was one of several regions that showed increased activation from smoking cues in comparison with neutral cues, as well as increased activation in smokers versus nonsmokers for smoking minus neutral cues [82]. The results suggested that there can be a decreased brain response to smoking cues after VR-CET. Further studies will be required to examine effects of VR-CET on brain function, subjective craving, and smoking behaviors, and whether VR-CET can become an effective and specific treatment for smoking addiction.

3.2. Cerebral palsy

Cerebral palsy is a disorder of brain development that includes dysfunction of movement, posture, perception, communication, and other cognitive functions. Virtual reality therapies are being developed to treat cerebral palsy [83]. The use of fMRI to assess effects of virtual reality therapy has been described for a single case of a child with hemiparetic cerebral palsy [84]. The child had a right hemiparesis with encephalomalacia in the left temperoparietal cortex and corona radiata. Virtual reality therapy was given to improve function in the right shoulder and elbow. FMRI of right elbow flexion-extension movement was conducted before and after therapy. Results included observation of bilateral activation of the primary sensorimotor cortices before virtual reality therapy but only contralateral activation after treatment, indicating that the brain was capable of neuroplastic changes in cerebral palsy. The investigators noted that further studies would be required to compare virtual reality to other forms of rehabilitation to determine whether effects were specific or not to use of a virtual reality therapy.

3.3. Depression

Depression is one of the most common medical disorders and an increasing global health problem. It has been associated with many complex changes in brain structure and function. Virtual reality tasks have contributed to understanding the brain in depression. For example, MEG has been used to compare brain function of depressed versus healthy individuals during a virtual Morris water navigation task [85]. Depressed individuals showed decreased navigational performance of virtual Morris water tasks in comparison with healthy controls. In comparison with healthy individuals depressed individuals also showed decreased regional theta (4-8 Hz) activity in multiple regions, including the hippocampus and parahippocampal regions and lateral prefrontal cortices, and increased theta activity in posterior cortical regions and cerebellum. Results suggested that "dysfunction of right anterior hippocampus and parahippocampal cortices may underlie this deficit (impaired spatial navigation)" [85, p.836] in depressed individuals.

Given the abnormal findings associated with depression the question has arisen whether abnormalities are normalized after treatment. Hviid et al. [86] used virtual reality and MRI to assess long-term hippocampal structure and function in individuals in remission from depression. The participants comprised individuals in remission from depression and healthy controls, all of whom had been originally recruited for a PET/Depression project and then followed for 8 years. Baseline PET studies had shown that depressed participants had increased regional cerebral blood flow (rCBF) in the right hippocampus during rest [87]. The depressed participants had received several kinds of pharmacologic treatments for depression. Hippocampal volume at 8 year follow-up was assessed with structural MRI. Hippocampal function was assessed with a virtual reality navigation task in a shopping mall, the Counter-Strike Citymall Task (CSCT). A performance measure of CSCT, i.e. precision of navigation, had been shown to correlate with PET regional cerebral blood flow in the right hippocampus and, therefore, this task was used as a "right hippocampus-dependent task" [86, p.179]. Results showed that there were no significant differences between individuals in remission from depression and healthy controls for performance of the virtual reality precision of navigation task or MRI hippocampal volumes. Results suggested that individuals with depression in remission can show results similar to healthy individuals for some measures of hippocampal volume and function.

3.4. Epilepsy

Individuals with epilepsy may have seizures in unpredictable circumstances and, therefore, can face limitations in their activities of daily life. An important daily activity for many people is driving. Because of the potential hazards of a seizure during driving many people with epilepsy face the prospect of being restricted from driving. To better understand how driving may be impaired during a seizure Yang et al. [88] used EEG to monitor seizure activity in individuals with epilepsy while they played a virtual reality driving simulation game. Seizure activity and "driving" impairment were able to be observed and showed the potential value of virtual reality driving simulation for patient assessments.

Some cases of temporal lobe epilepsy are treated with surgical resection of the anterior temporal lobe when seizures cannot be managed medically. However, surgery comes with high risk for cognitive dysfunction. For example, the temporal lobe supports memory and this may become impaired after surgery. Thus it is helpful to have assessments that could indicate the relative risks of memory loss versus seizure relief. One presurgical measure to assess postsurgical cognitive risk is the degree of lateralization of memory function. FMRI of a memory task for object location in a virtual environment has been used to assess medial temporal lobe function prior to surgery [89]. Increased lateralization of presurgical hippocampal activation during the virtual reality task towards the ipsilesional side correlated with decline in verbal memory after surgery.

3.5. Movement disorders

3.5.1. Huntington's disease

Huntington's disease is a genetic neurological disorder that shows progressive impairment of movement, cognition, and mood and other psychological functions. In early stages of the disorder it manifests with relatively selective impairment of the caudate. The basal ganglia (caudate) and hippocampus are important regions that have been related to two systems for navigational memory, with evidence for both competitive and non-competitive or compensatory interactions between the caudate and the hippocampus. To further assess possible interactions between the caudate and hippocampus Voermans and colleagues [90] selected early stage Huntington's disease as a model of caudate dysfunction and predicted that individuals with early stage Huntington's disease and caudate dysfunction would show compensatory hippocampal function and near normal performance of a navigational task. FMRI was used to assess brain function in individuals with early Huntington's disease versus healthy controls during a navigational memory task employing a virtual home. A key measure was route recognition performance. Results included that right caudate activity associated with route recognition correlated negatively with disease severity while bilateral hippocampal activity correlated positively with disease severity. Further, healthy controls showed greater activity in the right caudate associated with route recognition than observed in individuals with Huntington's disease, while individuals with Huntington's disease showed greater activity in the right hippocampus and parahippocampus associated with route recognition than observed in healthy controls. Results indicated that the hippocampus could compensate for impaired caudate function in individuals with early Huntington's disease. This compensatory function could explain observations of relatively normal navigational performance in these individuals.

3.5.2. Parkinson's disease

Parkinson's disease is a movement disorder with many symptoms including tremor, bradykinesia, rigidity, and cognitive and mood changes. About 50% of individuals with Parkinson's disease exhibit spontaneous "freezing" while trying to move. A single case fMRI study identified the neural mechanisms involved in freezing while a patient with Parkinson's disease used foot pedals to "walk" in a virtual reality environment, both on and off dopaminergic medication [91]. Different brain activation and deactivation patterns were observed for different walking tasks and freezing episodes. For example, freezing episodes occurring when the individual was off medications showed increased activations in the "pSMA (pre supplementary motor area), motor cortices, DLPFC, VLPFC, (dorsolateral and ventrolateral prefrontal cortices) and posterior parietal regions… deactivation within the frontopolar cortices and precuneus" [91, p.809]. The study demonstrated the ability to obtain knowledge of functional brain activation patterns related to walking and freezing by employing a virtual reality protocol.

3.6. Pain

Pain can sometimes be unremitting and treatment resistant even to maximal doses of opioid medications. Thus alternative or adjunctive treatments are required. The use of virtual reality as an adjunct treatment for control of pain has been proposed. One hypothesis is that virtual reality will engage attention and therefore distract or decrease attention to pain [75]. Virtual reality approaches to treatment of pain have been assessed with fMRI using normal volunteers. In one study normal individuals subjected to thermal pain showed activations in the anterior cingulate, insula, thalamus, and primary and secondary somatosensory cortex regions that decreased during immersion in a virtual reality protocol (throwing snowballs in SnowWorld) [75, 76]. In another study fMRI was conducted to assess effects of no treatment, virtual reality, opioid, and combined virtual reality and opioid treatments of thermal pain in normal individuals [77]. Both virtual reality and opioid treatments given separately decreased brain activation in the insula and thalamus, and virtual reality treatment also decreased activation in the primary somatosensory cortex. Combined virtual reality and opioid treatment led to further decreases in brain activations in the above regions as well as decreased activation in the secondary somatosensory cortex. Additional studies, such as brain imaging comparisons employing other kinds of distractors and treatment combinations, are required to better understand the neural mechanisms specific to use of virtual reality distractors to augment analgesia.

A protocol for an fMRI study of a pain syndrome – fibromyalgia – has also been published that would examine effects of a virtual reality exposure therapy [92]. Therapy would comprise virtual reality exposure to exercises that could induce pain catastrophizing, with fMRI conducted before and after treatment.

3.7. Post-Traumatic Stress Disorder (PTSD)

Post-traumatic stress disorder (often combined with traumatic brain injury) affects many service members returning from deployment in conflict and war zones. Although symptoms of PTSD in early stages may be a normal response to trauma, chronic PTSD is diagnosed when recovery fails to continue. Cognitive behavioral therapy that includes imagined reliving of the traumatic experience, such as Prolonged Exposure (PE) imaginal therapy, has been developed and shown benefits as a treatment for PTSD [93]. However, virtual reality exposure therapy (VRET) has been developed as an alternative treatment that may have advantages over PE imaginal treatments, including the possibility for "shared experience" of the battlefield with the therapist, preference as a treatment by some individuals because it uses computers, etc. [93, p.130]. Roy et al. [94] have begun to employ VRET as a treatment for PTSD in service members and assess its effects in a small number of participants using fMRI of an Affective Stroop test. Participants were randomly assigned to receive VRET or PE treatments. FMRI results for VRET and PE participants were pooled because of the small number (eight) of total fMRI participants. Results before treatment showed increased activations in the amygdala and lateral prefrontal cortex, as observed in other studies of PTSD and consistent with hypervigilance and negative arousal, and in the subcallosal gyrus as observed for depression. There were also deactivations in the anterior cingulate region. FMRI after treatment indicated normalized responses in these regions.

3.8. Schizophrenia

Schizophrenia can be a highly debilitating psychiatric disorder with many complex psychological, cognitive, and behavioral symptoms. To advance understanding of the neurocognitive underpinnings of schizophrenia investigators have drawn from cognitive behavioral paradigms employed in animal studies. One such paradigm is the Morris water task described above. This is a task that has often been used to study spatial navigation, learning, and memory in nonhuman species, for which virtual variants exist that allow MRI studies of human brain function. Several brain regions involved in performance of MWT have also been implicated in schizophrenia. Thus Folley et al. [95] examined individuals with schizophrenia using structural MRI and fMRI during performance of a novel virtual

Morris water task to assess brain morphometry, neural circuits derived from independent component analysis, and regional brain activations. In comparison with normal controls, individuals with schizophrenia revealed differences in several distributed brain networks engaged during the tasks. They also demonstrated a lack of association between hippocampal activation and gray matter concentration that was observed in normal controls. The study demonstrated the value of virtual reality paradigms and MRI for translational research from animal studies.

3.9. Traumatic Brain Injury (TBI)

A concussion is a mild traumatic brain injury (MTBI) with mild symptoms of brain dysfunction. It can be one of the sequelae of sports injuries. Although symptoms of brain dysfunction after MTBI have usually been temporary there has been increasing interest in the possibility of residual brain impairment in otherwise asymptomatic persons. Evidence of brain pathology has, however, been difficult to demonstrate. Recently, fMRI of a virtual reality task was used to probe for impaired brain function in asymptomatic individuals with a recent history (within 30 days) of MTBI [96]. The virtual reality task was designed to probe spatial memory functioning. It comprised moving through a virtual corridor to find a virtual room. Behavioral measures showed that individuals with MTBI were able to perform the virtual reality navigation task similar to normal individuals. However, fMRI revealed that individuals with MTBI showed increased cluster size of activations in the parietal cortex, dorsolateral prefrontal cortex, and hippocampus indicating increased activation in these regions. These results provided fMRI evidence for regional brain impairment in individuals who appeared normal based on behavioral measures. The possibility that "efficient performance of navigation task … may give rise to hyperactivation of focal clusters and to recruitment of additional cerebral resources" [96, p.352] was one explanation given for the increased activations. It was also noted that further studies would be required to assess whether the findings were transient early phase changes or would be observed long-term.

More serious traumatic brain injuries can lead to a variety of cognitive and social dysfunctions. A virtual reality interpersonal paradigm has been employed to assess and social functioning in individuals with a history of TBI in conjunction with structural MRI studies [97]. The main task consisted of observing virtual reality presentations of people in conflict. Structural MRI was conducted for measurements of cortical thickness. Participants with TBI showed impaired social performance, with more impulsive and self-centered results. Cortical thickness of several brain regions correlated with task performance, including the orbitofrontal cortex with its role in motivation and reward.

3.10. Tumors

Surgery to remove tumors of the brain is accompanied by high risk for impaired function that depends on location of the tumor. Qiu et al. [98] have described the use of a virtual reality environment to display MRI results to the surgeon for presurgical planning for resection of gliomas located near cortical and subcortical motor pathways. Structural 3D

MRI was conducted to image the tumor and primary motor cortex. DTI was conducted for tractography of the pyramidal tracts (white matter tracts of myelinated axons of pyramidal cells that go from the primary motor cortex to the spinal cord and brain stem). The structural MRI and DTI results were then used for a stereoscopic 3D virtual reality display of the tumor, motor cortex, and pyramidal tracts for pre-surgical planning and surgery simulations. The virtual reality system allowed the surgeon to be able to visualize both the brain tumor and pyramidal tracts from multiple directions and planes. A segmentation tool was used to virtually rehearse tumor resection and compare different surgical approaches. This virtual reality 3D stereoscopic visualization of the brain provided increased information for presurgical planning and showed promise for improving surgical outcomes.

3.11. Vascular diseases

Study of cognition, emotion, and behavior in individuals with brain lesions has been an important method to elucidate neuroanatomical substrates of normal brain function. This approach was employed by Weniger et al. [99] to help identify the neuroanatomical substrates of egocentric navigation as introduced above. The investigators examined individuals with unilateral parietal lobe infarctions or hemorrhagic lesions using structural MRI. They assessed performance of two virtual reality navigation tasks in both normal individuals and those with parietal lobe lesions. The tasks were navigation in a virtual maze to assess egocentric navigation or navigation in a virtual park to assess allocentric navigation. Results showed that individuals with parietal cortical lesions had impaired navigation on the virtual maze task but normal performance on the virtual park task. Performance on the virtual maze task in individuals with parietal lesions also correlated with increasing size of the right precuneus. Results provided evidence for a role of the parietal cortex in egocentric navigation.

4. Brain-computer interfaces, real-time neuroimaging, and virtual reality

Virtual reality technology also plays an important role in enhancing brain-computer interfaces (BCI) that are uniquely enabled by real-time neuroimaging, specifically EEG and fMRI. The integration of virtual reality, BCI and real-time imaging has shown promise for training healthy adults and treating neurological and psychiatric disorders. Combined virtual reality and brain imaging may see its most sophisticated technical development in future applications of real-time fMRI biofeedback employing virtual reality environments. Overall, although combined use of virtual reality and brain imaging technologies for application to clinical disorders and enhancement of normal levels of functioning is still in early stages of development, many innovative approaches have appeared that herald the possibility of important future contributions to understanding and treatment of many clinical disorders.

4.1. EEG-based approach

EEG monitoring during virtual reality protocols has been used for training purposes in normal persons. These studies may have implications for future development of cognitive, behavioral, and psychological treatments of clinical disorders. As an example of training in normal persons, training of actors has been enhanced by using an EEG based sensory motor rhythm (SMR) neurofeedback method with eyes-open training in a virtual reality auditorium performance environment [100]. Assessments of training outcome included ratings of Hamlet performances on the Globe Theatre stage, London. Results showed higher ratings for acting performance using SMR with the virtual reality environment versus a 2D computer screen.

EEG monitoring can take place in very fast time scales (milliseconds) with relatively good mobility and comfort and has become the most important modality in brain-computer interfaces. A novel study of a single individual with tetraplegia and muscular dystrophy was conducted in which BCI with EEG monitoring of the sensorimotor cortex was developed to allow the individual to use his motor intention to control an avatar in an internet virtual environment [101]. The individual "successfully walked and chatted with other virtual users while using the BCI at home" [101, p.3]. EEG results also showed changes in event-related synchronizations and desynchronizations over the 5 month period of the study that suggested cortical plasticity. The possibility of BCI in conjunction with virtual reality as a therapeutic regimen has also been discussed for walking rehabilitation and treatment of autism [102, 103].

4.2. fMRI-based approach

Functional MRI studies may be performed with a real-time and interactive manner, enabling participants to adaptively modify their cognitive and emotional processing strategies according to dynamic brain activities measured with fMRI. This real-time fMRI based biofeedback scheme makes it possible to enhance subjects' ability to control and modulate their central nervous system, and is expected to be valuable for learning and various therapeutic applications such as cognitive rehabilitation. The dynamic and vivid information provided by virtual reality technologies, not surprisingly, can directly improve the performance of real-time fMRI based biofeedback, in comparison with 2D visual feedback used in most conventional fMRI experiments, as summarized in this section.

Real-time biofeedback of brain activity information has been previously demonstrated with EEG, which measure changes of the electric fields associated with neuronal activation at high temporal resolution (on the order of milliseconds). The measured information can be fed back to subjects for behavioral changes [104-106]. A limitation of EEG and EEG-based biofeedback is that the spatial-resolution and spatial accuracy of functional mapping are limited. Furthermore, it is difficult to reliably measure neuronal activities in brain regions away from the skull using EEG.

As compared with EEG, fMRI provides imaging data at a much higher spatial resolution, and can image the whole brain at approximately equal sensitivity and accuracy. Because of these advantages, real-time fMRI based biofeedback is becoming an important research topic [107, 108]. Note that the data acquired with fMRI have lower temporal resolution (on the order of seconds) than EEG. Nevertheless, the biofeedback with information updated every few seconds is adequate for most behavioral paradigms.

The real-time fMRI based biofeedback is schematically illustrated in Figure, using an experimental design of fear control as an example. A subject is lying in the MRI scanner, watching a video showing a roller coaster scene while brain signals are dynamically acquired with an MRI system (panel A). The MRI signals are then transferred to a workstation (panel B) for real-time reconstruction of functional activation maps (panel C). Functional activation in pre-selected regions of interest, such as the amygdala (i.e., a region associated with fear: yellow circle in panel C), is quantified [109]. The level of amygdala activation is converted to a scale bar (panel D), and displayed in the computer screen (panel E) so that the subject is aware of his own level of fear quantified by functional MRI. The subject is then asked to suppress the fear based on the information provided visually and dynamically.

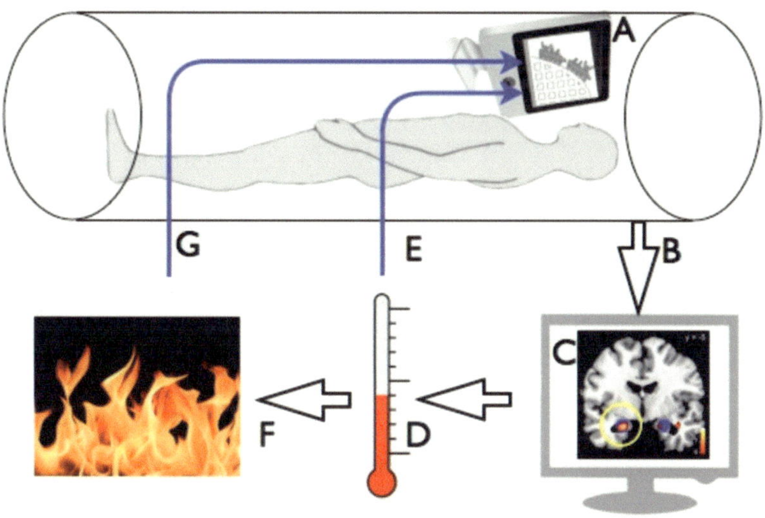

Figure 1. Schematic diagram of real-time fMRI biofeedback.

This real-time fMRI based biofeedback study can be improved by virtual reality technologies in several ways. First, if the roller coaster scene is displayed with 3D virtual reality through goggles instead of 2D viewing through a computer screen then the fear induced brain signals is expected to be significantly higher. Second, instead of showing a 2D thermometer-like scale bar in the computer screen, the biofeedback can be made more intense by employing 3D flames of different sizes to show the level of functional signals (panel F) stimulated by the virtual reality task [110].

In addition to the amygdala signal modulation that has been shown in previous real-time fMRI studies [109], other brain regions that have been investigated with real-time fMRI include somatomotor cortex [111-113], parahippocampal gyrus [114], the auditory cortex [115], the insular cortex [116], and the anterior cingulate cortex [112, 117].

Recent studies further showed that, in contrast to relying on fMRI signals in pre-selected regions of interest, the brain state derived from analyzing the overall activation patterns can be used in real-time fMRI based biofeedback without needing prior knowledge of activation regions [118]. It is expected that the brain state based biofeedback, without needing prior assumption on brain regions involved, should prove valuable for various types of virtual reality-based training in which multiple brain regions or even multiple neuronal connectivity networks are involved.

As summarized by deCharms [110], real-time fMRI biofeedback has several promising therapeutic applications, including treatment of neurological diseases and enhanced psychotherapy. For example, through training patients to modulate their central nervous system activities, the level of chronic pain can be reduced [117]. It appears that what a participant does internally while learning highly specific cognitive strategies for controlling pain using real-time fMRI is closely related to what the same participant might do while using other methods for learning cognitive control over pain, such as in cognitive behavioral therapy [110, p.725]. It is hoped that virtual reality will improve the efficacy of therapeutic procedures based on real-time fMRI biofeedback.

5. Conclusion

Many studies have now employed virtual reality paradigms in brain imaging studies performed with healthy individuals or those with clinical diseases. One of the most significant contributions of these studies has been to initiate brain imaging studies of some human behaviors that would have been difficult or impossible to assess with imaging methods such as MRI. Most often virtual reality paradigms have appeared as cognitive behavioral stimuli employed as tasks in brain imaging studies or as components of treatment protocols for which brain imaging is an important measure of treatment effects. There are also emerging possibilities for therapies with brain computer interfaces and real-time biofeedback at the neurological level that could employ complex virtual reality environments for which real-world counterparts would be impractical or impossible to use. We believe that with continued improvements in virtual reality and brain imaging technology this application holds considerable promise, both for theoretical frameworks of brain science and for translational applications.

Author details

Ying-hui Chou, David J. Madden, Allen W. Song and Nan-kuei Chen
Brain Imaging and Analysis Center, Duke University Medical Center, Durham, North Carolina, USA

Carol P. Weingarten and David J. Madden
Department of Psychiatry and Behavioral Sciences, Duke University Medical Center, Durham, North Carolina, USA

David J. Madden
Center for Cognitive Neuroscience, Duke University Medical Center, Durham, North Carolina, USA

Acknowledgement

This chapter was support by research grant R01-NS074045 from National Institute of Neurological Disorders and Stroke (NKC), and research grant R01 AG039684 (DJM) and training grant T32 AG000029 (YHC) from the National Institute on Aging.

6. References

[1] Burdea GC, Coiffet P (2003) Virtual Reality Technology. 2nd ed. Hoboken, NJ: John Wiley & Sons.

[2] Liang Z-P, Lauterbur PC (2000) Principles of Magnetic Resonance Imaging: A Signal Processing Perspective. New York: The Institute of Electrical and Electronics Engineers.

[3] Mori S (2007) Introduction to Diffusion Tensor Imaging. Boston: Elsevier.

[4] Hoekstra OS, Juweid ME, editors. (2011) Positron Emission Tomography. New York: Humana Press.

[5] Huettel SA, Song AW, McCarthy G (2009) Functional Magnetic Resonance Imaging. 2nd ed. Sunderland, Massachusetts: Sinauer.

[6] Lopes da Silva FH, Niedermeyer E, Schomer DL, editors. (2011) Niedermeyer's Electroencephalography : Basic Principles, Clinical Applications, and Related Fields. 6th ed. Philadelphia: Wolters Kluwer Health/Lippincott Williams & Wilkins.

[7] Sato S, editor. (1990) Magnetoencephalography. New York: Raven Press.

[8] Levitt TS, Lawton DT (1990) Qualitative Navigation for Mobile Robots. Artif Intell. 44: 305-360.

[9] Franz MO, Mallot HA (2000) Biomimetic Robot Navigation. Rob Auton Syst. 30: 133-153.

[10] Klatzky RL (1998) Allocentric and Egocentric Spatial Representations: Definitions, Distinctions, and Interconnections. In: Freksa C, Habel C, Wender KF, editors. Spatial Cognition. Heidelberg: Springer. pp. 1-17.

[11] Mou W, McNamara TP, Rump B, Xiao C (2006) Roles of Egocentric and Allocentric Spatial Representations in Locomotion and Reorientation. J Exp Psychol Learn Mem Cogn. 32: 1274-1290.

[12] Lawton CA (1994) Gender Differences in Way-Fining Strategies: Relationship to Spatial Ability and Spatial Anxiety. Sex Roles. 30: 765-779.

[13] Coluccia E, Louse G (2004) Gender Differences in Spatial Orientation: A Review. J Environ Psychol. 24: 329-340.

[14] Moffat SD (2009) Aging and Spatial Navigation: What Do We Know and Where Do We Go? Neuropsychol Rev. 19: 478-489.

[15] Galea LAM, Kimura D (1992) Sex Differences in Route-Learning. Pers Individ Differ. 14: 53-65.

[16] Sandstrom NJ, Kaufman J, Huettel SA (1998) Males and Females Use Different Distal Cues in a Virtual Environment Navigation Task. Brain Res Cogn Brain Res. 6: 351-360.

[17] Saucier DM, MacFadden A, Bell S, Elias LJ (2002) Are Sex Differences in Navigation Caused by Sexually Dimorphic Strategies or by Differences in the Ability to Use the Strategies? Behav Neurosci. 116: 403-410.

[18] Begega A, Cienfuegos S, Rubio S, Santin JL, Miranda R, Arias JL (2001) Effects of Ageing on Allocentric and Egocentric Spatial Strategies in the Wistar Rat. Behav Processes. 53: 75-85.

[19] Gallagher M, Pelleymounter MA (1988) Spatial Learning Deficits in Older Rats: A Model for Memory Decline in the Aged. Neurobiol Aging. 9: 549-556.

[20] Lukoyanov NV, Andrade JP, Dulce Madeira M, Paula-Barbosa MM (1999) Effects of Age and Sex on the Water Maze Performance and Hippocampal Cholinergic Fibers in Rats. Neurosci Lett. 269: 141-144.

[21] Moffat SD, Resnick SM (2002) Effects of Age on Virtual Environment Place Navigation and Allocentric Cognitive Mapping. Behav Neurosci. 116: 851-859.

[22] Aguirre GK, Detre JA, Alsop DC, D'Esposito M (1996) The Parahippocampus Subserves Topographical Learning in Man. Cereb Cortex. 6: 823-829.

[23] Bohbot VD, Lerch JL, Thorndycraft B, Iaria G, Zijdenbos AP (2007) Gray Matter Differences Correlate with Spontaneous Strategies in a Human Virtual Navigation Task. J Neurosci. 27: 10078-10083.

[24] Gron G, Wunderlich AP, Spitzer M, Tomczak R, Riepe MW (2000) Brain Activation During Human Navigation: Gender-Different Neural Networks as Substrate of Performance. Nat Neurosci. 3: 404-408.

[25] Iaria G, Petrides M, Dagher A, Pike B, Bohbot VD (2003) Cognitive Strategies Dependent on the Hippocampus and Caudate Nucleus in Human Navigation: Variability and Change with Practice. J Neurosci. 23: 5945-5952.

[26] Moffat SD, Elkins W, Resnick SM (2006) Age Differences in the Neural Systems Supporting Human Allocentric Spatial Navigation. Neurobiol Aging. 27: 965-972.

[27] Nowak NT, Resnick SM, Elkins W, Moffat SD (2011) Sex Differnces in Brain Activation During Virtual Navigation: A Functional MRI Study. 33rd Annual Meeting of the Cognitive Science Society; Boston, MA, USA.

[28] Hartley T, Maguire EA, Spiers HJ, Burgress N (2003) The Well-Worn Route and the Path Less Traveled: Distinct Neural Bases of Route Following and Wayfinding in Humans. Neuron. 37: 877-888.

[29] Maguire EA, Burgress N, Donnett JG, Frackowiak RSJ, Frith CD, O'Keefe J (1998) Knowing Where and Getting There: A Human Navigation Network. Science. 280: 921-924.

[30] Ramos-Loyo J, Sanchez-Loyo LM (2011) Gender Differences in EEG Coherent Activity before and after Training Navigation Skills in Virtual Environments. Hum Physiol. 37: 700-707.

[31] Pine DS, Grun J, Maguire EA, Burgress N, Zarahn E, Koda V, et al. (2002) Neurodevelopmental Aspects of Spatial Navigation: A Virtual Reality FMRI Study. Neuroimage. 15: 396-406.

[32] Morris RGM, Garrud P, Rawlins JNP, O'Keefe J (1982) Place Navigation Impaired in Rats with Hippocampal Lesions. Nature. 297: 681-683.

[33] Moffat SD, Kennedy KM, Rodrigue KM, Raz N (2007) Extrahippocampal Contributions to Age Differences in Human Spatial Navigation. Cereb Cortex. 17: 1274-1282.

[34] Sneider JT, Sava S, Rogowska J, Yurgelun-Todd DA (2011) A Preliminary Study of Sex Differences in Brain Activation During a Spatial Navigation Task in Healthy Adults. Percept Mot Skills. 113: 461-480.

[35] Driscoll I, Hamilton DA, Petropoulos H, Yeo RA, Brooks WM, Baumgartner RN, et al. (2003) The Aging Hippocampus: Cognitive, Biochemical and Structural Findings. Cereb Cortex. 13: 1344-1351.

[36] Maguire EA, Frackowiak RSJ, Frith CD (1997) Recalling Routes around London: Activation of the Right Hippocampus in Taxi Drivers. J Neurosci. 17: 7103-7110.

[37] Thier P, Andersen RA (1996) Electrical Microstimulation Suggests Two Different Forms of Representation of Head-Centered Space in the Intraparietal Sulcus of Rhesus Monkeys. Proc Natl Acad Sci USA. 93: 4962-4967.

[38] Maguire EA, Frith CD, Burgress N, Donnett JG, O'Keefe J (1998) Knowing Where Things Are Parahippocampal Involvement in Encoding Object Locations in Virtual Large-Scale Space. J Cogn Neurosci. 10: 61-76.

[39] Antonova E, Parslow D, Brammer M, Dawson GR, Jackson SHD, Morris RGM (2009) Age-Related Neural Activity During Allocentric Spatial Memory. Memory. 17: 125-143.

[40] Meulenbroek O, Petersson KM, Voermans N, Weber B, Fernandez G (2004) Age Differences in Neural Correlates of Route Encoding and Route Recognition. Neuroimage. 22: 1503-1514.

[41] O'Keefe J, Dostrovsky J (1971) The Hippocampus as a Spatial Map: Preliminary Evidence from Unit Activity in the Freely-Moving Rat. Brain Res. 34: 171-175.

[42] Astur RS, Taylor LB, Mamelak AN, Philpott L, Sutherland RJ (2002) Humans with Hippocampus Damage Display Severe Spatial Memory Impairments in a Virtual Morris Water Task. Behav Brain Res. 132: 77-84.

[43] Maguire EA (2001) The Retrosplenial Contribution to Human Navigation: A Review of Lesion and Neuroimaging Findings. Scand J Psychol. 42: 225-238.

[44] Rosenbaum RS, Ziegler M, Winocur G, Grady CL, Moscovitch M (2004) "I Have Often Walked Down This Street Before": FMRI Studies on the Hippocampus and Other Structures During Mental Navigation of an Old Environment. Hippocampus. 14: 826-835.

[45] Ghaem O, Mellet E, Crivello F, Tzourio N, Mazoyer B, Berthoz A, et al. (1997) Mental Navigation Along Memorized Routes Activates the Hippocampus, Precuneus, and Insula. Neuroreport. 8: 739-744.

[46] Mitchell MC (1985) Alcohol-Induced Impairment of Central Nervous System Function: Behavioral Skills Involved in Driving. J Stud Alcohol Suppl. 10: 109-116.

[47] Ma H, Rolka H, Mandl K, Buckeridge D, Fleischauer A, Pavlin J (2005) Implementation of Laboratory Order Data in Biosense Early Event Detection and Situation Awareness System. MMWR Morb Mortal Wkly Rep. 54 Suppl: 27-30.

[48] Strayer DL, Johnston WA (2001) Driven to Distraction: Dual-Task Studies of Simulated Driving and Conversing on a Cellular Telephone. Psychol Sci. 12: 462-466.

[49] Allen AJ, Meda SA, Skudlarski P, Calhoun VD, Astur R, Ruopp KC, et al. (2009) Effects of Alcohol on Performance on a Distraction Task During Simulated Driving. Alcohol Clin Exp Res. 33: 617-625.

[50] Calhoun VD, Carvalho K, Astur R, Pearlson GD (2005) Using Virtual Reality to Study Alcohol Intoxication Effects on the Neural Correlates of Simulated Driving. Appl Psychophysiol Biofeedback. 30: 285-306.

[51] Calhoun VD, Pekar JJ, McGinty VB, Adali T, Watson TD, Pearlson GD (2002) Different Activation Dynamics in Multiple Neural Systems During Simulated Driving. Hum Brain Mapp. 16: 158-167.

[52] Calhoun VD, Pekar JJ, Pearlson GD (2004) Alcohol Intoxication Effects on Simulated Driving: Exploring Alcohol-Dose Effects on Brain Activation Using Functional MRI. Neuropsychopharmacology. 29: 2097-2107.

[53] Meda SA, Calhoun VD, Astur RS, Turner BM, Ruopp K, Pearlson GD (2009) Alcohol Dose Effects on Brain Circuits During Simulated Driving: An FMRI Study. Hum Brain Mapp. 30: 1257-1270.

[54] Rzepecki-Smith CI, Meda SA, Calhoun VD, Stevens MC, Jafri MJ, Astur RS, et al. (2010) Disruptions in Functional Network Connectivity During Alcohol Intoxicated Driving. Alcohol Clin Exp Res. 34: 479-487.

[55] Walter H, Vetter SC, Grothe J, Wunderlich AP, Hahn S, Spitzer M (2001) The Neural Correlates of Driving. Neuroreport. 13: 1763-1767.

[56] Kan KYG (2010) Neural Correlates of Driving in a Virtual Reality Environment. Toronto, Canada: University of Toronto.

[57] Spiers HJ, Maguire EA (2007) Neural Substrates of Driving Behaviour. Neuroimage. 36: 245-255.

[58] Uchiyama Y, Ebe K, Kozato A, Okada T, Sadato N (2003) The Neural Substrates of Driving at a Safe Distance: A Functional MRI Study. Neurosci Lett. 352: 199-202.

[59] Mader M, Bresges A, Topal R, Busse A, Forsting M, Gizewski ER (2009) Simulated Car Driving in FMRI--Cerebral Activation Patterns Driving an Unfamiliar and a Familiar Route. Neurosci Lett. 464: 222-227.

[60] Calhoun VD, Pearlson GD (2012) A Selective Review of Simulated Driving Studies: Combining Naturalistic and Hybrid Paradigms, Analysis Approaches, and Future Directions. Neuroimage. 59: 25-35.

[61] Dumas G (2011) Towards a Two-Body Neuroscience. Commun Integr Biol. 4: 349-352.

[62] Sanger J, Lindenberger U, Muller V (2011) Interactive Brains, Social Minds. Commun Integr Biol. 4: 655-663.

[63] Pelphrey KA, Viola RJ, McCarthy G (2004) When Strangers Pass: Processing of Mutual and Averted Social Gaze in the Superior Temporal Sulcus. Psychol Sci. 15: 598-603.

[64] Mosconi MW, Mack PB, McCarthy G, Pelphrey KA (2005) Taking an "Intentional Stance" on Eye-Gaze Shifts: A Functional Neuroimaging Study of Social Perception in Children. Neuroimage. 27: 247-252.

[65] Schilbach L, Wohlschlaeger AM, Kraemer NC, Newen A, Shah NJ, Fink GR, et al. (2006) Being with Virtual Others: Neural Correlates of Social Interaction. Neuropsychologia. 44: 718-730.

[66] Wilms M, Schilbach L, Pfeiffer U, Bente G, Fink GR, Vogeley K (2010) It's in Your Eyes--Using Gaze-Contingent Stimuli to Create Truly Interactive Paradigms for Social Cognitive and Affective Neuroscience. Soc Cogn Affect Neurosci. 5: 98-107.

[67] Greicius MD, Krasnow B, Reiss AL, Menon V (2003) Functional Connectivity in the Resting Brain: A Network Analysis of the Default Mode Hypothesis. Proc Natl Acad Sci U S A. 100: 253-258.

[68] Newman-Norlund RD, Bosga J, Meulenbroek RG, Bekkering H (2008) Anatomical Substrates of Cooperative Joint-Action in a Continuous Motor Task: Virtual Lifting and Balancing. Neuroimage. 41: 169-177.

[69] Bohil CJ, Alicea B, Biocca FA (2011) Virtual Reality in Neuroscience Research and Therapy. Nat Rev Neurosci. 12: 752-762.

[70] Montague PR, Berns GS, Cohen JD, McClure SM, Pagnoni G, Dhamala M, et al. (2002) Hyperscanning: Simultaneous FMRI During Linked Social Interactions. Neuroimage. 16: 1159-1164.

[71] Cui X, Bryant DM, Reiss AL (2012) NIRS-Based Hyperscanning Reveals Increased Interpersonal Coherence in Superior Frontal Cortex During Cooperation. Neuroimage. 59: 2430-2437.

[72] Dumas G, Nadel J, Soussignan R, Martinerie J, Garnero L (2010) Inter-Brain Synchronization During Social Interaction. PLoS One. 5: e12166.

[73] Krill AL, Platek SM (2012) Working Together May Be Better: Activation of Reward Centers During a Cooperative Maze Task. PLoS One. 7: e30613.

[74] Moench T, Hollmann M, Grzeschik R, Mueller C, Luetzkendorf R, Baecke S, et al. (2008) Real-Time Classification of Activated Brain Areas for FMRI-Based Human-Brain-Interface. Medical Imaging: Physiology, Function, and Structure.

[75] Hoffman HG, Chambers GT, Meyer WJ, 3rd, Arceneaux LL, Russell WJ, Seibel EJ, et al. (2011) Virtual Reality as an Adjunctive Non-Pharmacologic Analgesic for Acute Burn Pain During Medical Procedures. Ann Behav Med. 41: 183-191.

[76] Hoffman HG, Richards TL, Bills AR, Van Oostrom T, Magula J, Seibel EJ, et al. (2006) Using FMRI to Study the Neural Correlates of Virtual Reality Analgesia. CNS spectrums. 11: 45-51.

[77] Hoffman HG, Richards TL, Van Oostrom T, Coda BA, Jensen MP, Blough DK, et al. (2007) The Analgesic Effects of Opioids and Immersive Virtual Reality Distraction: Evidence from Subjective and Functional Brain Imaging Assessments. Anesth Analg. 105: 1776-1783.

[78] Fischer E, Bulthoff HH, Logothetis NK, Bartels A (2012) Visual Motion Responses in the Posterior Cingulate Sulcus: A Comparison to V5/MT and MST. Cereb Cortex. 22: 865-876.

[79] Snow JC, Pettypiece CE, McAdam TD, McLean AD, Stroman PW, Goodale MA, et al. (2011) Bringing the Real World into the FMRI Scanner: Repetition Effects for Pictures Versus Real Objects. Sci Rep. 1: 130.

[80] Moon J, Lee JH (2009) Cue Exposure Treatment in a Virtual Environment to Reduce Nicotine Craving: A Functional MRI Study. Cyberpsychol Behav. 12: 43-45.

[81] Martin T, LaRowe S, Malcolm RJ (2010) Progress in Cue Exposure Therapy for the Treatment of Addictive Disorders: A Review Update. Open Addict J. 3: 92-101.

[82] Due DL, Huettel SA, Hall WG, Rubin DC (2002) Activation in Mesolimbic and Visuospatial Neural Circuits Elicited by Smoking Cues: Evidence from Functional Magnetic Resonance Imaging. Am J Psychiatry. 159: 954-960.

[83] Snider L, Majnemer A, Darsaklis V (2010) Virtual Reality as a Therapeutic Modality for Children with Cerebral Palsy. Dev Neurorehabil. 13: 120-128.

[84] You SH, Jang SH, Kim YH, Kwon YH, Barrow I, Hallett M (2005) Cortical Reorganization Induced by Virtual Reality Therapy in a Child with Hemiparetic Cerebral Palsy. Dev Med Child Neurol. 47: 628-635.

[85] Cornwell BR, Salvadore G, Colon-Rosario V, Latov DR, Holroyd T, Carver FW, et al. (2010) Abnormal Hippocampal Functioning and Impaired Spatial Navigation in Depressed Individuals: Evidence from Whole-Head Magnetoencephalography. Am J Psychiatry. 167: 836-844.

[86] Hviid LB, Ravnkilde B, Ahdidan J, Rosenberg R, Stodkilde-Jorgensen H, Videbech P (2010) Hippocampal Visuospatial Function and Volume in Remitted Depressed Patients: An 8-Year Follow-up Study. J Affect Disord. 125: 177-183.

[87] Videbech P, Ravnkilde B, Pedersen AR, Egander A, Landbo B, Rasmussen NA, et al. (2001) The Danish PET/Depression Project: PET Findings in Patients with Major Depression. Psychol Med. 31: 1147-1158.

[88] Yang L, Morland TB, Schmits K, Rawson E, Narasimhan P, Motelow JE, et al. (2010) A Prospective Study of Loss of Consciousness in Epilepsy Using Virtual Reality Driving Simulation and Other Video Games. Epilepsy Behav. 18: 238-246.

[89] Frings L, Wagner K, Halsband U, Schwarzwald R, Zentner J, Schulze-Bonhage A (2008) Lateralization of Hippocampal Activation Differs between Left and Right Temporal Lobe Epilepsy Patients and Correlates with Postsurgical Verbal Learning Decrement. Epilepsy Res. 78: 161-170.

[90] Voermans NC, Petersson KM, Daudey L, Weber B, Van Spaendonck KP, Kremer HP, et al. (2004) Interaction between the Human Hippocampus and the Caudate Nucleus During Route Recognition. Neuron. 43: 427-435.

[91] Shine JM, Ward PB, Naismith SL, Pearson M, Lewis SJ (2011) Utilising Functional MRI (FMRI) to Explore the Freezing Phenomenon in Parkinson's Disease. J Clin Neurosci. 18: 807-810.

[92] Morris LD, Grimmer-Somers KA, Spottiswoode B, Louw QA (2011) Virtual Reality Exposure Therapy as Treatment for Pain Catastrophizing in Fibromyalgia Patients: Proof-of-Concept Study (Study Protocol). BMC Musculoskelet Disord. 12: 85.

[93] Rothbaum BO, Rizzo AS, Difede J (2010) Virtual Reality Exposure Therapy for Combat-Related Posttraumatic Stress Disorder. Ann N Y Acad Sci. 1208: 126-132.

[94] Roy MJ, Francis J, Friedlander J, Banks-Williams L, Lande RG, Taylor P, et al. (2010) Improvement in Cerebral Function with Treatment of Posttraumatic Stress Disorder. Ann N Y Acad Sci. 1208: 142-149.

[95] Folley BS, Astur R, Jagannathan K, Calhoun VD, Pearlson GD (2010) Anomalous Neural Circuit Function in Schizophrenia During a Virtual Morris Water Task. Neuroimage. 49: 3373-3384.

[96] Slobounov SM, Zhang K, Pennell D, Ray W, Johnson B, Sebastianelli W (2010) Functional Abnormalities in Normally Appearing Athletes Following Mild Traumatic Brain Injury: A Functional MRI Study. Exp Brain Res. 202: 341-354.
[97] Hanten G, Cook L, Orsten K, Chapman SB, Li X, Wilde EA, et al. (2011) Effects of Traumatic Brain Injury on a Virtual Reality Social Problem Solving Task and Relations to Cortical Thickness in Adolescence. Neuropsychologia. 49: 486-497.
[98] Qiu TM, Zhang Y, Wu JS, Tang WJ, Zhao Y, Pan ZG, et al. (2010) Virtual Reality Presurgical Planning for Cerebral Gliomas Adjacent to Motor Pathways in an Integrated 3-D Stereoscopic Visualization of Structural MRI and DTI Tractography. Acta Neurochir (Wien). 152: 1847-1857.
[99] Weniger G, Ruhleder M, Lange C, Irle E (2012) Impaired Egocentric Memory and Reduced Somatosensory Cortex Size in Temporal Lobe Epilepsy with Hippocampal Sclerosis. Behav brain res. 227: 116-124.
[100] Gruzelier J, Inoue A, Smart R, Steed A, Steffert T (2010) Acting Performance and Flow State Enhanced with Sensory-Motor Rhythm Neurofeedback Comparing Ecologically Valid Immersive VR and Training Screen Scenarios. Neurosci Lett. 480: 112-116.
[101] Hashimoto Y, Ushiba J, Kimura A, Liu M, Tomita Y (2010) Change in Brain Activity through Virtual Reality-Based Brain-Machine Communication in a Chronic Tetraplegic Subject with Muscular Dystrophy. BMC Neurosci. 11: 117.
[102] Cheron G, Duvinage M, De Saedeleer C, Castermans T, Bengoetxea A, Petieau M, et al. (2012) From Spinal Central Pattern Generators to Cortical Network: Integrated BCI for Walking Rehabilitation. Neural Plast. 2012: 375148.
[103] Zhu H, Sun Y, Zeng J, Sun H (2011) Mirror Neural Training Induced by Virtual Reality in Brain-Computer Interfaces May Provide a Promising Approach for the Autism Therapy. Med Hypotheses. 76: 646-647.
[104] Lubar JF (1991) Discourse on the Development of EEG Diagnostics and Biofeedback for Attention-Deficit/Hyperactivity Disorders. Biofeedback Self Regul. 16: 201-225.
[105] Moore NC (2000) A Review of EEG Biofeedback Treatment of Anxiety Disorders. Clin Electroencephalogr. 31: 1-6.
[106] Linden M, Habib T, Radojevic V (1996) A Controlled Study of the Effects of EEG Biofeedback on Cognition and Behavior of Children with Attention Deficit Disorder and Learning Disabilities. Biofeedback Self Regul. 21: 35-49.
[107] Cox RW, Jesmanowicz A, Hyde JS (1995) Real-Time Functional Magnetic Resonance Imaging. Magn Reson Med. 33: 230-236.
[108] Cohen MS (2001) Real-Time Functional Magnetic Resonance Imaging. Methods. 25: 201-220.
[109] Posse S, Fitzgerald D, Gao K, Habel U, Rosenberg D, Moore GJ, et al. (2003) Real-Time FMRI of Temporolimbic Regions Detects Amygdala Activation During Single-Trial Self-Induced Sadness. Neuroimage. 18: 760-768.
[110] deCharms RC (2008) Applications of Real-Time FMRI. Nat Rev Neurosci. 9: 720-729.
[111] Posse S, Binkofski F, Schneider F, Gembris D, Frings W, Habel U, et al. (2001) A New Approach to Measure Single-Event Related Brain Activity Using Real-Time FMRI: Feasibility of Sensory, Motor, and Higher Cognitive Tasks. Hum Brain Mapp. 12: 25-41.

[112] Yoo SS, Fairneny T, Chen NK, Choo SE, Panych LP, Park H, et al. (2004) Brain-Computer Interface Using FMRI: Spatial Navigation by Thoughts. Neuroreport. 15: 1591-1595.

[113] deCharms RC, Christoff K, Glover GH, Pauly JM, Whitfield S, Gabrieli JD (2004) Learned Regulation of Spatially Localized Brain Activation Using Real-Time FMRI. Neuroimage. 21: 436-443.

[114] Weiskopf N, Scharnowski F, Veit R, Goebel R, Birbaumer N, Mathiak K (2004) Self-Regulation of Local Brain Activity Using Real-Time Functional Magnetic Resonance Imaging (FMRI). J Physiol Paris. 98: 357-373.

[115] Yoo SS, O'Leary HM, Fairneny T, Chen NK, Panych LP, Park H, et al. (2006) Increasing Cortical Activity in Auditory Areas through Neurofeedback Functional Magnetic Resonance Imaging. Neuroreport. 17: 1273-1278.

[116] Caria A, Veit R, Sitaram R, Lotze M, Weiskopf N, Grodd W, et al. (2007) Regulation of Anterior Insular Cortex Activity Using Real-Time FMRI. Neuroimage. 35: 1238-1246.

[117] deCharms RC, Maeda F, Glover GH, Ludlow D, Pauly JM, Soneji D, et al. (2005) Control over Brain Activation and Pain Learned by Using Real-Time Functional MRI. Proc Natl Acad Sci U S A. 102: 18626-18631.

[118] LaConte SM, Peltier SJ, Hu XP (2007) Real-Time FMRI Using Brain-State Classification. Hum Brain Mapp. 28: 1033-1044.

Chapter 11

Cybertherapy in Medicine – Experience at the Universidad Panamericana, IMSS and ISSSTE Mexico

José Luis Mosso, Gregorio T. Obrador, Brenda Wiederhold, Mark Wiederhold, Verónica Lara and Amador Santander

Additional information is available at the end of the chapter

http://dx.doi.org/10.5772/52349

1. Introduction

There are many reports of Virtual Reality analysis and clinical applications in Medicine since the end of the last century for many authors. In trauma head injuries [1], during chemotherapy in children [2], in burn wound care [3,4,5], lumbar puncture [6,], breast cancer [7], vein puncture [8], pain distraction [9,10,11,12], dental pain control [13], leg ulcer relief [14], night vision technology with robot control to treat bourn injuries using robot-like arm mounted VR goggles [15]. The first case reports related with VR and medical invasive procedures and surgery in hospitals was in 2004 with our group, beginning with upper gastrointestinal endoscopies and so on, until introduce VR in postoperative care unit of cardiac surgery [16,17].

Pain and anxiety in outpatients and inpatients is a regular symptom in hospitals. For pain are available medications and for anxiety also. Patients in; in-rooms, operating rooms and another different places, get treatments with medical and surgical procedures, and pain and anxiety are the meanly symptom. In this project we try to demonstrate that virtual reality is a complementary tool to reduce pain and anxiety in hospitals during medical procedures including surgical procedures. There are a lot reason and justifications to use VR in hospitals, as follows. *Neonatology* (0-28 days old) , there are newborns staying in unit cares since few days, weeks even months, growing without contact with the external world. The visit unit care is limited to parents in few hours a day. The psychological impact in growth and development in childhood is so hard, where neuro stimulation in a closed environment is a good alternative. In Infants (lower and higher infants) VR have been a good resource to reduce pain and anxiety in oncology. Hunter Hoffman has demonstrated in the benefits of

VR during medical rehabilitation in burned childhood. In *postoperative cardiac surgery unit care*, patients stay 3 days in different critical care units depending of their progress. In these units they stay under the influence of sedative drugs receiving treatments by vein catheters, gastric tube, etc. Their recovery is on beds with different degrees of limited position and the breathing relief is on bed also. *In ambulatory surgery (General surgery, endoscopic surgery, urology, dermatology, Orthopedic, angiology, pediatric surgery, obstetrics),* patients are awake, minutes, even hours awake during surgical procedures doing nothing in a same position on a surgical table. In the recovery they stay hours waiting for go home or to be hospitalized. Special group of ambulatory surgery, surgeon used night vision technology while patients navigate in VR environment. At the school of medicine from Universidad Panamericana we used night vision on animal models before the application on humans. *Gynecology and obstetrics.* Women during colposcopy are awake while gynecologist applies local anesthesia into the cervix to perform cervical conization with diathermal loop. The anxiety is present in a same gynecological position. *Gastrointestinal endoscopy.* Under local anesthesia, neurovegative response is present during and after each upper gastrointestinal exploration of esophagus, stomach and duodenum. Patients present plenty of saliva associated with shortness of. *Labor Room.* Pregnant patients without complications expected hours to complete cervical dilation in labor room to go into delivery room or operating room. During uterus contractions they increase their respiratory rate. *Pediatric.* Scholar children stay on bed many days in recovery. In their stay they distract with games and in some times with a TV only. *Epidural and Spinal block in anesthesia.* This is a relative fast procedure compared with the previous groups where patient in the fetal position they expect catheter installation in their back, place where they can´t see anything and anxiety is present despite local infiltration.

2. Methodology

The conditions to perform the following project are. Consent informed, full consciousness; agree to participate in the project; completeness vision, controlled cardiovascular and respiratory disease. Patients were afíliate at Instituto Mexicano del Seguro Social (IMSS) and Instituto de Seguridad y Servicios Sociales para los Trabajadores del Estado, (ISSSTE). In case of newborns, parents signed consent informed.

Technique. Before each procedure, nurse measured blood pressure, heart rate, breath rate (In the case of patients in unit cares, we considered arterial blood gases). Physician asks patients about his-her feelings (Anxiety) in a scale of 0 to 10 (where 0 is no pain and 10 is high pain). Physician installed the following equipment also: Laptop with 3 virtual scenarios (Enchanted forest, cliff final, icy city see figure 2) and connected to a vision goggles eMagin. When we used smart phones (Nokia N95 and iPhone G3) and PSP2 we used a mp3 video with a green valley scenario connected to goggles. Once set up the equipment, it was installed in the patient´s head. Physician begins the procedure and in the middle of the procedure, nurse measured vital signs again and physician repeats the same question about the feelings (Anxiety and pain). Head Mounted display is removed of the head´s patient at the end of each procedure and 15 minutes after, nurse made the third measure of vital signs and physician asks the feelings of the patient again. In the case of Night Vision Goggles (NV) for ambulatory surgery for total darkness was necessary to perform open surgery on rabbits at the school of Medicine at the Universidad Panamericana in 9 rabbits under general anesthesia before to make ambulatory surgeries on humans. NV equipment needs a scrub nurse to hold all equipment on the head of surgeon and first assistant after dressing with sterile clothing. Cables were connected beside the surgical table. Diagnosis of surgery group: big lipomas on head, arms, legs or abdomen, non complicated inguinal and umbilical hernias. Complicated and non complicated postoperative hernias in the abdominal wall. For Laparoscopic group: non acute cholecystitis, and inguinal hernias. With night vision technology group: big lipomas on arms, legs and non complicated inguinal hernia. Diagnosis for postoperative care unit of cardiac surgery: aortic and mitral valves rechange and coronary revascularization. Colposcopy group: cervicitis and premalignant lesions in cervix. Neonatology group: premature newborns in recovery with more than 5 days to 2 months of hospitalizations with multiple diagnosis such as necrotizing enterocolitis, sepsis, respiratory disorders. In the kidney transplantation only one patient in the recovery used VR. In obstetrics, patients with 48 weeks of pregnant with 5 to 10 of cervical dilation participate with normal fetus in labor room where epidural or spinal block were installed and we follow the same patients to the delivery room and to the operating rooms. In the group of endoscopy, diagnosis was hiatal hernia, peptic disease and colon disorder.

3. Results

The comparative measure of pain were made before, during and after each procedure in all groups using Virtual Reality and the control group, except in patients without any medical procedure such as the group of care unit of cardiac surgery and neonatology. The statistical method to measure pain used was the mean into the scale of 0 to 10 (Cero is no pain and 10 is high level of pain). We present statistical results in the 3 representative groups where pain and anxiety are high such as colposcopy (see table 3), ambulatory surgery (see table 4) and postoperative care unit of cardiac surgery (See table 5). With the results of these groups we can appreciate the possibilities to use this technique in another groups into the hospitals because we have been tested VR in a group seriously ills in critical care units in cardiac surgery and manipulations of tissues including abdominal cavity in ambulatory surgery group. In table 1 we measured mean pain (see figure 1) between before to during the procedure and between before to after (see table 1). In the colposcopy group that used VR the mean pain before the procedure was 7.5 and during was 5.35 with a difference of 2.15 that correspond of 28.66 % of reduction of pain. In the group of colposcopy that didn't use VR the mean pain before was 6.43 and during 6.78 with an increase and difference of 0.35 that correspond of 5.44 % of increasing of pain, in the same group considering mean pain before of 6.43 and 4.83 after, the difference was 1.6 that correspond of 24.88 % of reduction of pain. In all groups we demonstrated an important reduction of pain using VR during medical or surgical procedures (see table 2). In care unit of cardiac surgery group the mean pain before were 8 and after 30 was 3.64 with a difference of 4.36 that correspond to 54.5% of reduction of pain-anxiety (See table 2). In the group of surgery there were a mean of pain before of 5.7 and during the procedure 3.93 with a difference of 1.8 that correspond of 31.41 % of reduction on pain. Considering the mean pain before of 5.73 and 2.09 after, there is a difference of 3.64 that correspond to 63 % of reduction of pain. In the group of surgery that didn't use VR the pain mean were 5.57 before and during 5.19 with a difference of 0.48 that correspond to 6.8% of reduction, and considering the mean pain before of 5.57 and 3.52 after there is a difference of 2.05 that correspond to 36.80 % of reduction of pain. With these results we can appreciate the impact of distraction of VR to reduce anxiety and subsequently visceral and somatic pain in the case of surgery with regional anesthesia well placed.

$$\bar{x} = \frac{1}{n}\sum_{i=1}^{n} a_i = \frac{a_1 + a_2 + \cdots + a_n}{n}$$

Figure 1. Mean equation to obtain results

MEAN of pain and anxiety	Colposcopy with HMD	Colposcopy no HMD	Critical care unit of cardiac surgery	Ambulatory surgery. with HMD	Ambulatory surgery. No HMD
Before	7.5	6.43	8	5.73	5.57
During	5.35	6.78		3.93	5.19
After	2.7	4.83	3.64	2.09	3.52

Table 1. Comparative measure of mean pain before, during and after each procedure in 3 representative groups. Pain scale used 0-10. (0 means no pain and 10 high pain).

% of pain and anxiety Reduction	Colposcopy with HMD	Colposcopy no HMD	Critical care unit of cardiac surgery	Ambulatory surgery. with HMD	Ambulatory surgery. No HMD
Before - during	28.66 %	5.44 %	54.5 %	31.41 %	6.8 %
Before - after	64 %	24.88 %		63 %	36.80 %

Table 2. Pain and anxiety percentage reduction, between before and during procedures. Pain scale used 0-10. (0 means no pain and 10 higher pain).

As a result of patient selection, we suggest the following general criteria to use VR in Medicine: 1.- Patients with voluntary participation, and signed authorization with written consent informed, (Patients under 18 years of age need written informed consent from parents), 2.- Medical control of cardiovascular, respiratory and neurological diseases, and 3.- full consciousness. (See figures 2 to 12)

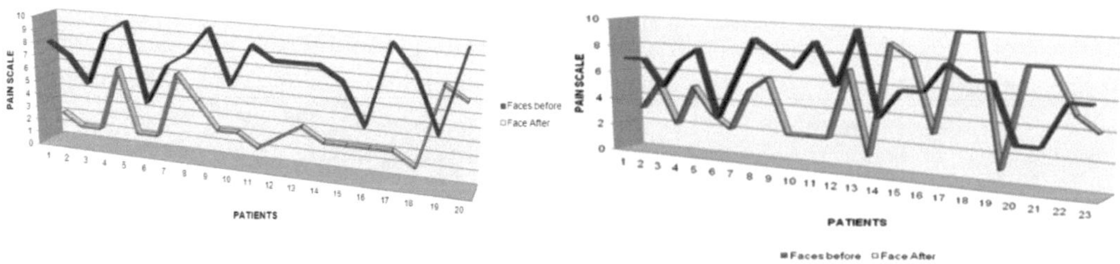

Table 3. Colposcopy. Group with VR (left) and control group (right)

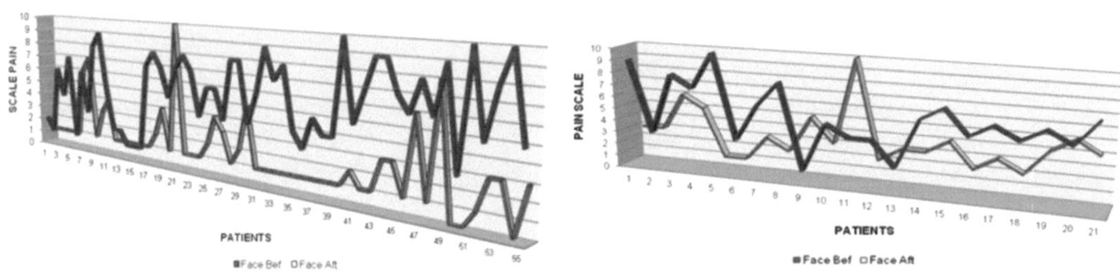

Table 4. Ambulatory surgery. Group with VR (left) and control group (right)

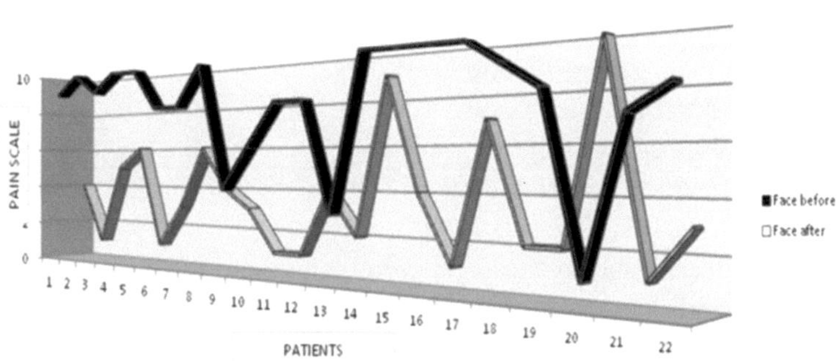

Table 5. Postoperative Care Unit of cardiac surgery

Medical complications. 2 patients of 22 in care unit of cardiac surgery group presented cardiac arrhythmias during the virtual navigation due to the slow effort on the bed, the treatment was the immediately disruption of VR exposure with total recovery. In ambulatory surgery group with local anesthesia, 2 patients presented nausea and vertigoes using VR and after 30 minutes patients become tired to carry goggles on their heads in the same position, for this reason we used a projector to display a screen in the roof of the operating room. In the subgroup of endoscopic surgery under regional anesthesia, 3 patients of 5, presented chest pain secondary of neumoperitoneum and it was necessary convert regional anesthesia to general (For this reason we don´t suggest VR for endoscopic surgery under regional anesthesia). (See table 6).

Area	Cases	Complications
Care unit of cardiac surgery	Care Unit	2 patients presented arrhythmias.
Surgery	General Surgery Endoscopic surgery Urology Angiology Dermatology Orthopedic Angiology Pediatric Obstetric	2 patients presented nausea and vertigoes, another one presented pain in peritoneum manipulation. 3 patients presented chest pain and difficulty breathing None None None None None None None None
Anesthesia	Epidural/spinal block	None
Obstetric	Labour room Childbirth room	None None
Endoscopy	Upper Endoscopy Colonoscopy	None None

Table 6. Complications

Technical limitations. In ambulatory surgery, patients need to move their hands to get more distraction; but they hands tied to the table to avoid contamination and in the other hand they become tired with the goggles on their heads and after 45 minutes they prefer change the projection of the scenario to the roof of the operating room. With NV we found limitation for surgeon's motions for the cables connected to the wall beside the table one meter away. We had distorted vision for the glasses used, and the vision was in black and white color. The surgical time was small longer than conventional surgical time. In neonatology group, newborns present 5 to 8 seconds of attention for VR. It was difficult to evaluate the benefits of this technology, this is the reason we don′t present statistical results. We suggest using another kind of device of VR to interact with these patients who have limited vision.

4. Conclusions

We found differences results according with age, gender, procedures, culture, origin of regions, diagnosis, prognosis, but in all groups we found reduction of pain and anxiety, one more than others, except the subgroup of endoscopic surgery, the only one we don′t suggest use VR. . The group that enjoyed better the virtual scenarios was patients between 5 to 14 years of age in pediatric group. The older patients demonstrated curiosity of new technology that had in their hands.

There where patients exposed with Virtual Reality in different conditions like age, sex, medical procedures and different physicians but all results show us reduction of anxiety and pain. The answers were different in each group because pathways of pain involved are two, somatic and visceral. Tissues and organs manipulated were different, for instance: Skin, fat, muscles, fascias, peritoneum, cervix, uterus, peritoneum, spine´s dura, ligaments, small and medium vessels, gallbladder, omentum, visceral peritoneum, parietal peritoneum, esophagus, stomach and colon´s mucous. On the other hand, Patient´s psyche played an important role for anxiety and pain. All patients required medications for local and regional anesthesia such as analgesics, sedatives, in different dosages. The high level of pain we saw in all groups of patients was those who underwent endoscopic surgery.

In the beginning of our experience, we used a Spider HMD game to distract patients under upper GI endoscopy in 2004. Thanks of these results we improved the devices introducing Virtual reality scenarios and a HMD developed by Brenda and Mark Wiederhold from the Virtual Reality Medical center from San Diego CA. Clinical applications were done at the Instituto de Seguridad y Servicios Sociales para los Trabajadores del Estado I.S.S.S.T.E. and at the Instituto Mexicano del Seguro Social IMSS, Public Health Centers in Mexico City. There were big and small groups because the results from critical care unit of cardiac surgery and ambulatory surgery allow us to conclude the usefulness of virtual reality to reduce pain and anxiety in inpatients without minimal complications and permit us to limit virtual applications in all areas of Medicine. If patients in critical care unit presented no more than 2 complications and in surgery we consider absolutely contraindicated in endoscopic surgery is obvious to use virtual reality without major risk in the rest of areas of medicine if we follow the absolute indications.

Virtual reality is a not invasive method, easy to use, easy to install and easy to carry and each day new generations of technology are lighter, smaller such as smartphones, PSP 2, iPADs that permit more interaction with users using friendly feedbacks such as tracking (Kinect), 3D images permit more immersion with Interactive screens that could be considered in future works because U screens, virtual cubes and cave could be an option for patients because these screens have demonstrated the utility in rehabilitation for the big space for body motions and for the stereoscopic immersion.

The best advantage for the healthcare institutions we found in the group for ambulatory surgery was to reduce the half of dosage of medication of fentanyl, and midazolam, and in many cases we avoid totally these medications in the intraoperative. This advantage represent an important saving for institutions always considering and appropriate selection of patients with absence of cardiovascular and respiratory disease that could permit physician works without concerns during procedures. With these results we can give the following advantages as: reduction of medication (Analgesic and in several cases, drugs to reduce anxiety as fentanyl and mydazolam), bed days and specially increase the wellbeing of patients. There were small groups of one or two patients that should increase the number of cases to get reliable results, however theirs results are a clear indicator of benefits of virtual reality. We didn't consider more areas of medicine, but we involve representative as surgical and clinical areas and both sex, males and females and different ages, newborns and adults. Technology developments permit use small device such as mobile phones to connect to head mounted display instead of laptops; they facilitate the handling of the virtual scenarios. 3D scenarios will permit more immersion in patients (see table 7). Display scenario on the wall is an option for navigations instead to use HMD which can be tiring for patients in prolonged time. Virtual reality is useful in hospital and we consider that in the future, virtual rooms could be integrated inside services, operating rooms, and waiting rooms.

Medicine Apps	Interface used	Virtual scenario	Area	Cases	Control group	Patients participants
Ambulatory Surgery	Laptop	Enchanted forest	Surgery	53	31	84
Ambulatory Surgery	Laptop & Night vision	Enchanted forest	Surgery	5	0	5
Ambulatory Surgery	Nokia	Green valley video	Surgery	23	0	23
Ambulatory Surgery	PSP2	Green valley video	Surgery	24	0	24
Ambulatory Surgery	X BOX	Video	Surgery	2	0	2
Ambulatory Surgery	iPhone G3	Green valley video	Surgery	2	0	2
Pediatric surgery	PSP2	Green valley video	Pediatric surgery	1	0	1
Recovery surgery	Laptop	Enchanted forest	Kidney transplantation	1	0	1
Surgical cleaning	Laptop	Enchanted forest	Surgery-Infectology	11	10	21
Epidural blockage	Laptop	Enchanted forest	Anesthesia	4	3	7
Labor room	Laptop	Enchanted forest	Obstetric			
Cesarean	Laptop	Enchanted forest	Obstetric	1	1	2
Delivery Room	Laptop	Enchanted forest	Obstetric	1	2	3
Neonatology Unit care	Laptop	Enchanted forest	Neonatology	5	0	5

Medicine Apps	Interface used	Virtual scenario	Area	Cases	Control group	Patients participants
Colposcopy	Laptop	Enchanted forest	Ginecology	20	23	43
Upper GI Endoscopy	Spider HMD Game	Spider game	Endoscopy	200	200	400
Colonoscopy	Laptop	Enchanted forest	Endoscopy	3	0	3
Unit care	Laptop	Enchanted forest	Cardiac surgery	20	22	42
TOTAL PATIENTS				376	292	668

Table 7. Interfaces and VR scenarios used in each group as well as the total patients participants in the groups studied and controls.

Figure 2. Enchanted forest scenario used in the majority of procedures.

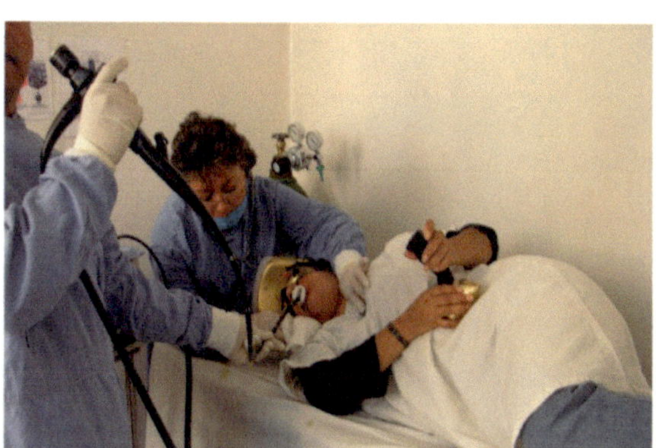

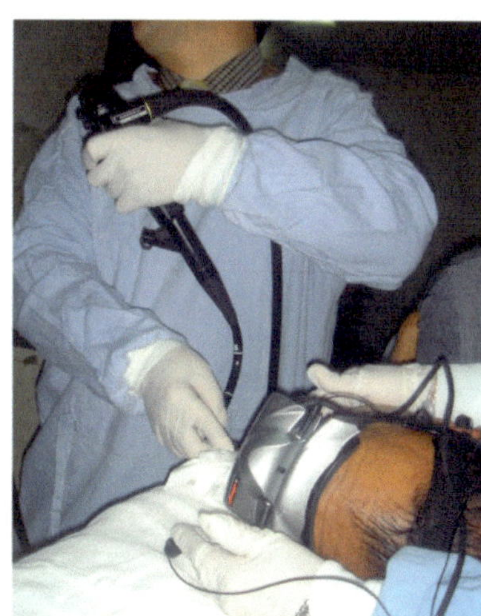

Figure 3. Upper Gastrointestinal Endoscopy.

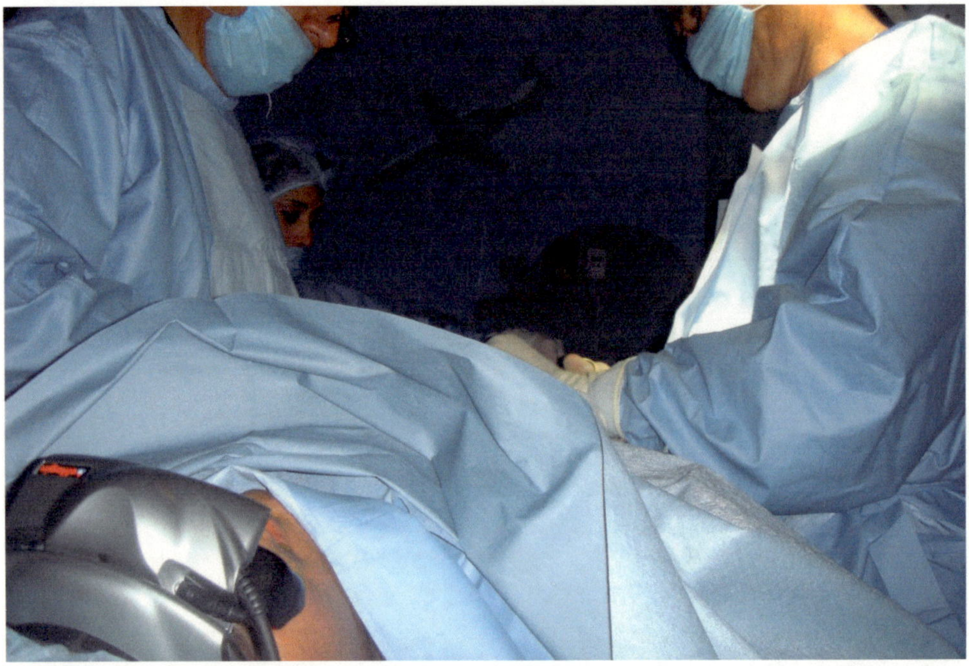

Figure 4. Ambulatory surgery with regional anesthesia

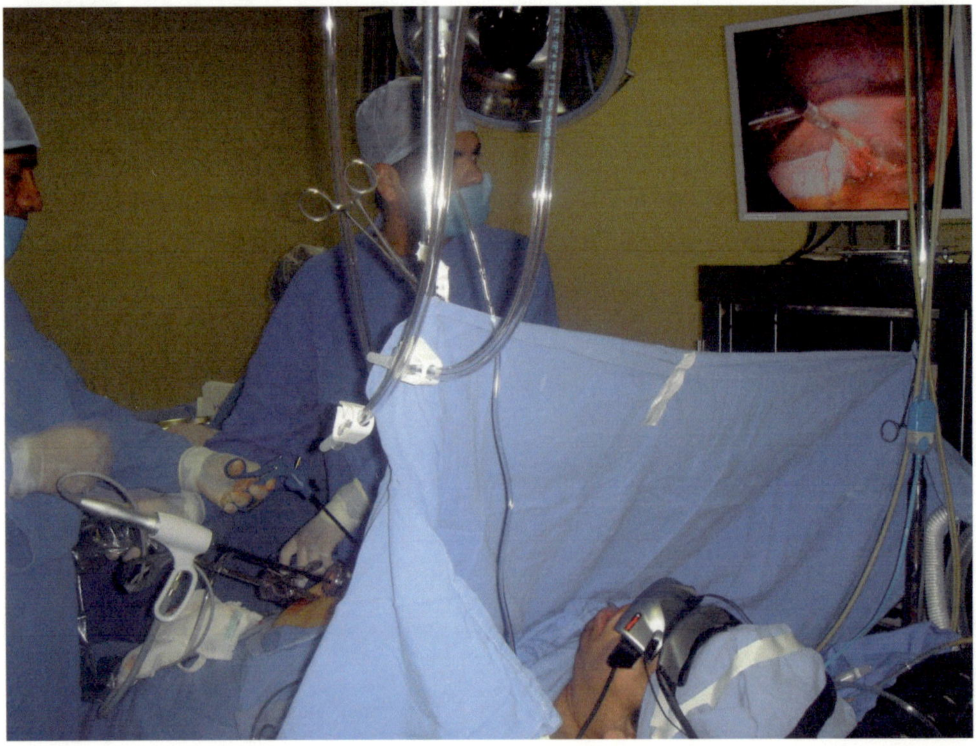

Figure 5. Endoscopic surgery with regional anesthesia

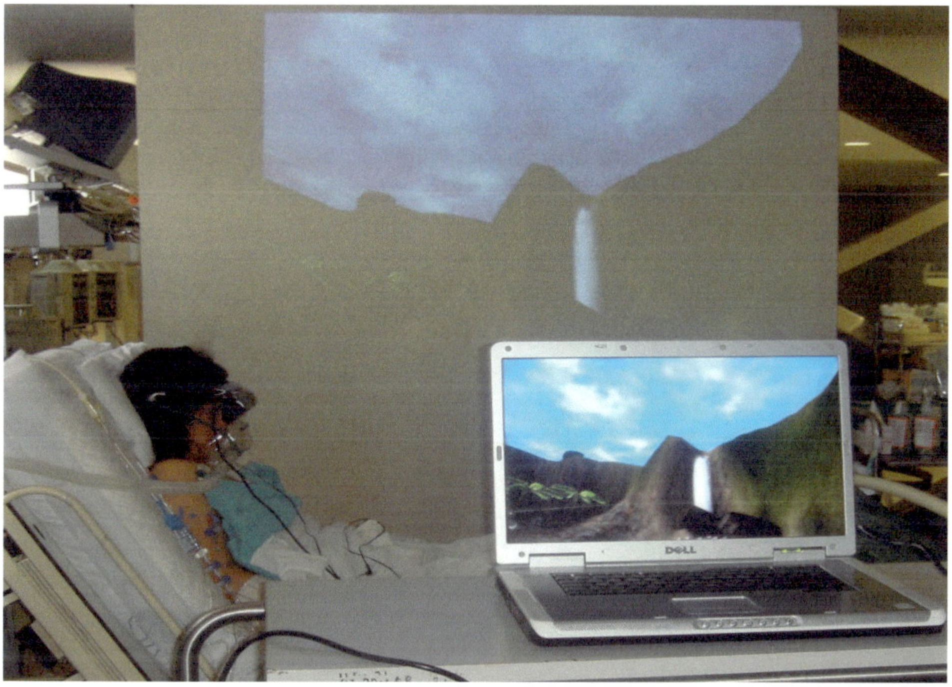

Figure 6. Care unit of cardiac surgery

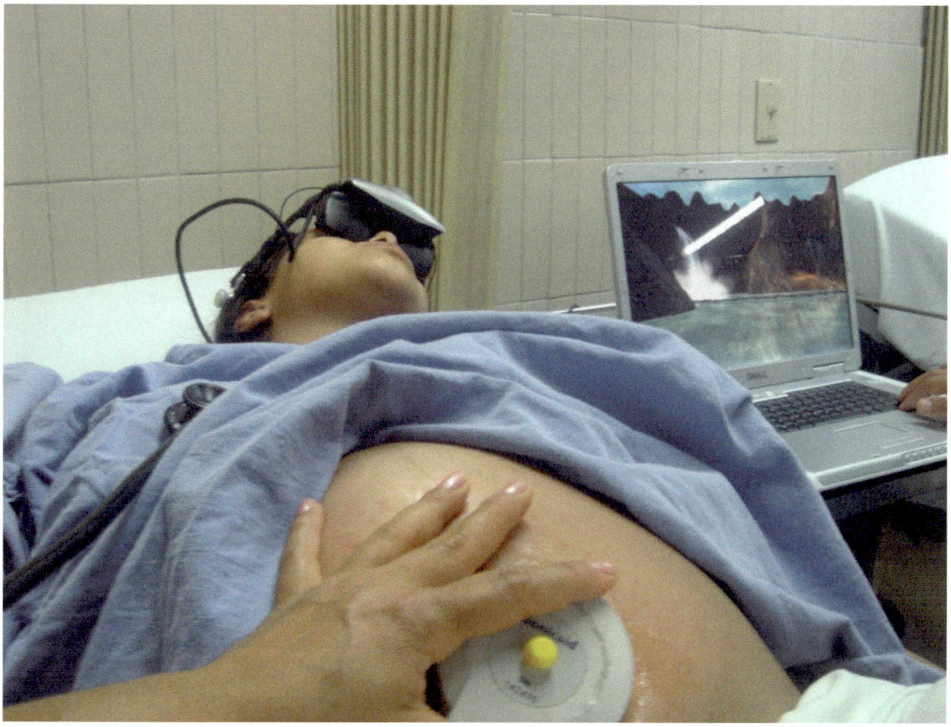

Figure 7. Labor room

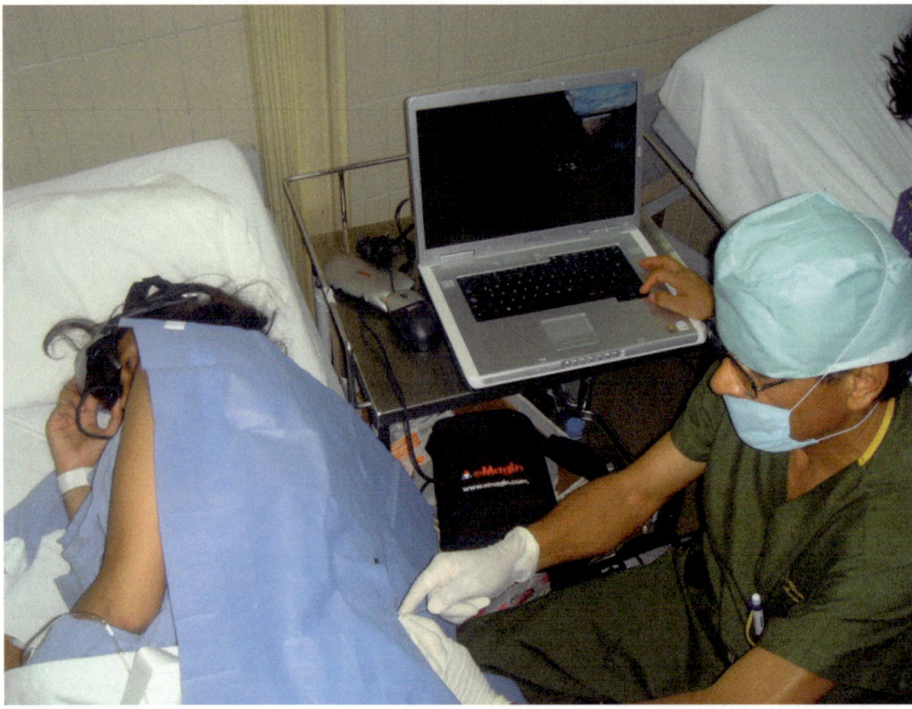

Figure 8. Anesthesia. Epidural/ spinal block

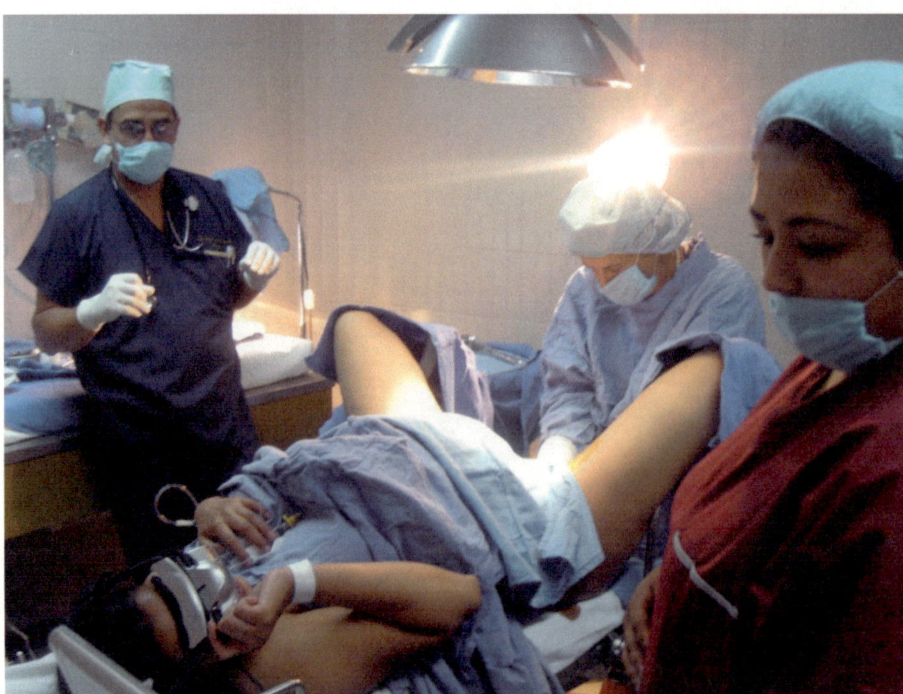

Figure 9. Delivery room

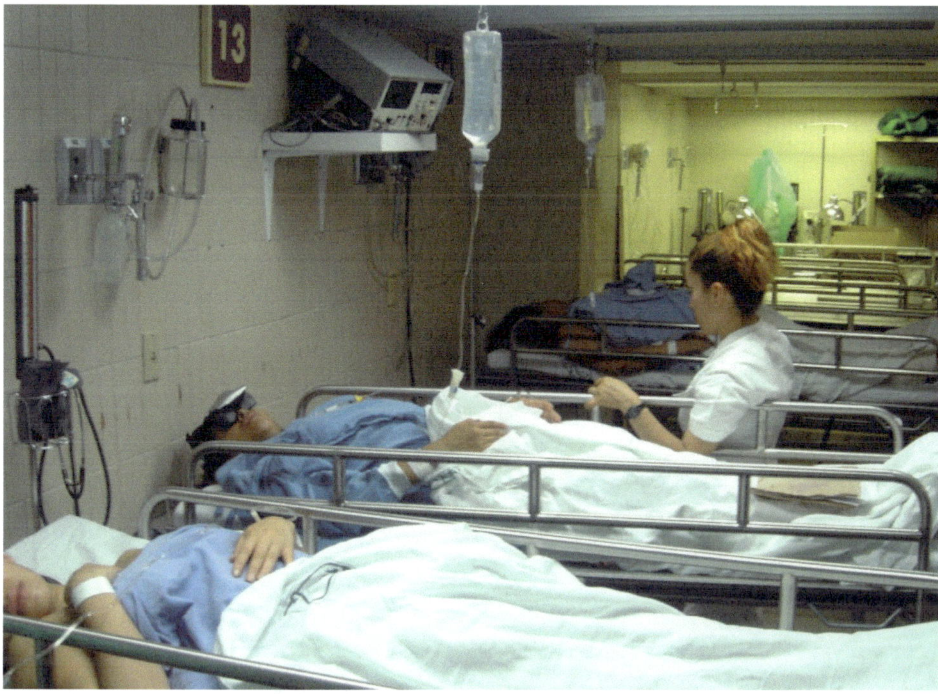

Figure 10. Recovery room. Post cesarean surgery

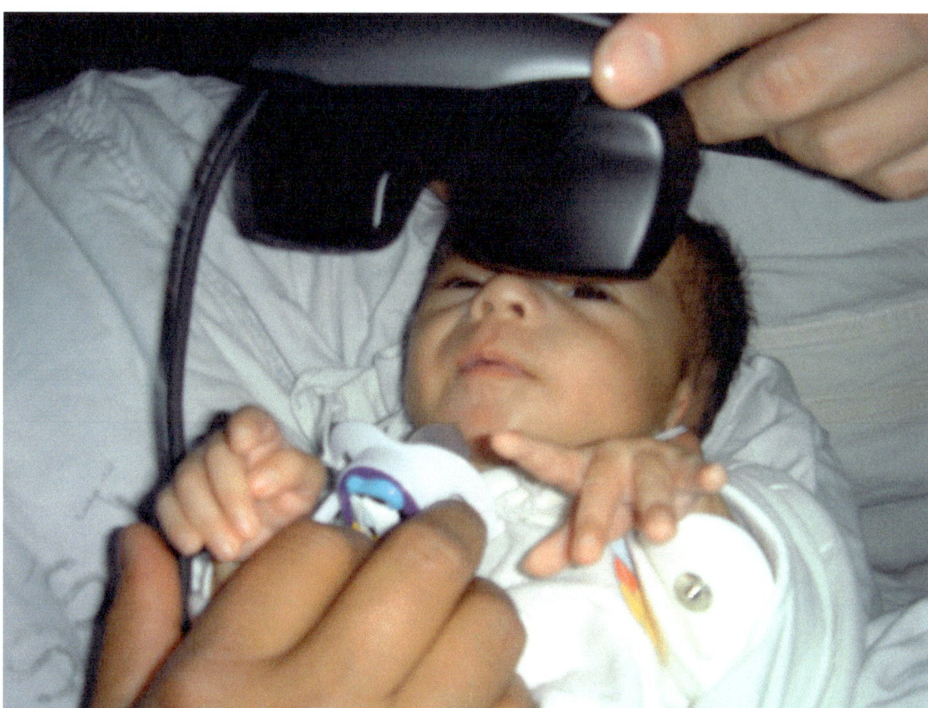

Figure 11. Neonatology care unit

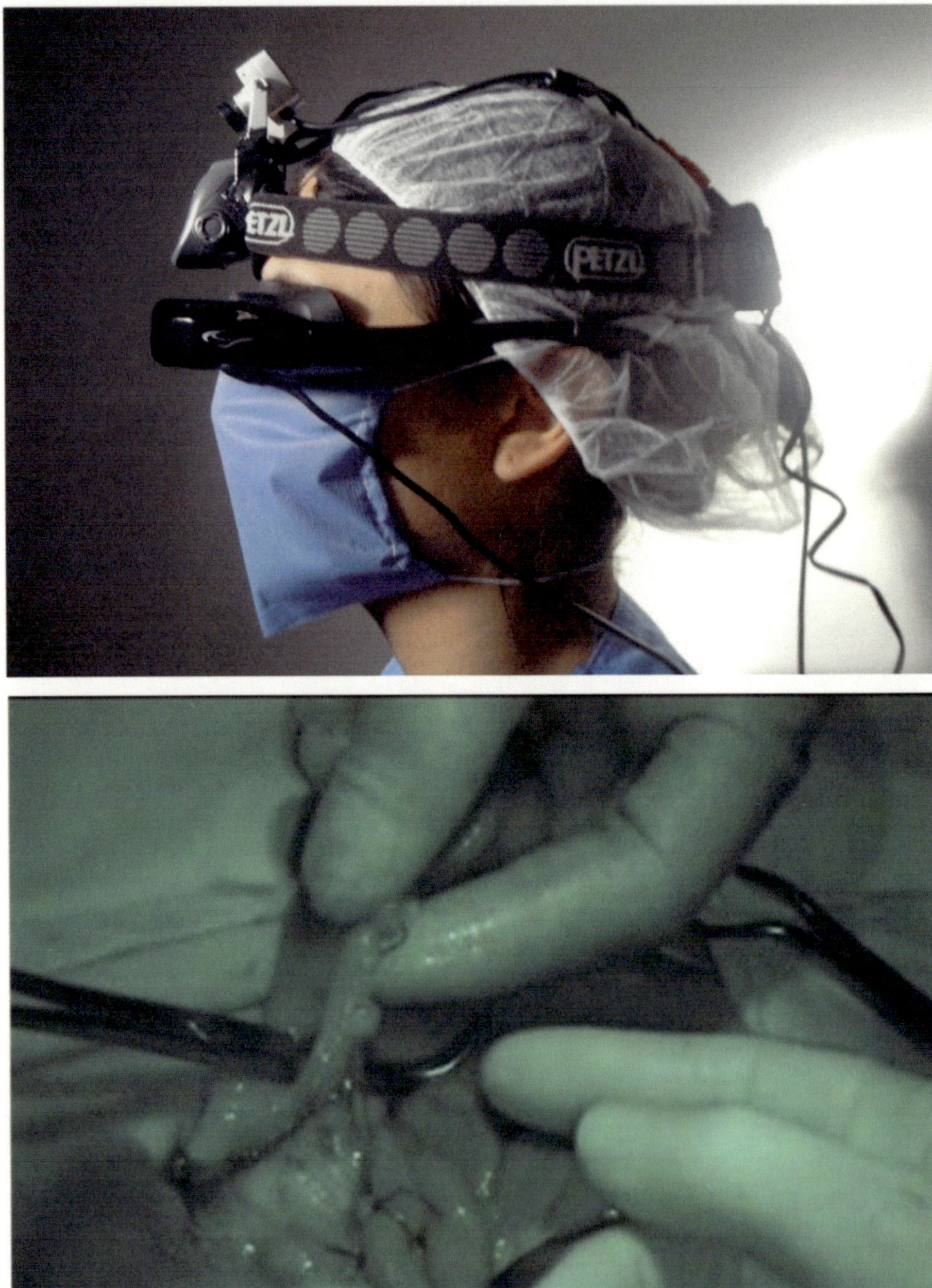

Figure 12. Night vision equipment (above). Appendectomy with night vision goggles. School of Medicine, Universidad Panamericana (Below)

Author details

José Luis Mosso
School of Medicine, Universidad Panamericana, Mexico
Regional Hospital No. 25 of the Instituto Mexicano del Seguro Social, IMSS, Mexico
Clínica Alberto Pisanty Ovadía ISSSTE, Mexico

Brenda Wiederhold and Mark Wiederhold
Virtual Reality Medical Center President, Interactive Media Institute, Mexico

Gregorio T. Obrador
School of Medicine Director, Universidad Panamericana, México

Verónica Lara
Centro Médico Nacional 20 de Noviembre ISSSTE, Hospital de Ginecología y Obstetricia Luis Castelazo Ayala No. 4, IMSS, Mexico

Amador Santander
UMAE, Centro Médico Nacional la Raza, IMSS, Mexico

5. References

[1] Rizzo, A.A. (1994). Virtual Reality applications for the cognitive rehabilitation of Persons with traumatic head injuries. In Murphy, H.J. (ed.), *Proceedings of the 2nd International Conference on Virtual Reality and Persons With Disabilities.* CSUN: Northridge.

[2] Schneider, S.M., Workman, M.L., *Effects of Virtual Reality on symptom distress in children receiving chemotherapy.* CyberPsychology and Behavior, 1999. 2(2): p. 125-134.

[3] Hoffman, H.G., Patterson, D.R., Carrougher, G.J, *Use of Virtual Reality for Adjunctive Treatment of Adult Burn Pain During Physical Therapy.* 2000.

[4] Hoffman, H.G., Doctor, J.N., Patterson, D.R., et al., *Virtual reality as an adjunctive pain control during burn wound care in adolescent patients.* Pain, 2000. 85(1-2): p. 305-309.

[5] Hoffman, H.G., Patterson, D.R., Carrougher, G.J., *Use of Virtual Reality for adjunctive treatment of adult burn pain during physical therapy: A controlled study.* The Clinical Journal of Pain, 2000. 16(3): p. 244-250. [11] Schneider, S.M., Workman, M.L., *Effects of Virtual Reality on symptom distress in children receiving chemotherapy.* CyberPsychology and Behavior, 1999. 2(2): p. 125-134.

[6] Schneider, S.M., Ellis, M., Coombs, W.T., et al., *Virtual Reality Intervention for Older Women with Breast Cancer.* CyberPsychology and Behavior, 2003. 6(3): p. 301-307.

[7] Sander W, Eshelman D, Steele J, Guzzetta CE. Effects of distraction using virtual reality glasses during lumbar punctures in adolescents with cancer. Oncol Nurs Forum 2002;29:E8–15.

[8] Reger GM. Effectiveness of virtual reality for attentional control to reduce children's pain during venipuncture. Paper presented at Piscataway NJ:Proceedings of the 2nd International Workshop in Virtual Rehabilitation, 62–67, 2003.

[9] Mark D. Wiederhold, MD, PhD, FACP, and Brenda K. Wiederhold, PhD, MBA, BCIA. *Virtual Reality and Interactive Simulation for Pain Distraction.* Pain Medicine. Volume 8, number S3. 2007 p.p. S182-S188.

[10] Brenda K. Wiederhold and Mark D. Wiederhold. *Managing Pain in Military Populations with Virtual Reality.*

[11] Wiederhold, M.D., & Wiederhold B.K. (2010). Virtual Reality and Interactive Simulation for Pain Distraction. CyberTherapy & Rehabilitation, 3(1), 14-19.

[12] Schneider, S.M. *Virtual Reality for the treatment of Breast Cancer.* in *CyberTherapy*. 2003. San Diego, CA: Interactive Media Institute.

[13] Hoffman, H.G., Garcia-Palacios, A., Patterson, D.R., et al., *The effectiveness of Virtual Reality for dental pain control: A case study.* CyberPsychology and Behavior, 2001. 4(4): p. 527-535.

[14] Tse, M.M.Y., Ng, J.K.F., and Chung, J.W.Y. *Visual stimulation as pain relief for Hong Kong Chinese patients with leg ulcers.* in *CyberTherapy*. 2003. San Diego, CA: Interactive Media Institute.

[15] Maani C.V., Hoffman H.G., Morrow M., Maiers A., Gaylord K., McGhee L.L., DeSocio P.A. *Virtual Reality pain control during burn wound debridement of combat-related burn injuries using robot-like arm mounted VR goggles.* J Trauma, 2011. 71(1 Suppl): S125-30.

[16] Mosso J.L., Gorini A., De La Cerda G., Obrador T., Almazan A., Mosso D., Nieto J.J., Riva G. *Virtual reality on mobile phones to reduce anxiety in outpatient surgery.* Stud Health Technol Inform, 2009. 142: 195-200.

[17] Mosso, J.L., Rizzo S., Wiederhold B., Lara V., Flores J., Espiritusanto E., Minor A., Santander A., Avila O., Balice O., Benavides B. *Cybertherapy—new applications for discomfort reductions. Surgical care unit of heart, neonatology care unit, transplant kidney care unit, delivery room-cesarean surgery and ambulatory surgery, 27 case reports.* Stud Health Technol Inform, 2007. 125:334-6.

Section 3

VR in Pedagogical Applications

Chapter 12

Using Augmented Reality Artifacts in Education and Cognitive Rehabilitation

Claudio Kirner, Christopher Shneider Cerqueira and Tereza Gonçalves Kirner

Additional information is available at the end of the chapter

http://dx.doi.org/10.5772/46416

1. Introduction

The first Augmented Reality Systems (ARS) were usually designed with basis on three main blocks, as is illustrated in Figure 1: Infrastructure Tracker Unit, Processing Unit, and Visual Unit. The Infrastructure Tracker was responsible for collecting data from the real world, sending them to the Processing Unit, which mixed the virtual content with the real content and sent the result to the Video Out module of the Visual Unit. Some designs used a Video In, to acquire required data for the Infrastructure Tracker Unit. The Visual Unit can be classified in two types of system, depending on the followed visualization technology:

1. Video see-through: It uses a Head-Mounted Display (HMD) that employs a video-mixing and displays the merged images on a closed-view HMD.
2. Optical see-through: It uses a HMD that employs optical combiners to merge the images within an open-view HMD

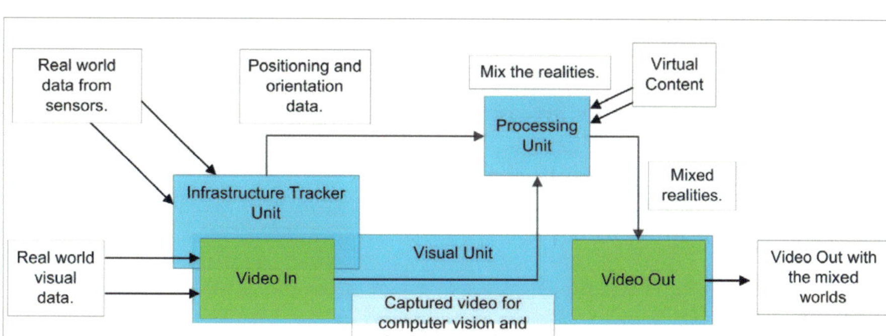

Figure 1. Augmented Reality Systems (ARS) standard design.

HMDs are currently the dominant display technology in the AR field. [5]. However they lack in several aspects, such as ergonomics, high prices and relatively low mobility due to their sizes and connectivity features. An additional problem involving HMD is the interaction with the real environment, which places virtual interactive zones to the user, making the collision with these zones hard due to the difficulty to interact with multiple points in different depths.

Alternative approachs to develop ARS involve the use of monitors and tablets. Monitors are used as an option for indirect view, since the user does not look directly into the mixed world. Tablets are used in direct view, since the user points the camera to the scene and looks directly into the mixed world. Both approaches still have difficulties in getting collision.

To make easy the collision actions, we developed series of artifacts, which help the user activity in the active zones, or int the active points by overlapping physical objects with virtual objects. The AR collision is implemented even if the user is not looking directly into the mixed world, but only into the artifact, providing another cognitive possibility as the user can use the tactile to collide. The cognitive function associated to the artifact empowered with augmented reality is based on an *Augmented Reality Cognitive Artifact* (ARCA).

This chapter presents the concepts and the technology involved with, emphasizing aspects of authoring and interaction on augmented reality applications based on multiple markers and multiple points. Moreover, the discusses augmented reality applications in the education and rehabilitation areas, which use artifacts aiming to overcome the main interaction problems. Finally, the last section concludes the chapter and presents future work.

1.1. Virtual, augmented and cross reality definitions

1.1.1. Virtual Reality (VR)

Virtual reality was the first three-dimensional interface option, allowing natural interaction using hands with virtual environment rendered on monitors, projections or through VR HMD. To interact with the virtual elements, it is necessary to have multimodal devices, such as VR gloves (with sensors and tracking capabilities), force-feedback devices, 3D mice, stereoscope glasses, etc.

Representative definitions of virtual reality are: "virtual reality is an advanced computer interface that has real-time simulation and interactions through multi-sensor channels" [6] and "virtual reality is a computer interface that allows the user to interact, in real time, in a computer generated three-dimensional world, using his senses through special devices" [19].

In virtual reality environments, the user sees the virtual world, through a window rendered on monitor screens or projection screens; or the user is inserted into the virtual world through HMD or projection rooms, called caves. When the user is totally inserted into the virtual world, through HMD, caves and multi-sensed devices, the virtual reality is called *Immersive* (Figure 2a). When the user is partially inserted into the virtual world, through monitor or equivalent, the virtual reality is called *Non-Immersive* (Figure 2b).

1.1.2. Augmented Reality (AR)

The AR definition has evolved with the technologic evolution. In early definitions, the real world augmentation was obtained only through visual elements [16]; however, with the development of audio and haptic interactions associated with spatial position in real-time, the AR concept has been extended.

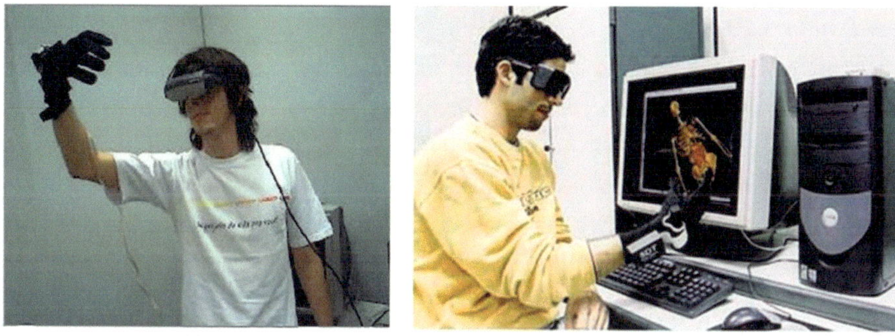

(a) (b)

Figure 2. Virtual reality examples. (a) Immersive VR environment; (b) Non-Immersive VR environment.

Azuma [2] [1] defines augmented reality as a system that allows the user to see the real world, with virtual objects superimposed upon or composed with the real world. That system has the following three characteristics: it combines real and virtual elements; it is interactive in real time; and it is registered in a three-dimensional (3D) way. Figure 3 shows some results of a real world augmentation with virtual elements.

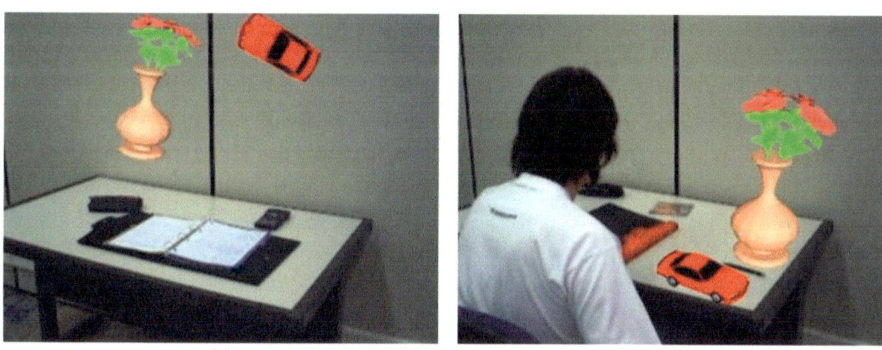

(a) (b)

Figure 3. Augmented reality examples. (a) Virtual objects are misplaced; (b) Virtual objects are correctly placed.

In addition, the development of audio and haptic interactions associated with spatial position in real-time, the AR concept has been expanded. In this context, a wider AR definition involves the real world empowered with virtual objects that consider several aspects, such as: visualization, audio, haptic, etc.

According to an updated definition "Augmented Reality is an interface based on computer generated information combination (static and dynamic images, spatial sounds and haptic sensations) with the real user environment, provided by technological devices and using natural interaction in the real world" [19].

A way to bring virtual information to the physical user environment is using a webcam, which captures a live stream of the real world, and tracks some features, allowing the computer to add virtual information to the real world. The result can be seen, heard and felt by monitors, projections, helmets and haptic devices, depending on the interaction devices that take part of the system.

1.1.3. Cross-Reality (CR)

Cross-reality involves a ubiquitous mixed reality environment that comes from the fusion of a network of sensors and actuators (which collects and sends data related to the real world) with shared virtual worlds, using an augmented reality interface, where the exchange of information is bidirectional between the real and the virtual world. [34] [17]

Cross-reality can be classified in two types: Non Overlapped and Overlapped.

- Non Overlapped Cross-Reality (NonOVER-CR). It is an AR environment with virtual elements, which are not overlapped with the real elements, using a network of sensors and actuators to acquire bidirectional communication between virtual and real objects. An example of this type of CR is a Seconf Life application that communicates with the real world [35].
- Overlapped Cross-Reality (OVER-CR). It is an AR environment with virtual elements overlapped with the real elements, using a network of sensors and actuators to acquire bidirectional communication between virtual and real objects. This type of CR was discussed by [27], who explained the concept of augmented space, overlaying layers of data over the physical space using technology. Keiichi Matsuda, in his thesis [29], describes some examples of physical space overlaid with dynamically changing multimedia information, localized for each user.

Figure 4 shows an example of NonOVER-CR and OVER-CR, developed with the basAR authoring tool (Section 4.2), where there are two buttons (one virtual and one real) and an overlapped lamp (LED and virtual light beam). The buttons are not overlapped, witch characterizes a NonOVER-CR example; additionally, the illumination characterizes an OVER-CR example. The interaction with the real or the virtual button activates or deactivates the overlapped lamp, showing the LED on and the virtual beam or the LED off without the virtual beam.

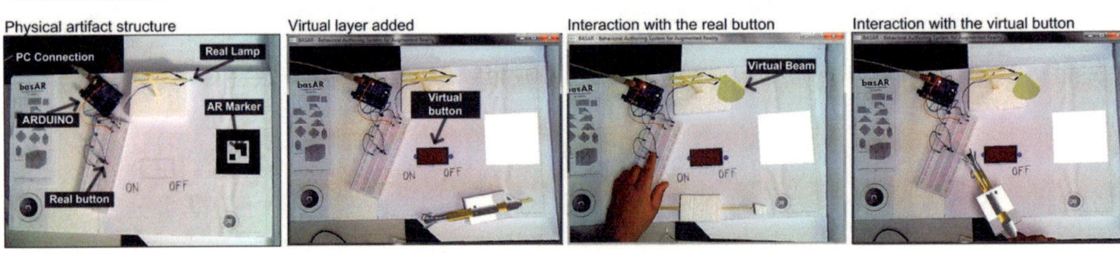

Figure 4. Cross-Reality example.

Figure 5 shows a CR representation obtained from the fusion of the augmented reality area (overlapping real and virtual environments) with ubiquitous computing, involving a network of real and virtual sensors and actuators. In this situation, interactions in the real environment reflect into the virtual environment and vice-versa. When the virtual and real sensors are overlapped, the interactions in the two worlds are concomitant. These results of this property could be useful in several types of applications, involving telepresence, collaborative and remote work in physical installations empowered by virtual elements, etc.

Cross-Reality makes possible to interact with remote equipments, as well as with complex experiments which are hard to simulate.

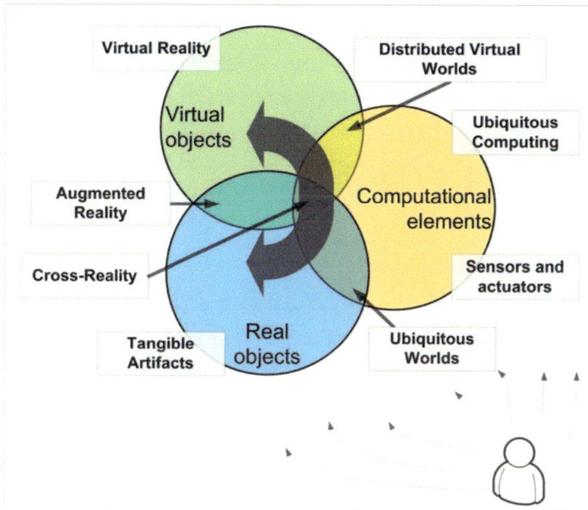

Figure 5. Cross-Reality, virtual and real interfaces.

2. Augmented reality artifacts

Humans actuating in the physical world frequently use artifacts as extension of their own knowledge and reasoning systems to support the remembering and processing of information [3] [32]. Classical examples are a shopping list and a string tied around a finger. In this way, artifacts that are used in cognitive applications, are named cognitive artifacts.

The term *cognitive artifact* was coined by Norman [31] and has different definitions, depending on the available technology and the type of application. An up-to-date definition, a cognitive artifact is a physical object or software application used to aid, enhance or improve thinking and reasoning.

Artifacts, including the cognitive ones, have significant potential to be implemented with augmented reality based on computer vision, once the prototype can present low cost and be easily distributed to interested users. Several interactive artifacts for rehabilitation are being developed, however most of them are applied in motor rehabilitation. There are few examples related to cognitive rehabilitation [4] [37] [12].

Artifacts based on Augmented Reality technology may fulfill the following requirements: [21]

- The artifact, for cogntive application have to involve muti-sensory perception, memory, attention, logic and motor control, in order to allow the preparation of cognitive exercises;
- The physical parts of the artifact has to be built with ordinary materials, involving a simple process, presenting availability and low cost;
- For this, it could be adopted materials such as Styrofoam, cardboard or wood, to implement the physical structure, tied with glue, always followed by instructions and templates;
- The logical parts of the artifact have to use augmented reality technology based on computer vision software. Authoring tools for rapid prototyping, using augmented reality can make easy the development of applications. A further section will present three possible authoring tools for these purposes;

- The user interactive actions on the artifact must be tangible and easy. This property, due to the coincident physical and virtual points, allows force feedback interactions, because when the user touches the interacting device (pointer) on the artifact, he feels the contact and the virtual action point is enabled. This characteristic is important because it gives more comfort to the user. When the points are placed into the 3D space, without physical association, they demand more ability and concentration from the user to collide the pointer with the virtual points.

Augmented Reality Artifacts Applications can be visualized with a projector or a HMD; however, using a computer monitor is cheaper, available and easy to operate. The artifact allows direct interaction with sound feedback, but the visualization will be indirect, when a monitor is used.

3. Authoring tools for AR artifact applications

In the last years, series of AR authoring tools were released to help users to develop spatial applications mediated by computer. Authoring augmented reality tools can be classified, according to their characteristics of programming and content design, in low and high level, considering the concepts abstraction and interface complexity incorporated in the tool.

Programming tools are based on basic or advanced libraries (basic or advanced ones), involving computer vision, registration, three-dimensional rendering, sounds, input/output and other functions. ARToolKit [16], MR [42], MX [11] and FLARToolKit are examples of low level programming tools. The development of applications, based on programming tools can be complex. Futhermore, authoring tools, templates and interfaces cover the development complexity and ease the steps to achieve the application abstraction, as illustrated in Figure 6

.

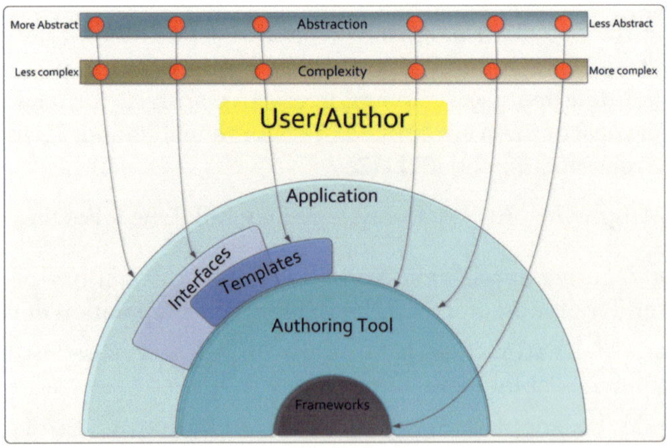

Figure 6. Complexity versus Abstraction in AR Applications Development.

ARToolKit is one of the first augmented reality programming tools that use marker registration and computer vision. In this tool, the developers need C/C++ programming skills to author the applications. A more recent tool, FLARToolKit is a wrapper from ARToolKit, developed with Action Script 3, the language from Adobe Flash environments. FLARToolKit has a distinguishing feature, which is to enable the creation of web-based augmented reality applications.

Content design tools are independent of a specific programming language, replacing it by the description of the virtual objects and their relationship with the real environment. IN this context, APRIL [24] is a low level example of this type of tool, which uses XML descriptions. IN the other had, high level content design tools use graphical user interfaces to represent the descriptions and interactions, as it occurs in DART [26], AMIRE [13], ECT [15], ComposAR [39] and ARSFG [43].

High level content design should be more intuitive and suitable for non-programmers. These tools can support scripting and visual interfaces, new functionalities added by user and real time interpretation.

Our research is different from other AR authoring tools, since it considers the following characteristics.

- A level of abstraction that covers the framework (ARToolKit or FLARToolkit);
- Authoring AR applications depends on editing configuration files and tangible operations;
- There are different authoring levels, depending on the skills of the developer;
- Authoring can use tangible operations, editing configuration files and mouse and keyboard support; however the end users can interact with the AR application using only one or two markers.

Authoring AR application basically depends on: structure of the AR environment; data structure and folders that support the tool; authoring interface; configuration tasks, action commands, system commands, and utilization procedures that support the end-user to navigate and interact with the augmented environment.

3.1. ARAS-NP

To make easy the development of AR applications with those elements, we developed the authoring tool ARAS-NP (Augmented Real Authoring System for Non-Programmers). It includes authoring and utilization characteristics, besides additional features related to a shared remote use, which enables user collaboration.

ARToolKit is the core of ARAS-NP and additional functionalities were programmed with C/C++. The software, user manual and applications of ARAS-NP are freely distributed by the authors [18].

Augmented reality involves more than superimpose virtual objects and annotations over the real world. Thus, the augmented world (Figure 7), as considered in this work, presents real and virtual objects, such as: interactive objects, which can change in certain situations; animated objects, which can be activated; visible or invisible objects, which vanish or appear in certain cases; visible or invisible points, which can be activated or deactivated; etc.

Moreover, the augmented reality environment can be modified after the initial authoring , for example, by adding, changing and deleting points and virtual objects.

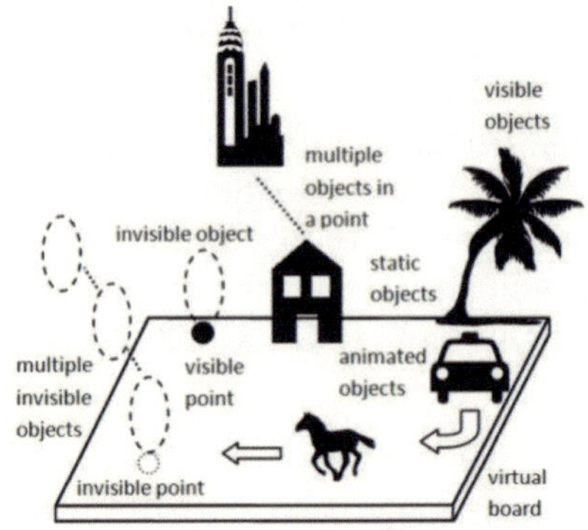

Figure 7. Augmented World.

The data structure of the augmented reality environment to be authored comprises reference markers, which have associated virtual boards, and their respective elements (points, virtual objects and sounds) that appear on the board, according to Figure 8. These elements must be placed in folders that the developer needs to manipulate in order to create the augmented reality environment.

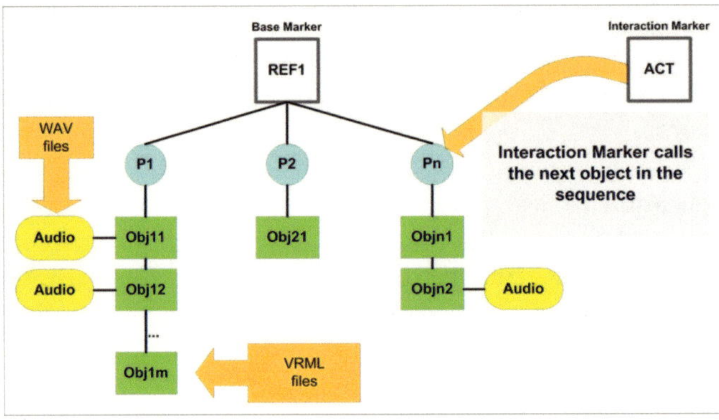

Figure 8. ARAS-NP Data Structure.

3.2. basAR

The basAR (Behavioral Authoring System for Augmented Reality) is an evolution of the ARAS-NP, once it uses the same AR framework ARToolKit as its core. Its configuration is based on description files and it follows the same approach of using action points, differently of other authoring tools that create behavior and interactivity based on marker position, orientation and proximity. The software, user manual and applications of basAR are freely distributed by the authors [8].

The basAR data structure is organized according to Figure 9.

The basAR tool involves a multi-layer approach, with the following features:

- Infrastructure: It defines the correlation between the real and virtual worlds, such as markers and their properties;

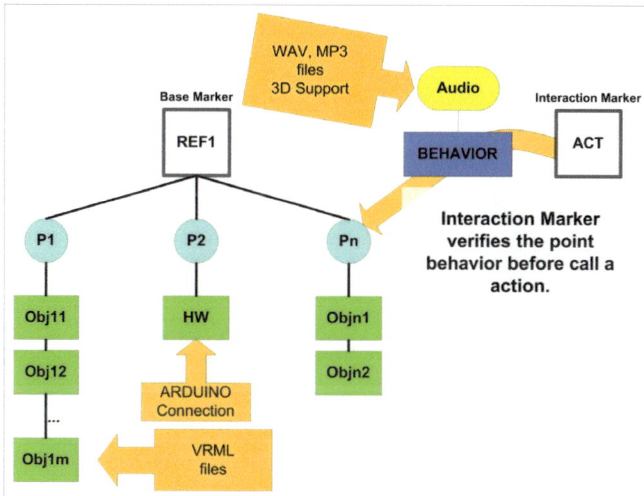

Figure 9. basAR Data Structure.

- Structure: It defines the virtual points layer and where they are located;
- Content: It defines the models, sounds, etc. that are used to create the application abstraction;
- Behavior: It defines how the augmented layer handles the feedbacks from the user interaction. The basAR behavior is structured by commands that describe dynamically the application; those commands are grouped on a language called basAR-AL (basAR Authoring Language) [8];
- Acting: It defines how the user interacts with the structure layer;
- Cross-Reality: It defines the keywords used by the behavior layer to communicate with the hardware.

3.3. FLARAS

FLARAS (Flash Augmented Reality Authoring System) is an augmented reality authoring tool based in the same action point approach of ARAS-NP; futhermore, it represents an evolution, mainly due to the graphic interface and because it allows to develop application to be hosted on the Internet and played on any computer that has Adobe Flash Player. This is an important advance, since most technologies are going forward the Web applications and the cloud computing. The software, user manual and applications of FLARAS are freely distributed by the authors [41].

The FLARAS data structure is organized according to Figure 10 .

4. Augmented reality applications for education

The main advantages to use ARS for educational purposes are: [7]

- Students are more motivated, because they live an experience proposed by the application and the use of a new technology;

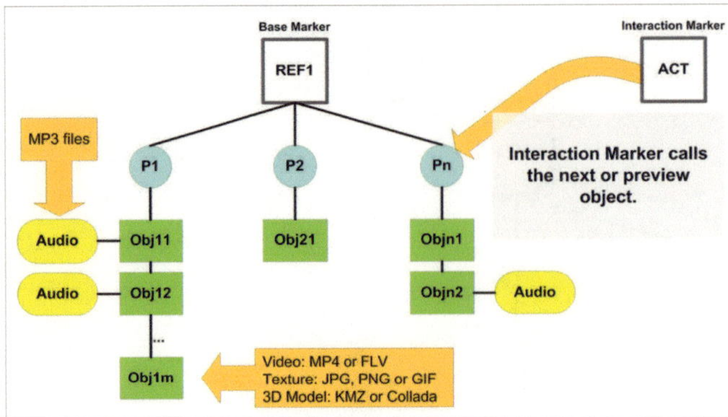

Figure 10. FLARAS Data Structure.

- AR can illustrate processes and characteristics that are not usually viewed by the user;
- AR allows detailed visualization and objects animation;
- AR allows micro and macro visualizations that cannot be seen with naked eyes, as well as proposes different view angles to understand the subject;
- AR allows interactive virtual learning using virtual experiments;
- AR allows the students to recreate the experiments out of the school environment;
- The students become more active due to the interactive application characteristics;
- AR encourages creativity, improving the experience;
- AR provides equal opportunities to different cultural students;
- AR helps to teach computational and peripheral skills.

AR technology has an strong appeal to the constructivism, where the students control their own learning [10]. AR environments allow the students to explore objects, perform tasks, learn concepts and develop skills. Using AR educational applications, each student can look for its own interests, in its own speed and need, which better suits to its individual characteristics. For example, in a historical place, using an AR application, each student can define its own discovery way [14].

This section shows some educational examples developed with different authoring tools, such as:

- ARAS-NP: AR books, Spatial Tutor, Q&A Applications, Perspective learning;
- basAR: Geometry teach and learning application;
- FLARToolKit: Electromagnetism teach and learning application.

4.1. AR books

The AR books comprise applications that have been much disseminated in the last years [14]. When a person looks to an AR books, it seems as any other book. However, when the user puts the books in front of a computer with a webcam, 3D objects, sounds, animation, extra explanations and several interactive elements seem to jump from the pages. These resources are added to the book to motivate the student to explore the presented theme, supporting the learning process.

Some examples of AR Books are the GeoAR [36] and the SpaceAR [33].

The GeoAR is an AR book to teach geometry subjects related to the main geometric shapes. Figure 11a presents the page of the square in the GeoAR, showing the marker, and some explanations and formulas. Figures 11b and 11c shows the book with the AR layer of the sphere and cube square pages.

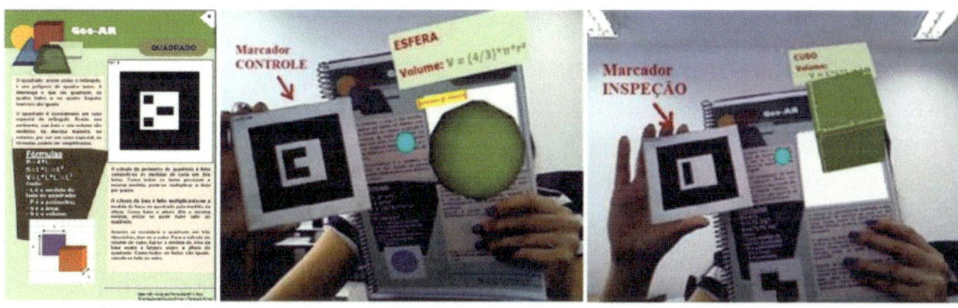

(a) (b) (c)

Figure 11. GeoAR examples. (a) Example page, (b) Sphere page and (c) Square page.

Another example of AR Book is the SpaceAR. It has information about the Solar System, and its pages guide the user into new discoveries of the objects that orbit the Sun. Figures 12a, 12b and 12c illustrate the use of the book, with the Sun and its information and a rotating animation.

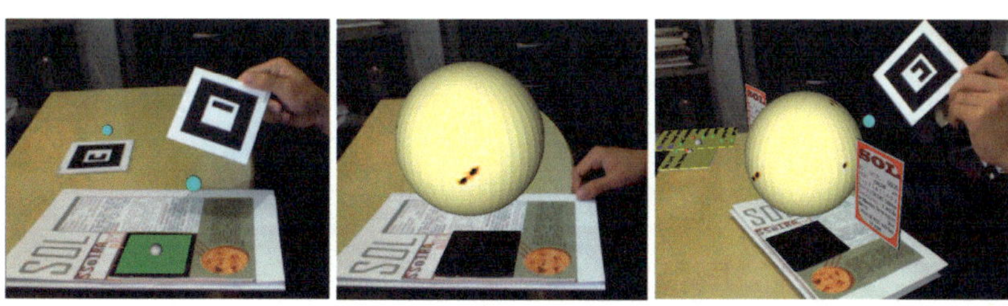

(a) (b) (c)

Figure 12. SpaceAR examples. (a) How to use the book, (b) Sun information page and (c) Sun animation.

4.2. AR spatial tutor to explore multimedia and three-dimensional environments

The AR Spatial Tutor aims at creating interaction with panels and mockups using AR, to expose 3D objects, annotations, sounds and animations.

This tutor is based on the ARAS-NP tool and includes two physical artifacts to show the tutor use. The first version is based on a Photographic Panel representing the Itaipú Hoover, in Brazil (Figure 13a). It has some action points located on the panel, which, when are clicked with the interaction artifact, they show annotations (Figure 13b), sounds and explanations. Those points can have multiple information elements that allow the expansion of contents or the fulfilling of different types of users.

Figure 13. AR Spatial Tutor - Multimedia. (a) Photographic Panel and (b) AR annotations.

The second version of the AR Spatial Tutor is based on a mockup of the same Itaipú Hoover, made from Styrofoam. An AR layer paints the Styrofoam and place the action points. A simple look at the mockup shows a static artifact, without interaction, which could be no attractive to students or users. However, when the AR layer is placed, the mockup is empowered with dynamic content, motivating its use. Figure 14a shows the mockup without the AR layer and Figure 14b shows the AR layer added.

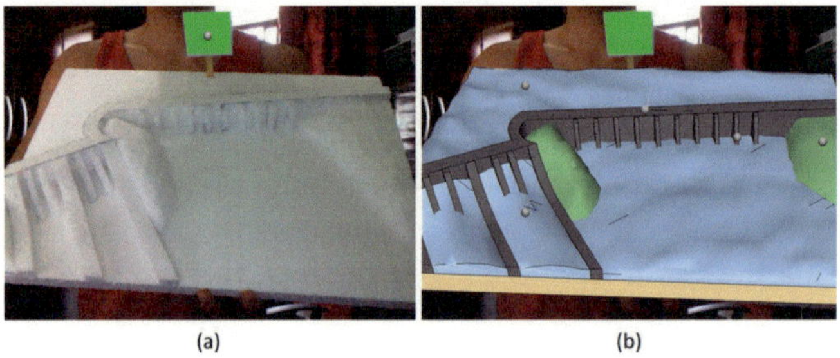

Figure 14. AR Spatial Tutor - Multimedia. (a) Mockup without AR layer and (b) with AR layer.

4.3. Q&A-AR game

The Q&A-AR educational spatial game is a multiplayer car racing game based on questions and answers, which works in an augmented reality environment [20].

The game Q&A-AR fulfills the following requirements:

- The game must have educational potential involving several themes of study using texts, illustrations and sounds;
- The physical parts of the game must be made with ordinary materials and process in order to have availability and low cost;
- The logical parts of the game must use augmented reality technology based on computer vision;
- The interactive actions to be executed on the game must be tangible and easy;
- The information related to the game (questions, answers, instructions) must be easily customized by teachers;
- The user interface of the game must consider usability factors, such as easy to understand, easy to learn and easy to use.

The game uses a series of artifacts, including nonmoving artifacts, and moving parts. The nonmoving artifact contains two perpendicular planes in order to present the game information to the user. The vertical plane contains the reference marker, which is used to superimpose the virtual information on the artifact. The horizontal plane presents the race path with ten cells, and a textual area for questions and answers.

The moving parts are composed by the player cars, the dices and an interaction pointer with a marker.

The virtual structure is composed of virtual buttons that overlap the printed buttons, the virtual cell buttons. To perform the interaction, the player only needs to touch the physical pointer on the printed button or on the top of the cars placed on the path cells.

Figures 15a, 15b, 15c and 15d present the nonmoving parts, the moving parts, the activation of a question and the answer elements, and the augmented reality environment of the game, respectively.

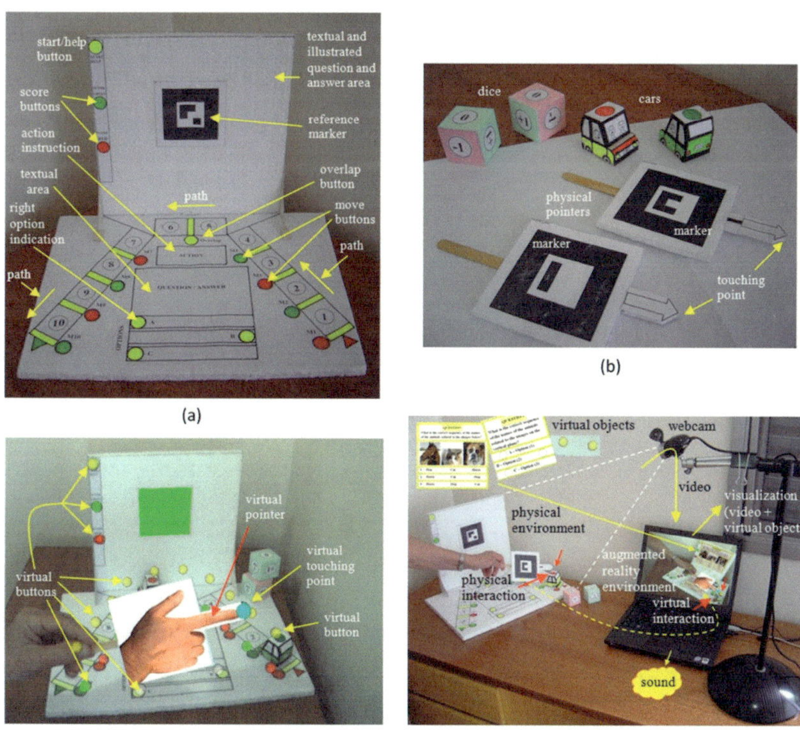

Figure 15. Q&A-AR Game. (a),(b),(c) and (d) present the nonmoving parts, the moving parts, the activation of a question and the answer elements, and the augmented reality environment of the game, respectively.

The goal of the game is to reach the end of the path first. The cars run over the path on the horizontal plane driven by moving information, involving dice, forward and backward movement indicated by buttons or by result of the player performance, and answering to the questions presented by the activation of the virtual path cell buttons.

4.4. Perspective learning

To see and describe real and imaginary three-dimensional scenes from the observer's viewpoint is an intuitive activity for non-impaired people; However, it is difficult and even impossible for congenitally blind people, once it involves abstract concepts for them, such as: perspective, depth planes, occlusion, etc. This project, supported by an augmented reality tool, helps blind people to understand, describe and convert three-dimensional scenes in two-dimensional embossed representations, like painting. To understand how the blind people can acquire those concepts, we developed an augmented reality application, working as an audio spatial tutor to make the perspective learning process easy [23] [38].

Figure 16 presents some developed ARCAs for perspective learning application.

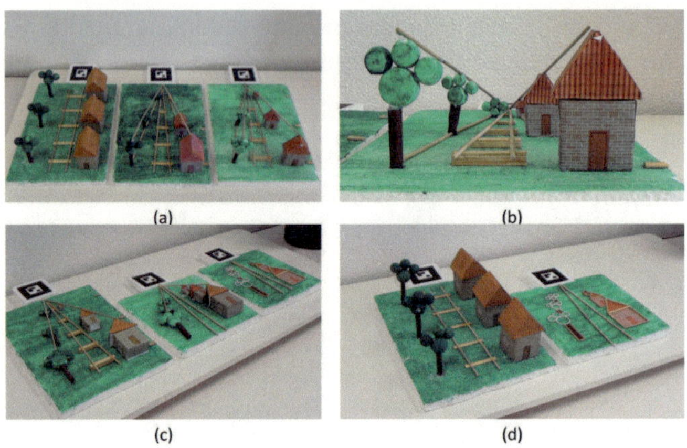

Figure 16. (a),(b),(c) and (d) are ARCAs used in this Perspective Learning, respectively.

4.5. Geometry learning

The development of spatial skills involves a critical understanding, when students start learning three-dimensional objects. In order to help this achievement, the teachers usually employ woodcraft artifacts and several orthographic and axonometric projections inside books [28]. A way to improve the learning of three-dimensional shapes is content based on AR. This application is used to teach polygon extrusion and revolution math concepts, using the authoring tool basAR to create the interactive application, where the student chooses the type of polygon and then apply the movement. Figure 17a shows the three possible polygons (circle, cube and triangle). When a polygon is chosen, it shows the two possibilities (extrusion or revolution), according to the Figure 17b. Figures 17c and 17b show the extrusion and revolution results of a circle.

This application can be found on the Internet [9].

4.6. Electromagnetism with augmented reality

Some concepts of electromagnetism, as they are relatively abstract, require more effort from the students to be understood. With the intention to offer an alternative way, that would be more interactive and dynamic, the MiniLabElectroMag-AR (Mini Laboratory of

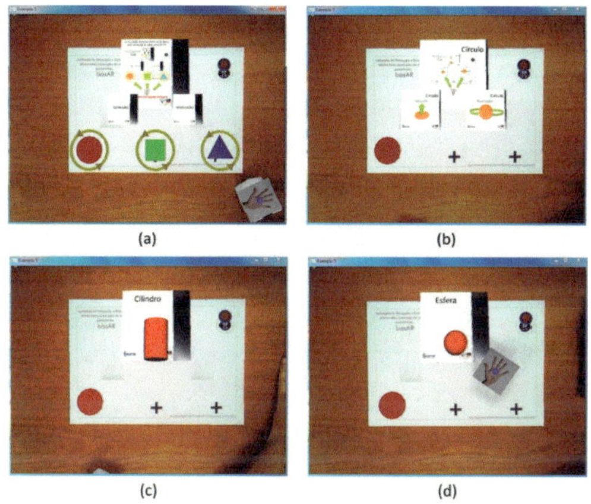

Figure 17. (a) Polygons choices; (b) user selected the circle; (c) extrusion result; (d) revolution result .

Electromagnetism with Augmented Reality) was developed. The purpose of this application is to work as a simple laboratory for experiments about electromagnetism, allowing, for example, that students explore in a practical way some basic concepts, such as electric currents, electric circuits, that inducted magnetic field generated by the flow of electric current on a straight wire, and also the simulation of the Oersted's experiment.

Figure 18a shows the artifact with the lamp, battery and switch elements draw. The Figure 18b shows the same artifact with the AR layer, with the virtual elements superimposed.

Figure 18c shows compass deflection due to the magnetic induction of the wire. Figure 18d shows two students collaboratively exploring the experience.

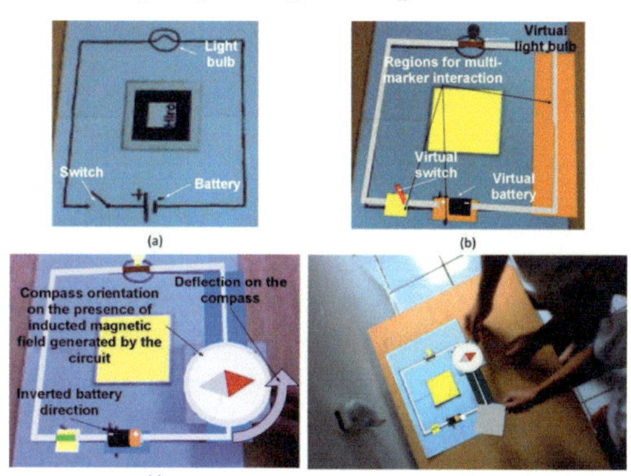

Figure 18. (a) Artifact with drawn elements; (b) Artifact with AR layer; (c) and (d) Oersted's experiments.

This application can be found on the Internet [40].

5. Augmented reality applications for cognitive rehabilitation

Nowadays, with the technological evolution, cognitive rehabilitation is using interactive artifacts, such as software applications (based on multimedia and virtual reality) and physical objects controlled by computer (PDAs, tablets, cellular phones, specific devices with GPS, accelerometers and other technological resources, etc.). Those artifacts are part of technology for cognitive rehabilitation and can help disabled people presenting traumatic brain injury, stroke, learning disabilities and multiple sclerosis. Besides, they have some potential to aid people with dementia, autism spectrum disorders and mental retardation [30].

The cognitive artifacts used for retraining and development of cognitive skills explores the following aspects: temporal and spatial orientation; attention, concentration and calculation; language understanding and speaking; understanding of social cues; judgment and abstraction; immediate recall, recent and remote memory; organization; planning and problem solving; mental processing speed; multi-sensory processing (visual, auditory and motor); self-control and self-confidence.

With recent technological trends, rehabilitation patients are getting access to advanced interactive devices with interesting features, such as highly technological, highly interactive and multi-sensory ones. Nevertheless, those devices present some disadvantages, such as: complex using, difficulty to convert the rehabilitation training to real-life benefits, low or medium availability, medium or high cost, medium or high dexterity demanding, etc. To overcome such problems, it is important to use assistive devices, presenting simplicity as their main feature [25].

We discuss next the development of interactive cognitive artifacts and their applications for retraining and improvement of cognitive skills, aiming at satisfying the main characteristics desired in a modern cognitive device, such as: low cost, easy customization, user-friendly interface, multi-sensory input/output, low dexterity demanding, etc.

The rehabilitation examples, presented next, are:

1. ARAS-NP: Artifact-AR
2. basAR: dGames-VI Memory Game and dGames-inclusive AR Pong

5.1. Artifact-AR

The Artifact-AR was implemented as a 3D structure built with three perpendicular planes, so that each one contains nine cells that can be virtually colored or has spatial colored virtual "coins" activated (Fig. 19). Besides, on the upper side, there is a plane extension used to accommodate the application marker and control buttons and to receive visual information like pictures and texts. The user interacts with the physical artifact, hears the auditory information from computer loudspeaker and visualizes effects (video of the physical artifact expanded with virtual information) on a monitor placed in front of him/her.

The visualization can also be obtained by a projector or an augmented reality HMD, although using a computer monitor is cheap, available and easy to operate. The artifact allows direct interaction with sound feedback; on the other hand, if a monitor or a projector is utilized, the visualization will be indirect, and, if a HMD is utilized, the visualization will be direct.

We initially developed two cognitive applications exploring identification, memorization, comparison and association of pictures, patterns and sounds. An application considers pre-built patterns whereas the other one allows the assembly of patterns through the interaction with each cell individually. Figure 19a presents examples where the user can select a picture on the right side and/or a pattern composed of virtual embossed "coins" on the perpendicular planes, comparing or associating pictures and patterns. Figure 19b presents an example of picture containing one plane pattern to serve as reference to be replicated by the user through the activation of cells on a plane.

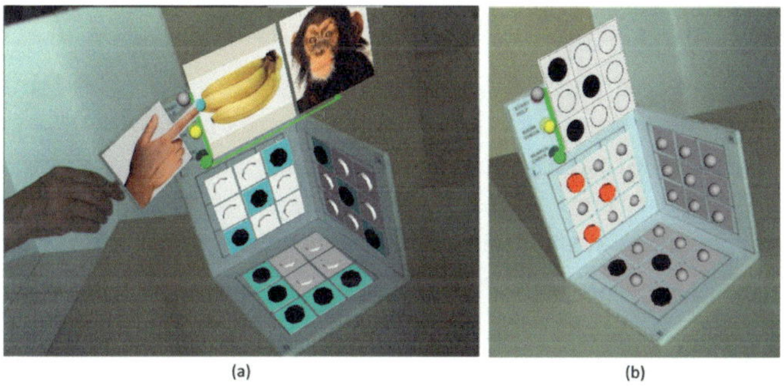

Figure 19. (a) Comparison and association of pictures and/or patterns; (b) Replication of patterns.

Although that application was implemented originally with ARAS-NP, it is being converted with FLARAS to work on the Internet.

The final project using FLARAS will be available on the Internet [22]

5.2. dGames-VI memory game

This project presents a solution to exercise cognitive skills, as association and memory , based on a classic memory card game, using a simple artifact, enhanced with AR. In this application, a therapist can setup several maps, with different characteristics and levels.

This artifact (Figure 20a) was developed using blended tactile and audition sense allowing its use for visual-impaired people. However, as it can also show images (Figure 20b), it can be used by non-visual impaired people as a memory game, or in classes activities, to teach word association, languages, scene associations, etc.

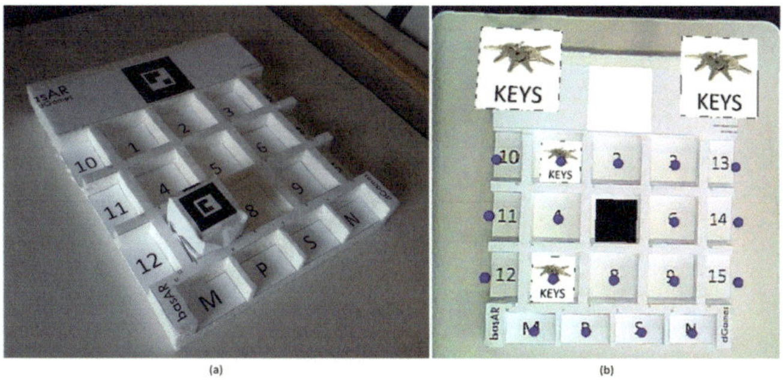

Figure 20. (a) Artifact design; (b) Artifact with AR layer.

An example of this artifact, applied as an inclusive memory card game, is given as follows (Figure 21):

1. The therapist builds several maps, with different characteristics, levels, etc.;

2. The therapist setups the environment with the artifact and the webcam/computer, adopting the AR required software;
3. The system creates the first option, asking to start the activity. The user chooses the next option. The system issues a start sound and shows the covered map.
4. The user chooses a first card (hole). The system issues the card sound and shows its image;
5. The user chooses a second card. The system issues the card sound and its image. If the pair of sounds (and/or images) related to the two cards matches, the system verifies if the amout of pairs on the map is completed. If it is completed, the system issues a game over sound and releases the next map, retrieving to step 3; otherwise it continues. If the pair does not match, it issues a mistaken sound, closing both selected cards and enabling the user to go to step 4.

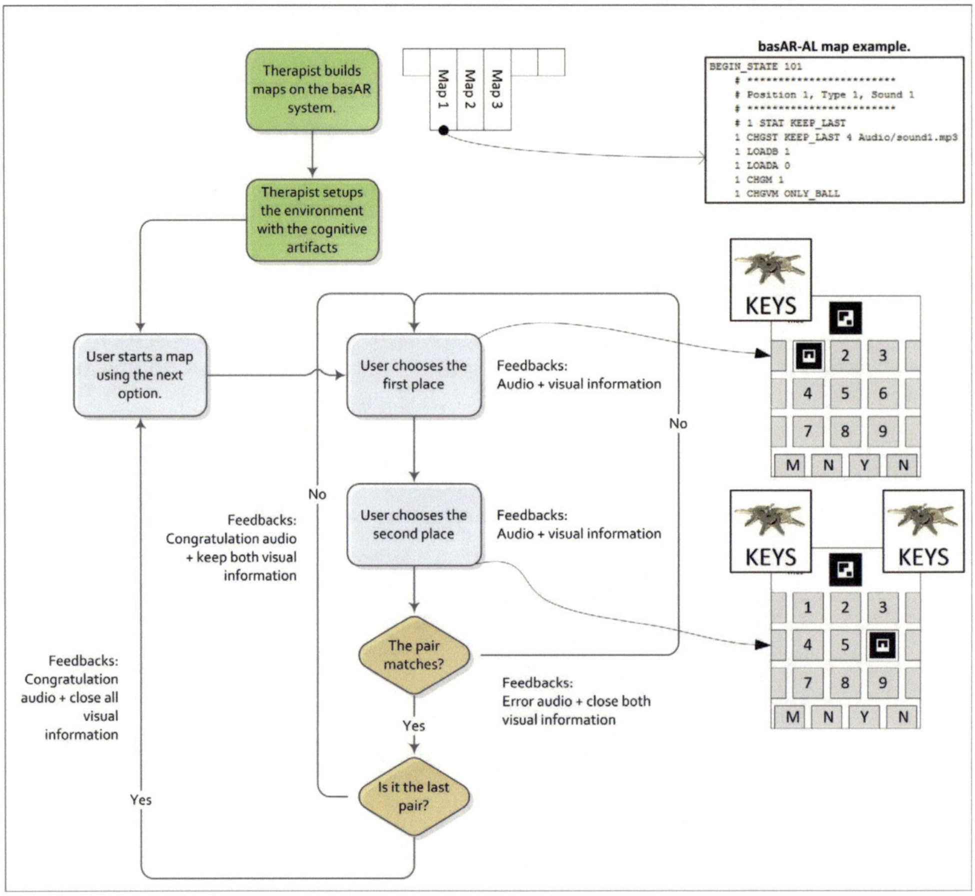

Figure 21. dGames-VI Memory Game Diagram.

5.3. dGames-inclusive AR pong

This project presents a solution to exercise spatial association of a 3D audio stimulus with its corresponding motor feedback. It was inspired on the Ping-Pong game, using a low cost and easily built Artifact enhanced with an AR layer provided by the basAR authoring tool. In this application, a blind people can play against the computer or against other player who is not necessarily blind as well. The game can has a therapeutic and, in this last case, the therapist can set exercises sequences to evaluate the patient.

Figure 22a shows the Styrofoam artifact, with the AR layer, from the camera view (Figure 22b). Figure 22c shows a therapeutic setup, where, in the right artifact (Therapist Artifact), the therapist can set the sequences and the speed in the top three spaces on his artifact grid.

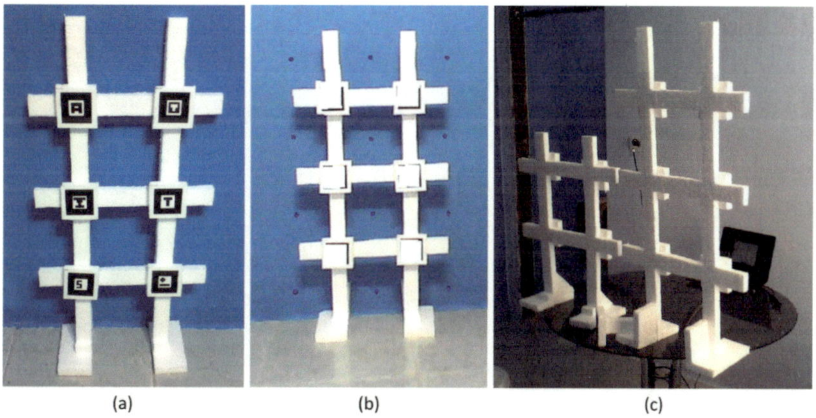

Figure 22. (a) Artifact design; (b) Artifact with AR layer; (c) Application setup.

The AR software layer provides a 3D audio placement, as Figure 23a, the horizontal placement is performed by the stereo balance; the vertical placement is performed by the frequency modulation, in which a higher pitch indicates a higher height and a lower pitch indicates a lower height; the deep placement is associated with the volume, in which higher volumes indicate that the object is near of the user. Figure 23b shows how to control the ball speed. Five stages, with four time intervals control the ball speed, that is, with high time intervals the speed is decreased, and to get fast speed the time intervals are decreased.

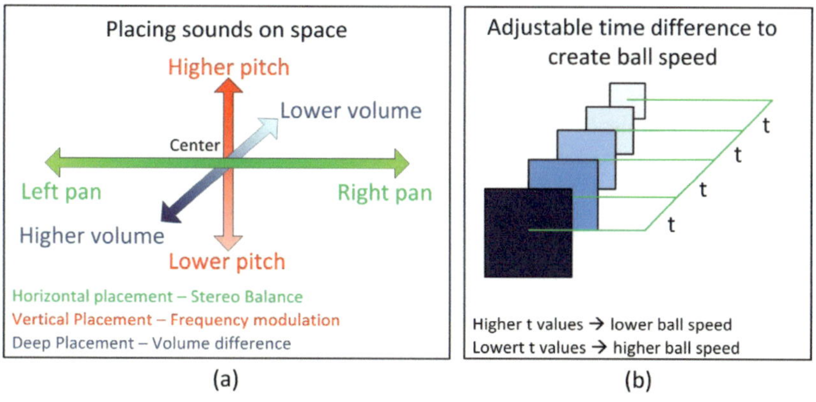

Figure 23. (a) Sound placement into a 3D space; (b) Adjusting the time intervals to control ball speed.

The inclusion of 3D placement in the artifact, enables the augmented reality properties. In this sense, each artifact cell has a deep placement (Figure 24a), to create each cell ball movement.

The vertical and horizontal placements are interlaced, to create nine possible combinations of pan and possible pitch (Figure 24b) so that the user can find the correct cell by its sound.

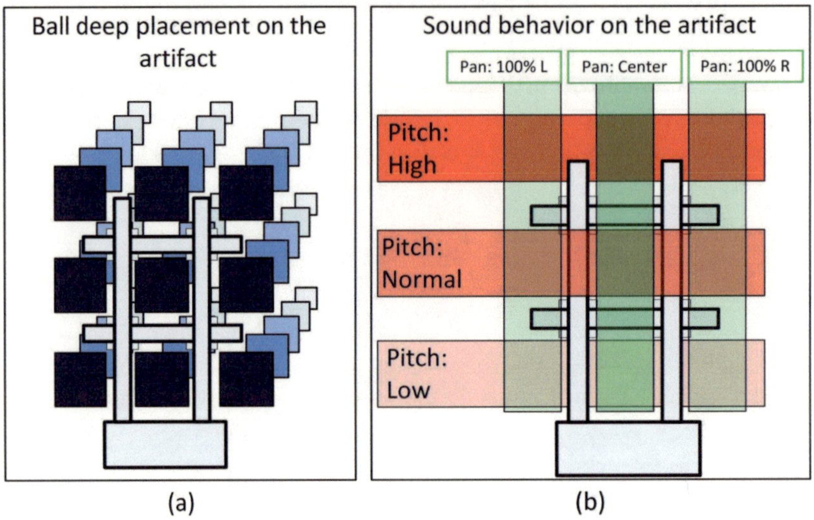

Figure 24. (a) Deep placement; (b) Sound behavior.

6. Conclusion

In this chapter, we presented the concepts and technology related to augmented reality applications, as well as authoring tools and developed applications in the areas of education and cognitive rehabilitation.

When using augmented reality systems based on multiple points instead of on multiple markers, we notice that the collision of the interaction device with the virtual points was hard to accomplish, due to the spatial positioning. To solve this problem, an augmented reality artifact was created, to place the virtual points over real points in a physical structure. The augmented reality artifacts solved the problem with the spatial collision in multiple depths. This allowed the development of several augmented reality applications for educational and cognitive rehabilitation purposes, once the artifact empowered with smart augmented reality reactions would provide significant support to students and patients, at low cost.

The fast prototyping of the application solutions with Styrofoam, cardboards, and easily found materials allows the creation of artifacts enriched with the augmented reality layer, which can be easily distributed and used. Even with the artifact weaknesses, it seems to be a very interesting option to be applied. Evaluation tests confirmed important strengths of using the artifacts, such as low cost, availability, user-friendly interfaces, multi-sensory, tangible interaction, non-demanding dexterity, etc.

The proposed authoring tools have distinct characteristics with respect to other approaches which are based on marker relation behavior, as we use action point interactions. This option allows the use of a minimum amount of marker, instead of a pack of markers to drive an application.

The authors believe that augmented reality artifacts have high potential to be applied in educational and cognitive rehabilitation applications, due to the specific potentiality provided by the augmented reality and the three-dimensional artifact features.

As future work, we are evolving the authoring tools, aiming at generating more powerful and easy-friendly versions and exploring the integration of web applications with online augmented reality applications implemented with the FLARAS authoring tool. Besides, we are developing cognitive and motor rehabilitation games, using the basAR authoring tool, and studying the use of cross-reality in innovative applications that could effectively contribute to the educational and rehabilitation areas.

Acknowledgments

This research was partially funded by Brazilian Agencies CNPq (Grants #558842/2009-7 and #559912/2010-2) and FAPEMIG (Grant #APQ-03643-10).

Author details

Claudio Kirner
UNIFEI - Universidade Federal de Itajubá, Brazil

Christopher Shneider Cerqueira
INPE - Instituto Nacional de Pesquisas Espaciais, Brazil

Tereza Gonçalves Kirner
UNIFEI - Universidade Federal de Itajubá, Brazil

7. References

[1] Azuma, R., Baillot, Y., Behringer, R., Feiner, S., Julier, S. & MacIntyre, B. [2001]. Recent advances in augmented reality, *IEEE Comput. Graph. Appl.* 21(6): 34–47. URL: *http://dx.doi.org/10.1109/38.963459*

[2] Azuma, R. T. [1997]. A survey of augmented reality, *Presence: Teleoperators and Virtual Environments* 6(4): 355–385.

[3] Bang, M. & Timpka, T. [2003]. Cognitive tools in medical teamwork: the spatial arrangement of patient records., *Methods Inf Med* 42(4): 331–6.

[4] Beato, N., Mapes, D. P., Hughes, C. E., Fidopiastis, C. & Smith, E. [2009]. Evaluating the potential of cognitive rehabilitation with mixed reality, *Proceedings of the 3rd International Conference on Virtual and Mixed Reality: Held as Part of HCI International 2009*, VMR '09, Springer-Verlag, Berlin, Heidelberg, pp. 522–531.

[5] Bimber, O. & Raskar, R. [2004]. *Spatial Augmented Reality - Merging Real and Virtual Worlds*, 1 edn, A K Peters, Wellesley, Massachusetts, USA.

[6] Burdea, G. C. & Coiffet, P. [2003]. *Virtual Reality Technology*, 2 edn, John Wiley & Sons, Inc., New York, NY, USA.

[7] Cardoso, A. & Lamonier Jr., E. [2009]. *Virtual and Augmented Applications - from Aplicações de Realidade Virtual e Aumentada*, SBC, chapter AR and VR Educational and Training Applications - from Aplicações de RV e RA na Educação e Treinamento, pp. 29–54.

[8] Cerqueira, C. S. & Kirner, C. [2012a]. basar, online.
URL: *http://www.cscerqueira.com/basar*

[9] Cerqueira, C. S. & Kirner, C. [2012b]. Geometry learning, online.
URL: *http://www.cscerqueira.com/basar/projects/005_geometry/*

[10] Chen, S.-J. [2007]. Instructional Design Strategies for Intensive Online Courses: An Objectivist-Constructivist Blended Approach, *Journal of Online Interactive Learning* 6(1).

[11] Dias, J. M. S., Monteiro, L., Santos, P., Silvestre, R. & Bastos, R. [2003]. Developing and authoring mixed reality with mx toolkit, *IEE Review* pp. 18–26.

[12] Grasielle, A., Correa, D., Assis, G. A. D. & Nascimento, M. [2007]. Genvirtual : An augmented reality musical game for cognitive and motor rehabilitation object . wt astiuatn uosiber ernative bengiuse rality inter todcollati st cala . pitc depends on, *Virtual Reality* pp. 1–6.
URL: *http://ieeexplore.ieee.org/lpdocs/epic03/wrapper.htm?arnumber=4362120*

[13] Grimm, P., Agc, F., Overview, I., Haller, M., Reimann, C., Paelke, V., Paderborn, U. & Zauner, J. [2002]. Amire - authoring mixed reality.
URL: *http://www.amire.net/*

[14] Hamilton, K. E. [2011]. Augmented reality in education, online.

[15] Hampshire, A., Seichter, H., Grasset, R. & Billinghurst, M. [2006]. Augmented reality authoring: generic context from programmer to designer, *Proceedings of the 18th Australia conference on Computer-Human Interaction: Design: Activities, Artefacts and Environments*, OZCHI '06, ACM, New York, NY, USA, pp. 409–412.
URL: *http://doi.acm.org/10.1145/1228175.1228259*

[16] Kato, H. & Billinghurst, M. [1999]. Marker tracking and hmd calibration for a video-based augmented reality conferencing system, *Proceedings of the 2nd IEEE and ACM International Workshop on Augmented Reality*, IWAR '99, IEEE Computer Society, Washington, DC, USA.
URL: *http://dl.acm.org/citation.cfm?id=857202.858134*

[17] Kim, M., Gak, H. J. & Pyo, C. S. [2009]. Practical rfid + sensor convergence toward context-aware x-reality, *Proceedings of the 2nd International Conference on Interaction Sciences: Information Technology, Culture and Human*, ICIS '09, ACM, New York, NY, USA, pp. 1049–1055.
URL: *http://doi.acm.org/10.1145/1655925.1656115*

[18] Kirner, C. [2011a]. Aras-np: Augmented reality authoring system for non-programmers, online.
URL: *http://www.ckirner.com/sacra*

[19] Kirner, C. [2011b]. *Tendências e Técnicas em Realidade Virtual e Aumentada*, SBC, chapter Prototipagem Rápida de Aplicações Interativas de Realidade Aumentada.

[20] Kirner, C. & Kirner, T. G. [2011a]. Development of an educational spatial game using an augmented reality authoring tool, *International Journal of Computer Information Systems and Industrial Management Applications*, Vol. 3, MIR Labs, pp. 602–611.

[21] Kirner, C. & Kirner, T. G. [2011b]. Development of an interactive artifact for cognitive rehabilitation based on augmented reality, *Virtual Rehabilitation (ICVR), 2011 International Conference on*, pp. 1 – 7.

[22] Kirner, C. & Kirner, T. G. [2012]. Artifact-ar, online.
URL: *http://www.ckirner.com/ar/artifact-ar/*

[23] Kirner, C., Kirner, T. G., Mataya, R. S. & Valente, J. A. [Sept. 2010]. Using augmented reality to support the understanding of three-dimensional concepts by blind people, *in* J. S. P M Sharkey (ed.), *WProc. 8th Intl Conf. on Disability, Virtual Reality and Assoc. Technologies*, pp. 41–50.

[24] Ledermann, F. & Schmalstieg, D. [2005]. April a high-level framework for creating augmented reality presentations, *Proceedings of the 2005 IEEE Conference 2005 on Virtual Reality*, VR '05, IEEE Computer Society, Washington, DC, USA, pp. 187–194.
URL: *http://dx.doi.org/10.1109/VR.2005.8*

[25] Lopresti, E. F., Mihailidis, A. & Kirsch, N. [2004]. Assistive technology for cognitive rehabilitation: State of the art, *Neuropsychological Rehabilitation* 14(1-2): 5–39.
URL: *http://www.tandfonline.com/doi/abs/10.1080/09602010343000101*

[26] MacIntyre, B., Gandy, M., Dow, S. & Bolter, J. D. [2004]. Dart: a toolkit for rapid design exploration of augmented reality experiences, *Proceedings of the 17th annual ACM symposium on User interface software and technology*, UIST '04, ACM, New York, NY, USA, pp. 197–206.
URL: *http://doi.acm.org/10.1145/1029632.1029669*

[27] Manovich, L. [2006]. The poetics of augmented space, *Visual Communication* 5(2): 219–240.
URL: *http://vcj.sagepub.com/cgi/content/abstract/5/2/219*

[28] Martín-Gutiérrez, J., Luís Saorín, J., Contero, M., Alcañiz, M., Pérez-López, D. C. & Ortega, M. [2010]. Education: Design and validation of an augmented book for spatial abilities development in engineering students, *Comput. Graph.* 34(1): 77–91.
URL: *http://dx.doi.org/10.1016/j.cag.2009.11.003*

[29] Matsuda, K. [2010]. *Domestic city - the dislocated home in augmented space*, Master's thesis, UCL - London's Global University.
URL: *http://www.keiichimatsuda.com/thesis.php*

[30] Morganti, F. [2004]. Virtual interaction in cognitive neuropsychology., *Studies in health technology and informatics* 99: 55–70.
URL: *http://view.ncbi.nlm.nih.gov/pubmed/15295146*

[31] Norman, D. A. [1991]. Designing interaction, Cambridge University Press, New York, NY, USA, chapter Cognitive artifacts, pp. 17–38.
URL: *http://dl.acm.org/citation.cfm?id=120352.120354*

[32] Norman, D. A. [1992]. Design principles for cognitive artifacts, *Research in Engineering Design* 4: 43–50. 10.1007/BF02032391.
URL: *http://dx.doi.org/10.1007/BF02032391*

[33] Okawa, E. S., Kirner, T. G. & Kirner, C. [2012]. Spacear, online.
URL: *http://www.ckirner.com/sacra/aplica/sol-ra/*

[34] Paradiso, J. A. & Landay, J. A. [2009]. Guest editors' introduction: Cross-reality environments, *IEEE Pervasive Computing* 8(3): 14–15.
URL: *http://dx.doi.org/10.1109/MPRV.2009.47*

[35] Reilly, D., Tang, A., Wu, A., Echenique, A., Massey, J., Mathiasen, N., Mazalek, A. & Edwards, W. K. [2011]. Organic uis and cross-reality spaces, *Workshop on Organic User Interfaces*, pp. 23–26.

[36] Reis, F., Kirner, T. G. & Kirner, C. [2012]. Geoar, online.
URL: *http://www.fernandamaria.com.br/geoar/*

[37] Richard, E., Billaudeau, V., Richard, P. & Gaudin, G. [2007]. Augmented reality for rehabilitation of cognitive disabled children: A preliminary study, *2007 Virtual Rehabilitation*, pp. 102–108.

[38] Saúde, L. M. S., Kirner, T. G. & Kirner, C. [2012]. Perspective learning system with augmented reality from sistema de aprendizagem de perspectiva com realidade aumentada, online.
URL: *http://ckirner.com/eventos/jornada2011/lara.html*

[39] Seichter, H., Looser, J. & Billinghurst, M. [2008]. Composar: An intuitive tool for authoring ar applications, *Proceedings of the 7th IEEE/ACM International Symposium on Mixed and Augmented Reality*, ISMAR '08, IEEE Computer Society, Washington, DC, USA, pp. 177–178.
URL: *http://dx.doi.org/10.1109/ISMAR.2008.4637354*

[40] Souza, R. C. & Kirner, C. [2012]. Minilabeletromag-ra, online.
URL: *http://ckirner.com/apoio/eletromag/*

[41] Souza, R. C., Moreira, H. C. F. & Kirner, C. [2011]. Flaras: Flash aumented reality authoring system, online.
URL: *http://www.ckirner.com/flaras*

[42] Uchiyama, S., Takemoto, K., Satoh, K., Yamamoto, H. & Tamura, H. [2002]. Mr platform: A basic body on which mixed reality applications are built, *Proceedings of the 1st International Symposium on Mixed and Augmented Reality*, ISMAR '02, IEEE Computer Society, Washington, DC, USA, pp. 246–.
URL: *http://dl.acm.org/citation.cfm?id=850976.854992*

[43] Yao, Y., Wu, D. & Liu, Y. [2009]. Collabrative education ui in augmented reality from remote to local, *Proceedings of the 2009 First International Workshop on Education Technology and Computer Science - Volume 02*, ETCS '09, IEEE Computer Society, Washington, DC, USA, pp. 670–673.
URL: *http://dx.doi.org/10.1109/ETCS.2009.409*

Chapter 13

Virtual Environments for Children and Teens

Jamshid Beheshti

Additional information is available at the end of the chapter

http://dx.doi.org/10.5772/51628

1. Introduction

Today's children and teens are technology savvy. Information and communication technology (ICT) is the prevalent mode of communication among them, with 75 percent of 12 to 17 year olds owning cell phones, and text messaging at an incredible rate of more than 3000 messages per month [1]. This new generation is referred to as 'net savvy' youth [2], Google generation [3], and generation M, for Media, MySpace, or the Millenials [4] among other titles. The most common term, however, is the *digital natives*; those born after 1989, who may process information "different from their predecessors" [5]. The digital natives do much more than text messaging; they *live* in the digital world. Whereas most adults are attached to their physical material artifacts, cherishing their hardcopy books and DVDs, the digital natives live in digital worlds where they own virtual artifacts, which are more than just digital music and streaming movies. They use social networks (SNS) to create personal spaces to store artifacts such as currency and familial possession as a means of self-expression [6].

Digital natives often are well versed in using computer games, most of which now take the form of simulations, such as those designed for the X-Box 360 and the Nintendo Wii. According to a UK survey, teens' reliance on the gaming console to surf the web has increased significantly [7]. Recent reports also suggest that virtual environments or worlds are one of the most popular modes of interaction on the web [8]. The total number of users registered for virtual world sites is now more than one billion, of which the largest demographic group is between the ages of 10 and 15 [9]. The young generation uses the gaming console to access the internet and social networks, and therefore is most likely to want to use the virtual reality technology for more than gaming.

Computer simulations can take various technological forms, including virtual reality, augmented reality, and virtual environments. In the taxonomy of virtual reality, Milgram and Kishino [10] identify a continuum that connects real environments to virtual environments. They define augmented reality as a display where real environments are augmented with virtual objects. VEs provide a computer-generated experience obtained by and through an interface that engages one or more of the user's senses, and almost always includes the visual sense [11].

By far the vast majority of VEs have been developed for entertainment and gaming. *World of Warcraft* is perhaps the best known and most used VE designed for adult gamers, while Habbo Hotel (www.habbo.com) is one of the most subscribed among the more than 250 virtual worlds constructed for teens. Habbo Hotel founded in 2000, is an example of a thriving VE, boasting to be the "world's largest online community" with more than 200 million registered characters. Other examples of gaming virtual environments for children and teens include Club penguin, (http://www.clubpenguin.com/), which is the most popular virtual world among children aged 7 to 10 in Europe [12], Pet Society (http://www.petsociety.com/), Secret Builder (http://secretbuilders.com/home.html), and Whyville (http://www.whyville.net/smmk/nice). Beals and Bers mention many other VEs for children and provide detailed statistics of their usage [13]. Many corporations have combined their virtual worlds with real children's toys to increase their traffic and therefore their revenues, such as Webkinz, Bratz dolls and Barbie dolls. Beals and Bers conclude that to some extent, parents may perceive these virtual spaces as safer havens than real brick and mortar buildings and encourage their children to use them.

The focus of this chapter is on the design and application of virtual environments (VEs) for children and teenagers by reviewing examples of current systems in education, health, and information settings, design methodologies, and engagement.

2. Education

Virtual environments can provide an interactive, stimulating learning environment beyond gaming. These environments sometimes referred to as virtual learning environment (VLE) [14], or Educational Virtual Environments (EVE) [15] have increasingly gained recognition among educators. Although the application of computer simulations in an educational context has raised some controversy, especially from those educators and developmental psychologists who have questioned the appropriateness of "virtual" experiences for children [see for example 16], a MediaWise report summarizing the findings of a number of research studies, states: "Video games are natural teachers. Children find them highly motivating; by virtue of their interactive nature, children are actively engaged with them; they provide repeated practice; and they include rewards for skilful play. These facts make it likely that video games could have large effects, some of which are intended by game designers, and some of which may not be intended." [17]. Roussos et al [18] point to evidence that immersion and presence can have a strong motivational impact on the users. As Mumtaz [19] argues, young people like computer applications, which often produce engagement and delight in learning. VEs present opportunities "to experience environments which, for reasons of time, distance, scale, and safety, would not otherwise be available to many young children ..." [18, p.247]. Students who performed poorly in the classroom seem to benefit most from VEs [20]. Educational simulations can produce engagement through elements such as "embodiment, cultural embedding, personalised maps, interactive artifacts, dynamic environments, mood, and contextual tasks." [21]. Dalgarno & Lee [22] outline the potential learning benefits of VLEs by identifying the unique characteristics of the 3D environment that may impact students in an educational setting. Immersive technologies and particularly immersive presence are shown to enhance education by allowing multiple perspectives or points of view, through 'situated experience', i.e., simulated field and laboratory work, and transfer of real world experiences [23].

Findings in cognitive research also point to the potential of educational simulations. Drawing upon recent research, theorists have advanced the notion of "constructionism" (a term coined by Papert in [24], and closely connected to Piaget's "constructivism"), contending that knowledge is not deposited by the teacher into the student – what Freire [25] termed "banking" – but rather constructed in the mind of the learner. The use of technology and particularly games in education is important in integrating new materials in a formal learning environment, as gaming is the preferred activity of children in the age range of four to seven [26]. Such activity seems to increase self-esteem [27] and perhaps motivation towards learning.

As Gee [28] has shown, VEs are especially effective in allowing students to experience new worlds where they can develop resources for problem solving and, ultimately, view the environment as a design space that can be engaged and changed. Or as Bass [29] put it, researchers should search for "the critical and productive affinities" between "materials, methods, and epistemology on the one hand, and the inherent structure and capabilities of interactive technologies, on the other." Mikropoulos and Natsis conducted a ten-year review of empirical research on virtual environments in education [15]. They surmised that constructivism is the hallmark of virtual learning environments, and includes seven principles:

"1. Provide multiple representations of reality – avoid oversimplification of instruction by representing the natural complexity of the world
2. Focus on knowledge construction not reproduction
3. Present authentic tasks (contextualizing rather than abstracting instruction)
4. Provide real world, case based learning environments, rather than predetermined instructional sequences
5. Foster reflective practice
6. Enable context, and content, dependent knowledge construction
7. Support collaborative construction of knowledge through social negotiation, not competition among learners for recognition." (p. 771)

In general, education-focused virtual environments support learning in a variety of fields such as architecture, language training, and archaeology. However, the application of VEs in learning is becoming increasingly prevalent, particularly in science and mathematics education. Mikropoulos and Natsis investigated 53 studies on the application of VEs in education, of which 40 referred to science and mathematics [15]. More recent studies include the Mediterranean Sea, for example, which is modeled to teach students about ecology in an interactive setting [30]. A total of 48 students from two grade six classes were assigned randomly to two groups, 'traditional' and 'virtual' classes, in the Mediterranean Sea experiment. While the results of the experiment did not show a significant difference between the two classes in the learning outcome as measured in the limited pre and post tests, significant results were observed in the students' evaluations of engagement and 'enjoyment' for the VE 'class'. To teach geographical concepts, Tuzun et al [31] created an environment, whereby students collaborate to collect artifacts in different continents. The researchers used quantitative and qualitative methods including pre and post tests, observations, interviews, and open-ended questions to gather data on 24 students divided equally with 12 boys and 12 girls from fourth and fifth grades. Data consisted of achievement and motivation test scores as well as open-ended questions on students' knowledge of the subject matter. All the results showed statistically significant gains by students, which prompted the researcher to conclude that the virtual game-like environment supports learning in geography, whilst increasing students' motivation and makes learning 'fun'.

VEs have been also successfully utilized to represent 3D models of chemical compounds in chemistry classes [32], where the researchers used inexpensive webcams and open-source software to develop their system. They conducted a survey after using the system in a classroom, the results of which showed improved performance when solving problems related to 3D chemical structures. Virtual reality has been applied in mathematics education (for example see [33]). Another technology related to virtual environments is Augmented Reality (AR), which is a promising tool for schools to teach students about a variety of experiential subjects including material science [34].

The use of multi-user virtual environments (MUVEs) in education has been gaining momentum in recent years, with increased number of practical implementations that are reaching classrooms and students [35]. One of the few MUVEs designed for a wide range of purposes is the well-known Second Life (SL), launched in 2003 and intended for users 16 years and older. It has now about one million active followers. Communication in the SL virtual worlds is conducted through avatars, characters that can take the shape of human, animal, vegetable, or mineral. SL has been the subject of much research in education (for example see [36]). Summarizing the results of research, Hew and Cheung [37] conclude that in K-12 and higher education, virtual worlds are utilized to facilitate communication among students for simulation of real world trials and procedures, and for experiential enactments.

3. Health

VEs have been used in behavior modification and intervention. Reviewing the literature, Riva concludes that virtual reality has been used to "induce an illusory perception" of various body parts, as well as improve body image in patients with obesity and eating disorders[38]. Merry et al [39] reports on an experiment using a VE system, SPARX, as a self-help intervention agent to decrease and alleviate depression among young people. The software, designed by Metia Interactive (http://www.metia.co.nz/), utilizes a 3D fantasy game environment to teach skills to manage symptoms of depression. Young users can customise their avatars to travel to one or more of the seven Provinces, where they learn about 'hope', 'being active', 'dealing with emotions', and so on. In the experiment, the researchers allocated 85 volunteers to SPARX, and 85 to conventional treatment, with a follow up after three months. The volunteers' average age was 15.6 years. The results of the study showed that remission rates were significantly higher in the SPARX group than the conventional group, as measured on various psychological metrics.

Virtual reality and VEs have been used successfully to engage young people in behavior change to manage their weight, to encourage physical activity, and in rehabilitation settings. Exergaming is the term used for those applications that utilize sensory surface for exercising and active video gaming such as *Konami's Dance Dance Revolution* and other arcade games. Exergaming has been studied to quantify energy cost associated with playing active games and increase activities. In a recent study, researchers randomly assigned 108 students between the ages of 10 and 15 years to one of the three experimental groups [40]. The first group used the "Jackie Chan Studio Fitness Action Run", an exergaming environment produced by XaviX; the second group was assigned to the 1-mile run/walk, and the third group utilized the Progressive Aerobic Cardiovascular Endurance Run program. The rate of perceived exertion (RPE), which is measured by a word scale, where zero means 'not tired', and nine means 'very, very tired', was used as a metric. RPE has been shown to correlate strongly with oxygen uptake and heart rate. The results show that the RPE for 'Jackie Chan' exergame group was significantly lower than the RPE for the other groups. The researchers conclude that the exergame can be a potential substitute for other types of physical activities, where space and facilities are limited, and they may also encourage children and teens to participate in aerobic fitness programs, regardless of their body mass index (BMI). Other applications of VEs in health care include brain injuries, pediatric oncology, for assessing attention deficit disorder, and for autism (see for example [41], [42]).

4. Information retrieval

Teens use the Internet for online shopping, downloading music, and sharing their personal information and artifacts. However, 62 percent reported that they surf the Web for finding and retrieving news and information about current events, and 31 percent reported that they search for health, dieting, or physical fitness information [43]. Despite all their online activities, a panel of experts, meeting to discuss the information behavior and needs of the new generation of users, concluded that a research agenda is urgently required to investigate the "characteristics and preferences of this tech savvy group that surprisingly lacks basic skills in information evaluation and retrieval" [44]. A growing body of research shows that children and young adults when seeking information under imposed tasks such as school projects encounter many problems and challenges. In a comprehensive review of the literature Large [45] concludes that children encounter problems in selecting appropriate search terms, move too quickly through the web pages while spending little time reading the materials, and have difficulty judging the relevance of the retrieved pages. These are just a few challenges among scores of obstacles facing today's teenagers, who for most part may be information illiterate. In fact, studies show that when the youth enter institutions of higher education, they lack information seeking, retrieval and evaluation skills [46].

Rowland et al [3] observed that the new generation is "hungry for highly digested content," and their "information seeking behaviour can be characterised as being horizontal, bouncing, checking and viewing in nature." By horizontal they refer to the skimming activities of the youth, so prevalent in their information seeking, be it at home or at school. Nicholas et al [47] imposed four tasks (pre-piloted questions) on 138 participants, whose age ranged from children to adults. The results show that while the younger generation was faster in searching and retrieving the results, they viewed fewer pages and websites, and conducted fewer searches, resulting in less confidence in their answers to the task questions. One potential inference from many of these studies is that while the young generation is confident in its technology abilities, it lacks information and media literacy skills to make informed judgment on searching, selecting, authenticating, retrieving, organizing, synthesizing, and applying the information to create new knowledge.

In response to the youth's technological skills, and their inadequate competencies in information literacy, librarians have investigated ways to use virtual environments to assist young people in their information seeking behavior. Among the few VEs, Second Life is used in informational settings, with an archipelago allocated to Cybrary City, where more than 40 virtual libraries are situated. Second Life is also the most studied VE among researchers (for example [48]). In addition to Cybrary City, librarian avatars residing on Info Islands formed the Community Virtual Library in April 2006. Within a year of its launch, more than 6,500 virtual reference questions were answered by the Community [49]. Citing statistics by *Gartner Inc* that by 2011 some 80 percent of Internet users will be in virtual worlds [50], several public libraries purchased islands in Second Life to serve their younger clients [51].

Buckland [52] reports on two pilot studies in two Canadian universities, which acquired space on Cybrary City in Second Life. The purpose of the studies was to explore: a) the efficacy of virtual reference service, b) resources and training to offer the service, and c) the need for the service. The staff and volunteers were trained, and virtual reference services with avatars were implemented in Cybrary City for both universities. The result of the pilot studies showed that while Second Life residents could use Google for finding information to fulfill their needs, they preferred to ask librarian-avatars for assistance.

4.1. A virtual information environment

Beheshti and Large [53] designed and tested a virtual information environment for children, which embodies many of the concepts discussed in this chapter. The impetus for the project may be summed up as:

- Obstacles: Researchers have concluded that children and youth lack appropriate training and skills to transfer their information needs to effective search strategies (see for example [54]). The new generation spends little time evaluating the retrieved information, and expects instant gratification with the search results. Younger children also face spelling and typing challenges that add more obstacles to their information seeking process.

- Browsing: Research suggests that browsing may be a viable alternative to keyword searching for younger users, who might otherwise have problems in seeking information from the web, and whose behavior consists of skimming activities. Studies also suggest that browsing may be the preferred mode of searching, considering that both searching and browsing may produce equally valid and efficient results [55]. Children have a tendency to explore, and view a digital library as "a place to wander about looking for different kinds of information" [56]. Browsing can also produce valuable serendipitous discoveries.
- Visualization: Browsing is primarily a visual activity, and visual-based exploratory interfaces support search activities for learning and investigating [57].
- Metaphors: Many systems have incorporated familiar metaphors in their interfaces, such as the shopping cart and notebook metaphors. These everyday metaphors have made interfaces more "comprehensible and fun" [58]. A familiar metaphor takes advantage of artifacts and context affordances for a more natural and intuitive interaction [59]. Visual interfaces designed for children should be based on familiar metaphors.
- Libraries: Although young people are increasingly using digital information, they are still well acquainted with traditional libraries. The library metaphor has been used in experimental projects (see for example [60]). The book metaphor has also been used in online information systems, as a recognized artifact to assist users in search and navigation (see for example [61]). The Bookhouse project was an innovative retrieval system for fictional work in a public library, which used the library and bookshelf metaphor. Subjective evaluation by the users showed that the new interface was preferred to conventional means of retrieving fiction [62]. In another example a realistic virtual environment of an existing library building was built and user acceptance tests were performed. The authors reported that high school and university users quickly learned how to navigate in the virtual environment without any assistance, showed high user engagement, and expressed positive first impressions [63].

The library virtual environment was constructed using the metaphor of a physical library with rooms, bookcases and books. The user, just as in a physical library, can walk around the library, move among the bookcases, scan the titles of books that are arranged on the bookshelves, select individual books, and open them. Once the prototype was constructed, a combination of Bonded Design and Informant Design (see next section on *Spectrum of Design Methodologies*) was used to obtain feedback from children and young adults on the library.

After the initial construct and testing, the methodology was repeated for a second iteration. The system was tested again, the results of which led to the third iteration. Each iteration yielded feedback from children, which paved the way to new insights and recommendations. At each step, children made a number of recommendations for improving the system: navigational maps, search workstations, classification of 'books', limited personalization (for example color of walls and floors), and limited application of sound for a more realistic setting. Perhaps the most interesting recommendation was children's desire for presence of avatars, particularly a 'librarian', who would provide help and assistance on demand.

Figure 1. The Library virtual environment

The final product (Figure 1) was the library virtual environment, which was developed as an alternative interface for children's web portals. In this environment, users can utilize *search stations* situated in different locations of the library to conduct conventional keyword and term searches, the results of which are displayed as red dots on a plan of the library. The library contains about 1500 links to English-language websites on Canadian history deemed to be appropriate in content and language for elementary students. The database of links was initially created for *History Trek* (http://www.historytrek.ca), a children's portal on Canadian history [64]. Based on one of the suggestions made by children, all the websites were classified by the Dewey Decimal Classification (DDC) system to organize the collection similar to a typical public or school library.

Three focus group studies were conducted to evaluate and assess the efficacies of the library virtual environment. In the first study [65], eight children and teens between the ages of 11 and 16 participated in two focus groups to assess and evaluate the library virtual environment. In order to encourage the evaluation of the interface, participants completed four tasks, which were used in previous studies with a conventional interface. Similar qualitative methodologies were used in two other focus group studies [66, 67]. The results of these studies showed that the library virtual environment was an engaging alternative to conventional searching. Children reported that the virtual environment was much more 'fun'. The word "cool" was repeatedly used by all the children in the studies, regardless of their gender or age. One child stated: "If I have an option between Google and this [the library], I would use this. Lots more fun...it is different." Another youngster suggested that the library "would always be more interesting than Google." Children participating in the studies seemed to have a positive first impression, which is one of the crucial factors in measuring satisfaction and the desire of the user to continue working with the application. Norman [59] outlines three levels of design: visceral, behavioral and reflective. The visceral level is about the initial feelings a new product provokes, which may be independent of culture or experience and may have a significant impact on the success of the subsequent interactions. The library seems to be a fun and engaging virtual environment, where children and teens can spend time exploring, browsing, and scanning the digital information, the outcome of which may lead to more successful learning.

5. Spectrum of design methodologies

Although much has been written about the design criteria for technical aspects of virtual reality and virtual environments, only recently researchers have discussed other aspects of design such as content, aesthetics, and behavior. For example, Messinger et al [68] suggest that the "fundamental technical preconditions for a world to support education appear to be (a) realistic rendering, (b) expressive and behaviorally rich avatars, (c) high performance, and (d) easy-to-use tools for education providers to develop the materials necessary for their objectives." Bers [69] divides the design criteria for virtual worlds by age group: early childhood, elementary years, and high school age. She proposes the Positive Technological Development theoretical framework within which systems that can promote "engaging a young person in a good, healthy, and productive development trajectory" can be built. This framework consists of three components: Assets, such as caring, contribution, competence, etc.; Behaviors, which includes communication, collaboration, creativity, and content creation; and Context of practice. Technological tools bridge the gap between Assets and Behaviors, and learning culture, routines, and values may determine the Context of practice.

Until recently, research was conducted by adults with little input from the target audience, children and youth. As Hanna and her colleagues at Microsoft's Hardware Ergonomics and Usability Department commented: "Usability research with children has often been considered either too difficult to carry out with unruly subjects or not necessary for an audience that is satisfied with gratuitous animations and funny noises." [70] In the mid 1990s, a paradigm shift in usability studies began to take shape, when researchers began focusing on young users' participation in the actual design process with the hope that the resulting information retrieval systems meet their needs. As more researchers involved children in the system design process, a need for adopting conceptual models for children as designers arose [71]. There is no doubt that the role of children is critical in the design process as users of the system, as testers in usability studies, as informants in the design process, and as partners [72].

Depending on the role of the users, the models or methodologies applied in the design process may be divided into seven broad categories [71], from the least amount of user involvement to the most participation. Based on this spectrum, the user-centered design is at lowest level involvement, followed by Contextual Design, which utilizes ethnographic methods to determine how users work; Learner-Centered design, which is based on learning theories and assumes that everyone is a learner, and that learning cannot be detached from practice; Participatory Design, which utilizes the principle that users know how to improve their work and they are qualified to contribute to the development of new systems through their perceptions of technology; Informant Design, which views children (and students) as much more informed of learning practices than researchers; and Cooperative Inquiry methodology, which is developed for the design process involving younger participants, and where children are treated as equal partners in an intergenerational design team.

Large and his colleagues [73] adapted and incorporated a modified version of Cooperative Inquiry in their research. The new methodology was called Bonded Design, because of a natural bond between adults as design experts, and children as experts on being children. Large and his colleagues successfully utilized the Bonded Design methodology to create a portal on Canadian history for Grade 6 students [64].

6. Engagement

One of the most important factors for success of the gaming industry is engagement. For children and teens, engagement is a vital component of user experience (UX), which describes the totality of experience of the user and includes how easy a system is to learn, its efficiency, memorability, error management, user satisfaction [59], as well as holistic, aesthetic and hedonic factors, emotion and affect factors, and an experiential factors [74]. Laurel [75] defines engagement as a first-person experience involving playfulness, fun, and sensory integration that sustains a user's attention. Peters et al [76] suggest that engagement is perhaps the most important concept in human-computer interaction for the design of intelligent interfaces that are capable of adapting to users. For children and teens, engagement show "sustained behavioural involvement in learning activities accompanied by positive emotional tone." [77] On the other hand, youth may become bored, passive, and anxious in the learning environment when systems are not engaging. User engagement is a complex phenomenon which describes "how and why applications attract people to use them within a session and make interaction exciting and fun…" [78]

Among the many factors that make up user engagement is presence, which for children and youth is portrayed in virtual environments as a 2D virtual character or a 3D avatar. A primary concept of presence, aesthetics may be more influential in user preferences than is usability [59], and plays a crucial role in technology's overall attractiveness and its initial usage [79]. There is no doubt that interfaces with highly rated aesthetical appeal provide high overall user satisfaction [80]. Ngo et al [81] identified measures for aesthetic design, including: balance, 'the distribution of optical weight'; equilibrium, 'a midway centre of suspension'; symmetry, the 'axial duplication'; sequence, "arrangement of objects in a layout in a way that facilitates the movement of the eye through the information displayed" (p. 30); unity, the 'totality of elements that is visually all one piece'; simplicity, 'directness'; economy, careful use of display elements; and rhythm, 'regular patterns of changes'. In one of the few studies on visualizations for children, Large et al [82] found that many of these criteria were present in children's drawings of an ideal interface, suggesting that perhaps young people gravitate naturally to the principles of aesthetical design.

Aesthetics is correlated with the use of metaphors and with 'fun', which is a significant criterion in designing systems for children and teens [78]. Fun has been defined for system designers as: what is expected of the system and whether the system disappoints or satisfies, engagement in terms of time spent on the system, and endurability – what is remembered about the system and the desire to return to the system [83]. Virtual environments designed for children and youth, therefore, should adhere to basic principles of user experience and particularly user engagement.

7. Conclusion

Many systems have been developed for the technology savvy generation. Most of these systems are designed for entertainment, utilizing the affordances of virtual reality to entice the new generation, who live in the digital world. More recently, the application of computer simulation and virtual reality has gained momentum in education, particularly in science, mathematics, geography, architecture and archaeology, where students can delve in the virtual environments for experiential enactment. Multi-user virtual environments such as Second Life have been an appealing teaching and learning tool for both the educators and students. In health care, several systems have been designed to help young patients cope with and alleviate their symptoms, such as pain, distress, and obesity.

Digital information regardless of the format, whether it is text, images, audio or video, plays a crucial role in the young generation's lives. Yet, very few virtual reality applications are available for distributing this information. While children and teens utilize mobile technology and social networks for their everyday life information transfer they still depend heavily on Google and other search engines to find and retrieve information. In so doing, they encounter obstacles and face difficulties in expressing their needs in keywords and expressions, upon which the systems can efficiently and effectively act. In other words, although technology savvy, the young generation lacks certain information literacy skills.

The youth's reliance on technology for consuming and producing information requires the development of new tools for knowledge and information transfer. These tools may be developed by using methodologies that involve and include children and teens as equal partners on the design teams. Their input is invaluable; they are experts at being children and can contribute significantly to the design process. As one example, the library virtual environment was created through a combination of different methodologies as an alternative information retrieval technique for a children's portal.

The vast majority of studies on the application of VEs in education, health, and information retrieval, however, have observed one crucial factor: engagement. Today's new generation is conditioned and accustomed to active participation both in consumption and production of information and knowledge. Participation means engagement. Designing and developing any virtual environment must be engaging for children and youth to realize the full potential of the technology.

Author details

Jamshid Beheshti
Associate Professor, School of Information Studies, McGill University, Montreal, Canada

8. References

[1] Pew Research Center. Pew Internet & American Life Project. September 2009 Teens and Mobile Phones. (2011). http://www.pewinternet.org/Shared-Content/Data-Sets/2009/September-2009-Teens-and-Mobile.aspx. (Accessed April, 2012).

[2] Levin, D., J. Richardson, & S. Arafeh. Digital disconnect: students' perceptions and experiences with the Internet and education. In P. Baker & S. Rebelsky (Eds), Proceedings of ED-MEDIA, World Conference On Educational Multimedia, Hypermedia and Telecommunications. Norfolk, VA: Association for the Advancement of Computing in Education; 2002. 51-52.

[3] Rowlands, I., D. Nicholas, P. Williams, P. Huntington, M. Fieldhouse, B. Gunter, R. Withey, H. Jamali, T. Dobrowolski, C. Tenopir. The Google generation: the information behaviour of the researcher of the future. *A*slib Proceedings, 2008; 60(4) 290-310.

[4] Vie, S. Digital Divide 2.0: "Generation M" and Online Social Networking Sites in the Composition Classroom. Computers and Composition, 2008; 25 (1) 9-23.

[5] Prensky, M. Digital Natives, Digital Immigrants." On the Horizon 2001; 9 1-5.

[6] Odom, W., Zimmerman, J., Forlizzi, J. Teenagers and their virtual possessions: Design opportunities and issues. CHI 2011, Proceedings of the 2011 annual conference on Human factors in computing systems, May 7-12, 2011, Vancouver, B.C. New York: ACM. 1491-1500.

[7] Ofcom. UK children's media literacy, March 2010. Accessed March, 2012, http://stakeholders.ofcom.org.uk/market-data-research/media-literacy/archive/medlitpub/medlitpubrss/ukchildrensml11/

[8] Harris, A.L. & A. Rea. Web 2.0 and virtual world technologies: A growing impact on IS education. Journal of Information Systems Education 2009; 20 (2) 137–44.

[9] Watters, A. Number of virtual world users breaks 1 billion, roughly half under age 15. ReadWriteWeb. 2010. Accessed April, 2012, http://www.readwriteweb.com/archives/number_of_virtual_world_users_breaks_the_1_billion.php

[10] Milgram, P. & F. Kishino. A Taxonomy of mixed reality visual displays. IEICE Transactions on Information and Systems, E77 1994; (12) 1321-1329.

[11] Wilson, J. R. & M. D'Cruz. Virtual and interactive environments for work of the future. International Journal of Human-Computer Studies, 2006; 64 158–169.

[12] Digital Stats. Accessed March 2012. http://digital-stats.blogspot.ca/2010/06/club-penguin-is-most-popular-virtual.html

[13] Beals, L. & Bers, M. U. A Developmental lens for designing virtual worlds for children and youth. International Journal of Learning and Media 2009; 1(1) 51-65.

[14] Pan, Z., A. D. Cheok, H. Yang, J. Zhu, J. Shi. Virtual reality and mixed reality for virtual learning environments. Computers & Graphics 2006; 30 (1) 20-28.

[15] Mikropoulos, T. A., Natsis, A. Educational virtual environments: A ten-year review of empirical research (1999-2009). Computers & Education 2011; 56(3) 769-780.

[16] Brooks, F. Virtual Reality in Education: Promise and Reality panel statement. Proceedings of the IEEE Virtual Reality Annual International Symposium (VRAIS '98) 1998.

[17] Walsh, D., Gentile, D., Gieske, J., Walsh, M. & Chasco, E. *Ninth Annual MediWise Video Game Report Card*. Minneapolis, MN: National Institute on Media and the Family. 2004.

[18] Roussos, M., Johnson, A., Moher, T., Leigh, J., Vasilakis, C. & Barnes, C. Learning and Building Together in an Immersive Virtual World. Presence 1999; 8, 247-263.

[19] Mumtaz, S. 2001. Children's enjoyment and perception of computer use in the home and the school. Computers and Education 2001; 36, 347-362.

[20] Virvou, M. ,Katsionis, G. & Manos, K. Combining software games with education: Evaluation of its educational effectiveness. Educational Technology & Society 2005; 8, 54-65.

[21] Champion, E. Heritage Role Playing-History as an interactive digital game. In: Pisan, Y. Interactive Entertainment Workshop, Sydney, NSW, Australia, February 2004; 47-65.

[22] Dalgarno, B. & Lee, M.J.W. What are the learning affordances of 3-D virtual environments? British Journal of Educational Technology 2010; 41(1) 10-32.

[23] Dede, C. Immersive interfaces for engagement and learning. Science 2009; 323 (5910) 66-69.

[24] Harel, I. & Papert, S. Constructionism. Norwood, NJ: Ablex. 1991.

[25] Freire, P. Pedagogy of the Oppressed. New York: Herder & Herder. 1970.

[26] McKenney, S., Voogt, J. Technology and young children: How 4-7 year olds perceive their own use of computers. Computers in Human Behavior 2010; 26, 656-664.

[27] Miller, D., Robertson, D. 2010. Using a games console in the primary classroom: Effects of 'Brain Training' programme on computation and self-esteem. British Journal of Educational Technology 2010; 41(2) 242-255.

[28] Gee, J. P. What Games Have to Teach Us About Learning and Literacy. New York: Palgrave/Macmillan. 2003.

[29] Bass, R. The Garden in the Machine: The Impact of American Studies on New Technologies. 1997. Accessed March, 2012, http://www9.georgetown.edu/faculty/bassr/garden.html.

[30] Wrzesien, M., Raya, M. A. Learning in serious virtual worlds: Evaluation of learning effectiveness and appeal to students in the E-Junior project. Computers & Education 2010; 55, 178-187.

[31] Tüzün, H., Yılmaz-Soylu, M., Karakuş, T., İnal, Y., Kızılkaya, G. The effects of computer games on primary school students' achievement and motivation in geography learning. Computers & Education 2009; 52(1) 68-77.

[32] Nunez Redo, M., Arturo Quintana Torres, Ricardo Quiros, Inma Nunez Redo, Juan Carda Castello, and Emilio Camahort. New augmented reality applications: Inorganic chemistry education. In Teaching through Multi-User Virtual Environments: Applying dynamic elements to the modern classroom, by Giovanni Vincenti and James Braman, 365-386. Hershey, Pennsylvania: IGI Global. 2010.

[33] Yeh, Andy J. & Nason, Rodney A. Knowledge Building of 3D Geometry Concepts and Processes within a Virtual Reality Learning Environment. In World Conference on

Educational Multimedia, Hypermedia & Telecommunications, 21-26, June, 2004, Lugano, Switzerland.

[34] Tan, K. T. W., Lewis, E. M., Avis, N. J., Withers. P. J. Using augmented reality to promote an understanding of materials science to school children. In ACM SIGGRAPH ASIA 2008 educators programme (SIGGRAPH Asia '08). ACM, New York, NY.

[35] Vincenti, G., Braman, J. Multi-user virtual environments for the classroom: Practical approaches to teaching in virtual worlds. Hershey, Pennsylvania: IGI Global. 2011.

[36] Loureiro, A., Santos, A. Bettencourt, T. Virtual worlds as an extended classroom. In Lanyi, C. S. (ed.) Application of Virtual Reality. Rijeka: InTech; 2012, p. 89-108. DOI: 10.5772/34959.

[37] Hew, K.F. & W. S. Cheung. Immersive virtual worlds in K-12 and higher education. British Journal of Educational Technology 2010; 41 (1) 33-55.

[38] Riva, G. The Key to unlocking the virtual body: virtual reality in the treatment of obesity and eating disorder. Journal of Diabetes Science Technology 2011; 5(2) 283-292.

[39] Merry, S.N., Stasiak, K., Shepherd, M., Frampton, C., Fleming, T., Lucassen, M.F.G. The effectiveness of SPARX, a computerised self help intervention for adolescents seeking help for depression: randomised controlled non-inferiority trial. BMJ 2012; 344: e2598.

[40] Haddock, B., et al. Fitness assessment comparison between the "Jackie Chan Action Run" videogame, 1-mile run/walk, and PACER. Games for Health Journal. 2012; 1(3) 223-227.

[41] Mitchell, L., Ziviani, J., Oftedal, S., Boyd, R. The effect of virtual reality interventions on physical activity in children and adolescents with early brain injuries including cerebral palsy. Developmental Medicine & Child Neurology, 28 January, 2012. DOI: 10.1111/j.1469-8749.2011.04199.x.

[42] Anton, R., Opris, D., Dobrean, A., David, D., Rizzo, A. Virtual reality in rehabilitation of attention deficit / hyperactivity disorder: The instrument construction principles. Virtual Rehabilitation International Conference, June 29-July, 2009; 2, 59-64.

[43] Pew Internet & American Life Project. Trends data for teens. 2011. http://www.pewinternet.org/Static-Pages/Trend-Data-for-Teens/Online-Activites-Total.aspx

[44] Radford, M. L., Silipigni Connaway, L., Agosto, D.E., Cooper, L. Z., Reuter, K., Zhou, N. Behaviors and preferences of digital natives: information a research agenda. Proceedings of the American Society for Information Science and Technology 2008; 44(1) 2.

[45] Large, A. Children, Teens and the Web. In: Cronin, B. (ed.) Annual Review of Information Science and Technology, Volume 39, Medford: Information Today 2005; 347-392.

[46] Salisbury, F. & Karasmanis, S. Are they ready? Exploring student information literacy skills in the transition from secondary to tertiary education. Australian Academic & Research Libraries, March, 2011. http://findarticles.com/p/articles/mi_go2490/is_1_42/ai_n57776793/?tag=content;col1

[47] Nicholas, D., I. Rowlands, D. Clark, P. Williams. Google Generation II: web behaviour experiments with the BBC. Aslib Proceedings 2011; 63 (1) 28 – 45.

[48] Czarnecki, K. & Matt, G. Meet the New You: In Teen Second Life, Librarians Can Leap Tall Buildings in a Single Bound and Save Kids from Boring Assignments--All before Lunch. School Library Journal 2007; 53(1) 36.

[49] Alliance Library System. Trends report 2008, Accessed April 2012, www.alliancelibrarysystem.com/pdf08/TrendsReport2008.pdf

[50] Gartner, April 27, 2007. Accessed March, 2012, http://www.gartner.com/it/page.jsp?id=503861

[51] Czarnecki, K. Building community as a library in a 3-D environment. Australasian Public Libraries and Information Services 2008; 21(1) 25-7.

[52] Buckland, A. Save the time of the avatar: Canadian academic libraries using chat reference in multi-user virtual environments. The Reference Librarian 2009; 51(1) 12-30.

[53] Beheshti, J., Large, A. Preliminary Design Indicators to Desktop Virtual Reality Environments. ASIST 2006 Annual Meeting, Information Realities: Shaping the Digital Future for All. SIG HCI Research Symposium: Human-Computer Interaction in Information Intensive Environments. November 5, 2006, Austin, Texas.

[54] Bilal, D. Children's use of the Yahooligans! web search engine. III. Cognitive and physical behaviors on fully self-generated search tasks. Journal of the American Society for Information Science and Technology 2002; 53(13) 1170-1183.

[55] Druin, A., Bederson, B. B., Hourcade, J. P., Sherman, L., Revelle, G., Platner, M., and Weng, S. Designing a digital library for young children: an intergenerational partnership. In *Proc. of JCDL,* ACM Press 2001; 398-405.

[56] Druin, A. et al, Designing a digital library for young children: an intergenerational partnership 2001, p.404.

[57] Cubaud, P., Stokowski, P., and Topol, A. Binding browsing and reading activities in a 3D digital library. In *Proc. of JCDL,* ACM Press 2002; 281-282.

[58] Shneiderman, B. Designing for Fun: How can we design user interfaces to be more fun. Interactions 2004; 11(5), p. 49.

[59] Norman, D. A. Emotional design : why we love (or hate) everyday things. Basic Books, New York, NY, 2004.

[60] Rauber, A., Merkl, D. The SOMLib digital library system. Lecture Notes in Computer Science, 1999; 1696/1999, 852, 323-342.

[61] Card, S. K., Hong, L., Mackinlay, J. D., and Chi, E. H. 3Book: a scalable 3D virtual book. In *Proc. of CHI,* ACM Press 2004); 1095-1098.

[62] Pejtersen, A. M. The Book House: An icon based database system for fiction retrieval in public libraries. In B. Cronin (Ed.), The marketing of library and information services. London: ASLIB 1992; 2, 572–591.

[63] Christoffel, M. and Schmitt, B. Accessing libraries as easy as a game. Visual Interfaces to digital libraries: Lecture notes in computer science 2002; 2539, 25-38.

[64] Large, A., Beheshti, J., Nesset, V., Bowler, L. Designing web portals in intergenerational teams: Two prototype portals for elementary school students. Journal of the American Society for Information Science and Technology 2004; 55 (13) 1–15.

[65] Beheshti, J., Large, A., Julien, C.A. 2005. Designing A Virtual Reality Interface for Children's Web Portals. Data, Information, and Knowledge in a Networked World. Canadian Association for Information Science 2005 Annual Conference, June 2-4. London, Ontario, 2005.

[66] Beheshti, J., Large, A. Preliminary Design Indicators to Desktop Virtual Reality Environments. ASIST 2006 Annual Meeting, Information Realities: Shaping the Digital Future for All. SIG HCI Research Symposium: Human-Computer Interaction in Information Intensive Environments. November 5, 2006, Austin, Texas.

[67] Beheshti, J., Large, A., Clement, I., Tabatabaei, N. Evaluating the usability of a virtual reality information system for children. Information Sharing in a Fragmented World: Crossing Boundaries. Canadian Association for Information Science 2007 Annual Conference, May 10-12, Montreal, Quebec.

[68] Messinger, P.R., et al. Virtual worlds – past, present, and future: New directions in social computing. Decision Support Systems 2009; 47(3) 208.

[69] Bers, M. U. Designing digital experiences for positive youth development: from playpen to playground. New York: Oxford University Press, 2012. p. 10.

[70] Hanna, L., Risden, K., and Alexander, K. Guidelines for usability testing with children. Interactions 1997; 4(5), p. 4.

[71] Nesset, V. and Large, A. Children in the information technology design process: A review of theories and their applications. Library & Information Science Research 2004; 26(2) 140-161.

[72] Druin, A. The role of children in the design of new technology. Behaviour and Information Technology 2002; 21(1) 1-25.

[73] Large, A., V. Nesset, J. Beheshti, L. Bowler. 'Bonded design': A novel approach to intergenerational information technology design." Library and Information Science Research, 2006; 28 (1) 64-82.

[74] Hassenzahl, M. & N. Tractinsky. User experience - a research agenda. Behaviour & Information Technology 2006; 25(2) 91-97.

[75] Laurel, B. Computers as theatre. Reading, MA: Addison-Wesley 1993.

[76] Peters, C., G. Castellano, S. de Freitas. An exploration of user engagement in HCI. In Proceedings of the International Workshop on Affective-Aware Virtual Agents and Social Robots (AFFINE '09), Ginevra Castellano, Jean-Claude Martin, John Murray, Kostas Karpouzis, and Christopher Peters (Eds.). ACM, New York, 2009.

[77] Skinner, E. A. & M. J. Belmont. Motivation in the classroom: Reciprocal effects of teacher behavior and student engagement across the school year. Journal of Educational Psychology 1993; 85(4) p. 572.

[78] Sutcliffe, A. Designing for user engagement: Aesthetic and attractive user interfaces. In Carroll, J.M. (Series editor) Synthesis Lectures on Human-Centered Informatics. San Rafael, CA: Morgan Claypool, 2009.

[79] Cawthon, N., & Vande Moere, A. The effect of aesthetic on the usability of data visualization. Proceedings of the 11th International Conference on Information Visualization. Zurich, 2007; IEEE 637–648.

[80] Hartmann, J., Sutcliffe, A., De Angeli, A. Towards a theory of user judgment of aesthetics and user interface quality. ACM Transactions on Human–Computer Interaction, 2008; 15(4) 1–30.

[81] Ngo, D.C.L., Teo, L. S., Byrne, J.G. Modelling interface aesthetics. Information Sciences, 2003; 152, 25–46.

[82] Large, A., Beheshti, J., Tabatabaei, N., Nesset, V. Developing a visual taxonomy: Children's view on aesthetics. Journal of the American Society for Information Science and Technology, 2009; 60 (9) 1808-1822.

[83] Read, J.C., MacFarlane, S.J., Casey, C. Endurability, Engagement and Expectations: Measuring Children's Fun. Proceedings of Interaction Design and Children. Eindhoven: Shaker Publishing, 2002; 189 - 198.